FUNDAMENTALS OF BIOGEOGRAPHY

Fundamentals of Biogeography presents an engaging and comprehensive introduction to biogeography, explaining the ecology, geography, and history of animals and plants. Defining and explaining the nature of populations, communities and ecosystems, the book examines where different animals and plants live and how they came to be living there; investigates how populations grow, interact, and survive, and how communities are formed and change; and predicts the shape of communities in the twenty-first century.

Illustrated throughout with informative diagrams and attractive photos (many in colour), and including guides to further reading, chapter summaries, and an extensive glossary of key terms, *Fundamentals of Biogeography* clearly explains key concepts, life systems, and interactions. The book also tackles the most topical and controversial environmental and ethical concerns including: animal rights, species exploitation, habitat fragmentation, biodiversity, metapopulations, patchy landscapes, and chaos.

Fundamentals of Biogeography presents an appealing introduction for students and all those interested in gaining a deeper understanding of the key topics and debates within the fields of biogeography, ecology and the environment. Revealing how life has and is adapting to its biological and physical surroundings, Huggett stresses the role of ecological, geographical, historical and human factors in fashioning animal and plant distributions and raises important questions concerning how humans have altered Nature, and how biogeography can affect conservation practice.

Richard John Huggett is a Senior Lecturer in Geography at the University of Manchester

ROUTLEDGE FUNDAMENTALS OF PHYSICAL GEOGRAPHY SERIES

Series Editor: John Gerrard

This new series of focused, introductory textbooks presents comprehensive, up-to-date introductions to the fundamental concepts, natural processes and human/environmental impacts within each of the core physical geography sub-disciplines: Biogeography, Climatology, Hydrology, Geomorphology and Soils.

Uniformly designed, each volume includes student-friendly features: plentiful illustrations, boxed case studies, key concepts and summaries, further reading guides and a glossary.

FUNDAMENTALS OF BIOGEOGRAPHY

Richard John Huggett

Routledge Fundamentals of Physical Geography

London and New York

First published 1998
by Routledge
11 New Fetter Lane, London EC4P 4EE

Simultaneously published in the USA and Canada
by Routledge
29 West 35th Street, New York, NY 10001

Typeset in Garamond by Keystroke, Jacaranda Lodge, Wolverhampton
Printed and bound in Great Britain by The Bath Press

British Library Cataloguing in Publication Data
A catalogue record for this book is available from the British Library

Library of Congress Cataloging in Publication Data
Huggett, Richard J.
Fundamentals of biogeography/Richard John Huggett
p. cm. — (Routledge Fundamentals of Physical Geography Series)
Includes bibliographical references and index.
1. Biogeography. I. Title. II. Series
QH84.H84 1998
578'.09—dc21 97–38998

ISBN 0–415–15498–7 (hbk)
ISBN 0–415–15499–5 (pbk)

For my family

CONTENTS

SERIES EDITOR'S PREFACE

We are presently living in a time of unparalleled change and when concern for the environment has never been greater. Global warming and climate change, possible rising sea levels, deforestation, desertification and widespread soil erosion are just some of the issues of current concern. Although it is the role of human activity in such issues that is of most concern, this activity affects the operation of the natural processes that occur within the physical environment. Most of these processes and their effects are taught and researched within the academic discipline of physical geography. A knowledge and understanding of physical geography, and all it entails, is vitally important.

It is the aim of this *Fundamentals of Physical Geography Series* to provide, in five volumes, the fundamental nature of the physical processes that act on or just above the surface of the earth. The volumes in the series are *Climatology*, *Geomorphology*, *Biogeography*, *Hydrology* and *Soils*. The topics are treated in sufficient breadth and depth to provide the coverage expected in a *Fundamentals* series. Each volume leads into the topic by outlining the approach adopted. This is important because there may be several ways of approaching individual topics. Although each volume is complete in itself, there are many explicit and implicit references to the topics covered in the other volumes. Thus, the five volumes together provide a comprehensive insight into the totality that is Physical Geography.

The flexibility provided by separate volumes has been designed to meet the demand created by the variety of courses currently operating in higher education institutions. The advent of modular courses has meant that physical geography is now rarely taught, in its entirety, in an 'all-embracing' course but is generally split into its main components. This is also the case with many Advanced Level syllabuses. Thus students and teachers are being frustrated increasingly by lack of suitable books and are having to recommend texts of which only a small part might be relevant to their needs. Such texts also tend to lack the detail required. It is the aim of this series to provide individual volumes of sufficient breadth and depth to fulfil new demands. The volumes should also be of use to sixth form teachers where modular syllabuses are also becoming common.

Each volume has been written by higher education teachers with a wealth of experience in all aspects of the topics they cover and a proven ability in presenting information in a lively and interesting way. Each volume provides a comprehensive coverage of the subject matter using clear text divided into easily accessible sections and subsections. Tables, figures and photographs are used where appropriate as well as boxed case studies and summary notes. References to important previous studies and results are included but are used sparingly to avoid overloading the text. Suggestions for further reading are also provided. The main target readership is introductory level undergraduate students of physical geography or environmental science, but there will be much of interest to students from other disciplines and it is also hoped that sixth form teachers will be able to use the information that is provided in each volume.

John Gerrard, 1997

PLATES

COLOUR

BLACK AND WHITE

FIGURES

TABLES

BOXES

AUTHOR'S PREFACE

Biogeography means different things to different people. To biologists, it is traditionally the history and geography of animals (zoogeography) and plants (phytogeography). This historical biogeography explores the long-term evolution of life and the influence of continental drift, global climatic change, and other large-scale environmental factors. Its origins lie in seventeenth-century attempts to explain how the world was restocked by animals disembarking from Noah's ark. Its modern foundations were laid by Charles Darwin and Alfred Russel Wallace in the second half of the nineteenth century. The science of ecology, which studies communities and ecosystems, emerged as an independent study in the late nineteenth century. An ecological element then crept into traditional biogeography. It led to analytical and ecological biogeography. Analytical biogeography considers where organisms live today and how they disperse. Ecological biogeography looks at the relations between life and the environmental complex. It used to consider mainly present-day conditions, but has edged backwards into the Holocene and Pleistocene.

Physical geographers have a keen interest in biogeography. Indeed, some are specialist teachers in that field. Biogeography courses have been popular for many decades. They have no common focus, their content varying enormously according to the particular interests of the teacher. However, many courses show a preference for analytical and ecological biogeography, and many include human impacts as a major element. Biogeography is also becoming an important element in the growing number of degree programmes in environmental science. Biogeography courses in geography and environmental science departments are supported by a good range of fine textbooks. Popular works include *Biogeography: Natural and Cultural* (Simmons 1979), *Basic Biogeography* (Pears 1985), *Biogeography: A Study of Plants in the Ecosphere* (Tivy 1992), and *Biogeography: An Ecological and Evolutionary Approach* (Cox and Moore 1993), the last being in its fifth edition with a sixth in preparation.

As there is no dearth of excellent textbooks, why is it necessary to write a new one? There are at least four good reasons for doing so. First, all the popular texts, though they have been reissued as new editions, have a 1970s air about them. It is a long time since a basic biogeography text appeared that took a fresh, up-to-date, and geographically focused look at the subject. Second, human interaction with plants and animals is now a central theme in geography, in environmental science, and in environmental biology. Existing textbooks tackle this topic, but there is much more to be said about application of biogeographical and ecological ideas in ecosystem management. Third, novel ideas in ecology are guiding research in biogeography. It is difficult to read articles on ecological biogeography without meeting metapopulations, heterogeneous landscapes, and complexity. None of these topics is tackled in existing textbooks. They are difficult topics to study from research publications because they contain formidable theoretical aspects. Nevertheless, it is very important that students should be familiar with the basic ideas behind them. First-year and second-year undergraduates can handle them if they are presented in an informative and interesting way that avoids excessive mathematical formalism. Fourth, environmentalism in its glorious variety has mushroomed into a vast interdisciplinary juggernaut. It impinges on biogeography to such an extent that it would be inexcusably remiss not to let it feature in a substantial way. It is a facet of biogeography that geography students find fascinating. Without doubt, a biogeography textbook for the next millennium should include discussion of environmental and ethical concerns about such pressing issues as species exploitation, environmental degradation, and biodiversity. However, biogeography is a vast subject and all textbook writers adopt a somewhat individualistic viewpoint. This book is no exception. It stresses the role of ecological, geographical, historical, and human factors in fashioning animal and plant distributions.

I should like to thank many people who have made the completion of this book possible. Nick Scarle patiently drew all the diagrams. Sarah Lloyd at Routledge bravely took yet another Huggett book on board. Several people kindly provided me with photographs. Rob Whittaker and Chris Fastie read and improved the section on vegetation succession. Michael Bradford and other colleagues in the Geography Department at Manchester University did not interrupt my sabbatical semester too frequently. Derek Davenport again discussed all manner of ideas with me. And, as always, my wife and family lent their willing support.

Richard Huggett
Poynton
December 1997

Note

Every effort has been made to trace the owners of all copyright material. In a few cases, the copyright owners could not be traced. Apologies are offered to any copyright holders whose rights may have been unwittingly infringed. The copyright of photographs remains with the individuals who supplied them.

INTRODUCTION:
STUDYING BIOGEOGRAPHY

Biogeography deals with the geography, ecology, and history of life – where it lives, how it lives there, and how it came to live there. It has three main branches – analytical biogeography, ecological biogeography, and historical biogeography. **Historical biogeography** considers the influence of continental drift, global climatic change, and other large-scale environmental factors on the long-term evolution of life. **Ecological biogeography** looks at the relations between life and the environmental complex. **Analytical biogeography** examines where organisms live today and how they spread. It may be considered as a division of ecological biogeography.

This book explores the ecological and historical biogeography of animals and plants, and, in doing so, it considers human involvement in the living world. It is designed to lead students through the main areas of modern biogeographical investigation. Basic ideas are carefully explained using numerous examples from around the world. The chief points are summarized at the end of each chapter. Each chapter also has essay questions, which are designed to consolidate key ideas, and suggestions for further reading. The glossary will help students to understand technical terms.

Chapter 1 addresses the basic question: What is biogeography? Chapter 2 looks at how life copes with its biological and physical environments. It covers four areas. First, it examines the places in which organisms live, focusing on habitats (landscape elements, landscapes, and regions), habitat requirements (habitat generalists, habitat specialists), and ecological tolerance (limiting factors, tolerance range, ecological valency). Second, it discusses climatic factors – radiation and light, temperature, moisture (including snow), and climatic zones (ecozones, biomes, zonobiomes, and orobiomes). Third, it explains how substrate and soil, topography (altitude, aspect, inclination, insularity), and disturbance influence living things. Fourth, it looks at ways of living (ecological niches, ecological equivalents), life-forms, and autoecological accounts.

Chapter 3 investigates the distribution of organisms – where they live and how they came to live there. It deals with five topics. First, it illustrates the geographical patterns displayed by species (and higher-order units, such as families) – large and small, widespread and restricted (micro-endemics, endemics, pandemics, and cosmopolites), continuous and broken (evolutionary, jump dispersal, geological, and climatic disjunctions), relict groups, the geography of range sizes and shapes, patterns within geographical ranges (home range, territory, habitat selection), and limits to geographical distributions.

Second, it examines dispersal and range change, considering how organisms move (agents of dispersal, dispersal abilities, dispersal routes), dispersal in action, and dispersal in the past. Third, it explores the splitting of geographical ranges (vicariance), discussing the effects of continental breakup on various groups of organism (Triassic reptiles, Cenozoic land mammals, Pangaean plants, large and flightless birds, Greater Antillean insectivores). Fourth, it looks at the effects of continental fusion on life (faunal mixing and the Great American Interchange). Fifth, it discusses the ways in which humans aid and abet the spread of organisms, using examples from the animal kingdom (the American mink and muntjac deer in Britain, introduced predators on Pacific islands) and fungal kingdom (the chestnut blight in the eastern United States).

Chapter 4 tackles populations. It covers three areas. First, it describes the demography of single populations, explaining the nature of population growth (exponential, logistic, population irruptions, population crashes, and chaotic change), age and sex structure (life tables, survivorship curves, and cohort-survival models), and metapopulations (loose, tight, extinction-and-colonization, and mainland–island). Second, it outlines population survival strategies, examining opportunists and competitors (r-strategists and K-strategists); competitors, stress-tolerators, and ruderals; and migration strategies (fugitive, opportunist, and equilibrium). Third, it studies population exploitation and control, looking at overexploited populations (passenger pigeon, auks, northern fur seal, and American bison), controlled populations (Swedish beavers, New Zealand goats, and Marion island cats), and the role of metapopulation theory in conservation (northern spotted owl, common wallaroo or euro, Leadbeater's possum).

Chapter 5 surveys interacting populations. It develops five areas. First, it looks at ways of living together – protocooperation (animal–animal, plant–plant, and animal–plant protocooperation, mimicry), mutualism, and commensalism (cattle egrets and cattle, midge and mosquito larvae in pitcher-plant pools, mynas and king-crows, river otters and beavers). Second, it looks at ways of staying apart, discussing

types of competition (competitive exclusion, scramble competition, contest competition) and mechanisms for avoiding competition (resource partitioning, character displacement, spatial complexity). Third, it examines herbivory, looking at plant eaters (kinds of herbivore and how plants defend themselves against them) and plant–herbivore interactions (seed predation and grazing). Fourth, it examines carnivory, describing some aspects of flesh-eating (carnivorous specializations, prey switching, prey selection, carnivore communities), interactions between predators and their prey (predator–prey cycles, chaos in Finnish weasel and vole populations), and geographical effects (laboratory and mathematical experiments with spatially heterogeneous environments). Fifth, it shows how life is pitted against life to control populations, considering biological control (prickly-pear cactus in Australia, false ragweed in Australia, agricultural pests in the Mediterranean region), genetic control, and integrated pest management.

Chapter 6 delves into communities. It covers four topics. First, it describes the nature of communities and ecosystems, looking at a local ecosystem – the Northaw Great Wood, England – and the global ecosystem. Second, it discusses roles within communities, looking at community production (primary producers, primary production), community consumption (consumers, decomposers, and detritivores), and ecosystem turnover (biogeochemicals, biogeochemical cycles). Third, it considers food chains and food webs, exploring types of food web (grazing food chains, decomposer food chains, ecological pyramids), keystone species (keystone predators, keystone herbivores and omnivores, the effects of removing keystone species), and contaminated food webs (biological magnification, the long-range transport of radioactive isotopes and pesticides). Fourth, it examines biodiversity, probing species–area relationships (species–area curves, habitat diversity and area-alone hypotheses), diversity gradients and hot-spots, and diversity change (why diversity matters and how it can be safeguarded).

Chapter 7 is about community change. It expounds four topics. First, it explains the nature of equilibrium communities, looking at the climatic climax, balanced ecosystems, succession models (facilitation model, tolerance model, inhibition model, allogenic and autogenic changes), primary succession (Glacier Bay, Alaska, Krakatau Islands, Indonesia), and secondary succession (abandoned fields in Minnesota, ghost towns in the western Great Basin). Second, it explains the nature of disequilibrium communities, discussing multidirectional succession (Hawaiian montane rain forest, Glacier Bay again, Krakatau Islands again) and community impermanence (no-modern-analogue communities, disharmonious communities, chaotic communities). Third, it tackles land-cover transformation and its effect on communities, focusing on habitat fragmentation (the skipper butterfly in Britain, the malleefowl in Australia, the reticulated velvet gecko in Australia) and the loss of wetlands (bottomland forest in the Santee River floodplain in Georgia, coastal ecosystems in southern Louisiana). Fourth, it predicts the possible shape of communities in the twenty-first century, looking at

species under pressure and communities under pressure (prairie wetlands in North America, mires in the Prince Edward Islands, biome and ecotone shifts, various examples of forest change induced by global warming, lessons from ecological simulations of community change).

Chapter 8 turns to the human dimension of biogeography and ecology. It looks into three areas. First, it considers biorights, asking if organisms have rights (culling, the wildlife trade, zoos) and if all Nature has rights. Second, it discusses attitudes towards Nature, looking at brands of environmentalism (technocentric, ecocentric), the alignments of ecologists, and ecosystem management. Third, it elucidates the connection between conservation practice and biogeographical and ecological theory, discussing environmental exploitation and enduring equilibrium, balanced ecosystems, evolutionary disequilibrium, the edge of chaos, and the biodiversity bandwagon.

1

WHAT IS BIOGEOGRAPHY?

Biogeographers study the geography, ecology, and evolution of living things. This chapter covers:

- ecology – environmental constraints on living
- history and geography – time and space constraints on living

Biogeographers address a misleadingly simple question: why do organisms live where they do? Why is the speckled rangeland grasshopper confined to short-grass prairie and forest or brushland clearings containing small patches of bare ground? Why does the ring ouzel live in Norway, Sweden, the British Isles, and mountainous parts of central Europe, Turkey, and south-west Asia, but not in the intervening regions? Why do tapirs live only in South America and South-east Asia? Why do the nestor parrots – the kea and the kaka – live only in New Zealand? Why do pouched mammals (marsupials) live in Australia and the Americas, but not in Europe, Asia, Africa, or Antarctica? Why do different regions carry distinct assemblages of animals and plants?

Two groups of reasons are given in answer to such questions as these – ecological reasons and historical-cum-geographical reasons.

ECOLOGY

Ecological explanations for the distribution of organisms involve several interrelated ideas. First is the idea of **populations**, which is the subject of analytical biogeography. Each species has a characteristic life history, reproduction rate, behaviour, means of dispersal, and so on. These traits affect a population's response to the environment in which it lives. The second idea concerns this biological response to the **environment** and is the subject of ecological biogeography. A population responds to its physical surroundings (**abiotic environment**) and its living surroundings (**biotic environment**). Factors in the abiotic environment include such physical factors as temperature, light, soil, geology, topography, fire, water, water and air currents; and such chemical factors as oxygen levels, salt concentrations, the presence of toxins, and acidity. Factors in the biotic environment include competing species, parasites, diseases, predators, and humans. In short, each species can tolerate a range of environmental factors. It can only live where these factors lie within its tolerance limits.

Speckled rangeland grasshopper

This insect (*Arphia conspersa*) ranges from Alaska and northern Canada to northern Mexico, and from California to the Great Plains. It is found at less than 1,000 m elevation in the northern part of its range and up to 4,000 m in the southern part. Within this extensive latitudinal and altitudinal range, its distribution pattern is very patchy, owing to its decided preference for very specific habitats (e.g. Schennum and Willey 1979). It requires short-grass prairie, or forest and brushland openings, peppered with small pockets of bare ground. Narrow-leaved grasses provide the grasshopper's food source. The bare patches are needed for it to perform courtship rituals. These ecological and behavioural needs are not met by dense forest, tall grass meadows, or dry scrubland. Roadside meadows and old logged areas are suitable and are slowly being colonized. Moderately grazed pastures are also suitable and support large populations.

Even within suitable habitat, the grasshopper's distribution is limited by its low vagility (the ease with which it can spread). This is the result of complex social behaviour, rather than an inability to fly well. Females are fairly sedentary, at least in mountain areas, while males make mainly short, spontaneous flights within a limited area. The two sexes together form tightly knit population clusters within areas of suitable habitat. The clusters are held together by visual and acoustic communication displays.

Ring ouzel

The biogeography of most species may be explained by a mix of ecology and history. The ring ouzel or 'mountain blackbird', which goes by the undignified scientific name of *Turdus torquatus* (Box 1.1), lives in the cool temperate climatic zone, and in the alpine equivalent to the cool temperate zone on mountains (Figure 1.1). It likes cold climates. During the last ice age, the heart of its range was probably the Alps and Balkans. From here, it spread outwards into much of Europe, which was then colder than now. With climatic warming during the last 10,000 years, the ring ouzel has left much of its former range and survives only in places that are still relatively cold because of their high latitude or altitude. Even though it likes cold conditions, most ring ouzels migrate to less severe climates during winter. The north European populations move to the Mediterranean while the alpine populations move to lower altitudes.

HISTORY AND GEOGRAPHY

Historical-cum-geographical explanations for the distribution of organisms involve two basic ideas, both of which are the subject of historical biogeography. The first idea concerns **centres-of-origin** and **dispersal** from one place to another. It argues that species originate in a particular place and then spread to other parts of the globe, if they should be able and willing to do so. The second idea considers the importance of geological and climatic changes splitting a single population into two or more isolated groups. This idea is known as **vicariance** biogeography. These two basic biogeographical processes are seen in the following case studies.

Tapirs

The tapirs are close relatives of the horses and rhinoceroses. They form a family – the Tapiridae. There are four living species, one of which dwells in Southeast Asia and three in Central and South America (Plate 1.1). Their present distribution is thus broken and poses a problem for biogeographers. How do such closely related species come to live in geographically distant parts of the world? Finds of fossil tapirs help to answer this puzzle. Members of the tapir family were once far more widely distributed than at present (Figure 1.2). They are known to have lived in North America and Eurasia. The oldest fossils come from Europe. A logical conclusion is that the tapirs evolved in Europe, which was their centre of origin, and then dispersed east and west. The tapirs that went north-east reached North America and South America. The tapirs that chose a south-easterly dispersal route moved into South-east Asia. Subsequently, probably owing to climatic change, the tapirs in North America and the Eurasian homeland went extinct. The survivors at the tropical edges of the distribution spawned the present species. This explanation is plausible, though it is not watertight – it is always possible that somebody will dig up even older tapir remains from somewhere else. Historical biogeographers are dogged by the incompleteness of the fossil record, which means that they can never be fully confident about any hypothesis.

Box 1.1

WHAT'S IN A NAME?
CLASSIFYING ORGANISMS

Everyone knows that living things come in a glorious diversity of shapes and sizes. It is apparent, even to a casual observer, that organisms appear to fall into groups according to the similarities between them. No one is likely to mistake a bird for a beetle, a daisy for a hippopotamus. Zoologists and botanists classify organisms according to the similarities and differences between them. Currently, five great kingdoms are recognized – prokaryotae (monera), protoctista, plantae, fungi, and animalia. These chief subdivisions of the kingdoms are phyla. Each phylum represents a basic body plan that is quite distinct from other body plans. This is why it is fairly easy, with a little practice, to identify the phylum to which an unidentified organism belongs. Amazingly, new phyla are still being discovered (e.g. Funch and Kristensen 1995).

Organisms are classified hierarchically. Individuals are grouped into species, species into genera, genera into families, and so forth. Each species, genus, family, and higher-order formal group of organisms is called a **taxon** (plural taxa). Each level in the hierarchy is a taxonomic category. The following list shows the classification of the ring ouzel:

Kingdom:	Animalia (animals)
Phylum:	Chordata (chordates)
Subphylum:	Vertebrata (vertebrates)
Class:	Aves (birds)
Subclass:	Neornithes ('new birds')
Superorder:	Carinatae (typical flying birds)
Order:	Passeriformes (perching birds)
Suborder:	Oscines (song birds)
Family:	Muscicapidae (thrush family)
Subfamily:	Turdinae (thrushes, robins, and chats)
Genus:	*Turdus*
Species:	*Turdus torquatus*

Animal family names always end in -idae, and subfamilies in -inae; they may be less formally referred to by dropping the initial capital letter and using -ids as an ending, as in felids for members of the cat family. Plant family names end in -aceae or -ae. The genus (plural genera) is the first term of a binomial: genus plus species, as in *Turdus torquatus*. It is always capitalized and in italics. The species is the second term of a binomial. It is not capitalized in animal species, and is not normally capitalized in plant species, but is always italicized in both cases. The specific name signifies either the person who first described it, as in *Muntiacus reevesi*, Reeve's muntjac deer, or else some distinguishing feature of the species, as in *Calluna vulgaris*, the common (= vulgar) heather. If subspecies are recognized, they are denoted by the third term of a trinomial. For example, the common jay in western Europe is *Garrulus glandarius glandarius*, which would usually be shortened to *Garrulus g. glandarius*. The Japanese subspecies is *Garrulus glandarius japonicus*. In formal scientific writing, the author or authority of the name is indicated. So, the badger's full scientific name is *Meles meles* L., the L. indicating that the species was first described by Carolus Linnaeus (1707–78). The brown hare's formal name is *Lepus europaeus* Pallas, which shows that it was first described by Peter Simon Pallas (in 1778).

After its first appearance in a paper or book, the species name is usually abbreviated by reducing the generic term to a single capital letter. Thus, *Meles meles* becomes *M. meles*. This practice will not be adopted in the present book because it gives a rather 'stuffy' feel to the text. Likewise, the authorities will be omitted because they confer an even stuffier feel.

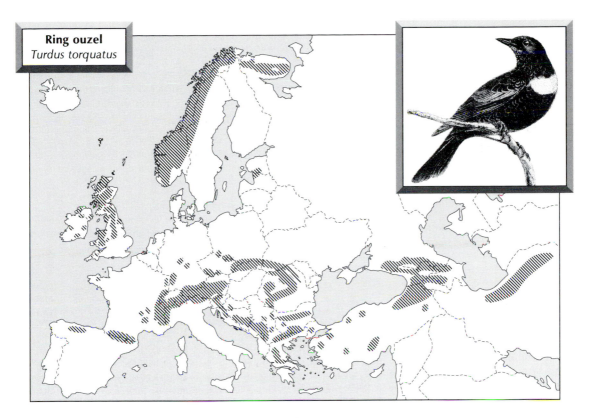

Ring ouzel
Turdus torquatus

Figure 1.1 The breeding distribution of the ring ouzel (*Turdus torquatus*).
Source: Map after Cramp (1988); picture from Saunders (1889)

Nestor parrots

The nestor parrots (Nestorinae) are endemic to New Zealand. There are two species – the kaka (*Nestor meridionalis*) and the kea (*Nestor notabilis*) (Plate 1.2). They are closely related and are probably descended from a 'proto-kaka' that reached New Zealand during the Tertiary period. Then, New Zealand was a single, forest-covered island. The proto-kaka became adapted to forest life. Late in the Tertiary period, the north and south parts of New Zealand split. North Island remained forested and the proto-kakas there continued to survive as forest parrots, feeding exclusively on vegetable matter and nesting in tree hollows. They eventually evolved into the modern kakas. South Island gradually lost its forests because mountains grew and climate changed. The

proto-kakas living on South Island adjusted to these changes by becoming 'mountain parrots', depending on alpine shrubs, insects, and even carrion for food. They forsook trees as breeding sites and turned to rock fissures. The changes in the South Island proto-kakas were so far-reaching that they became a new species – the kea. After the Ice Age, climatic amelioration promoted some reforestation of South Island. The kakas dispersed across the Cook Strait and colonized South Island. Interaction between North and South Island kaka populations is difficult across the 26 km of ocean. In consequence, the South Island kakas have become a subspecies. The kaka and the kea are now incapable of interbreeding and they continue to live side by side on South Island. The kea has never colonized North Island, probably because there is little suitable habitat there. The biogeography

Plate 1.1 Central American or Baird's tapir (*Tapirus bairdi*), Belize
Photograph by Pat Morris

of the nestor parrots thus involves adaptation to changing environmental conditions, dispersal, and vicariance events.

Marsupials

The pouched mammals or marsupials are now found in Australia, South America, and North America. Fossil forms are known from Eurasia, North Africa, and Antarctica. They are commonly assumed to have evolved in North America, where the oldest known marsupial fossils are found, from an ancestor that also sired the placental mammals. The marsupial and placental mammals split about 130 million years ago, early in the Cretaceous period. There are several rival explanations for the current distribution of marsupials (L. G. Marshall 1980).

The classic explanation of marsupial distribution was proposed before **continental drift** was accepted and assumed stationary continents. It argued that marsupials dispersed from a Cretaceous North American homeland to other continents. Some time in the Late Cretaceous period, marsupials hopped across islands linking North and South America. During the Eocene epoch, they moved into Asia and Europe across a land bridge spanning the Bering Sea between Alaska and Siberia. From there they spread into Europe and, using Indonesia as an embarkation point, into Australia. Several variations on this 'centre of origin followed by dispersal' hypothesis played out on a stationary land surface were forthcoming. The variations involved different centres of origin (South America or Antarctica) and different dispersal routes.

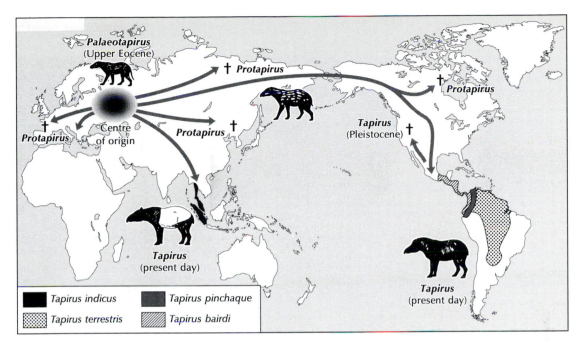

Figure 1.2 Tapirs: their origin, spread, and present distribution.
Source: After Rodríguez de la Fuente (1975)

Plate 1.2 Nestor parrots (a) Kaka (*Nestor meridionalis*) (b) Kea (*Nestor notabilis*)
Photographs by Pat Morris

As soon as it was accepted that the continents do drift, revised explanations of the marsupial history were suggested. Some of these new hypotheses still invoked centre-of-origin and dispersal. They were similar to the hypothesis developed for stationary continents, but they did not need to invoke fanciful land bridges between widely separated continents. Other hypotheses laid emphasis on the fragmentation of **Pangaea**, the Triassic supercontinent, and stressed vicariance events rather than dispersal over pre-existing barriers.

It now seems likely that, even if the first marsupials did appear in Mesozoic North America (and that is far from certain), they quickly became widely distributed over the connected land masses of South America, Antarctica, and Australia. The breakup of Pangaea, which started in earnest during the Mid Jurassic period, isolated the Mesozoic marsupials on South America and Australia and these two main branches then evolved independently. The American marsupials reached Europe, North Africa, and Asia in Palaeocene and Eocene times, but by the end of the Miocene epoch, North American and European marsupials were extinct. South American marsupials invaded North America in the Pleistocene epoch. Current explanations of marsupial biogeography thus call on dispersal and vicariance events.

Biogeographical regions

Different places house different kinds of animals and plants. This became apparent as the world was explored. In 1628, in his *The Anatomy of Melancholy*, Robert Burton wrote:

> Why doth Africa breed so many venomous beasts, Ireland none? Athens owls, Crete none? Why hath Daulis and Thebes no swallows (so Pausanias informeth us) as well as the rest of Greece, Ithaca no hares, Pontus [no] asses, Scythia [no] wine? Whence comes this variety of complexions, colours, plants, birds, beasts, metals, peculiar to almost every place?

> (Burton 1896 edn: vol. II, 50–1)

By the nineteenth century, it was abundantly clear that the land surface could be divided into several large **biogeographical regions**, each of which sup-

ports a distinct set of animals and a distinct set of plants. Using bird distributions, Philip L. Sclater (1858) recognized two basic divisions – the Old World and the New World. The Old World he divided into Europe and northern Asia, Africa south of the Sahara, India and southern Asia, and Australia and New Guinea. The New World he divided into North America and South America. Sclater's system was adopted by Alfred Russel Wallace (1876), with minor amendments, to provide a long-lasting nomenclature that survives today as the Sclater–Wallace scheme (Figure 1.3a). Six regions are recognized – Nearctic, Neotropical, Palaearctic, Ethiopian, Oriental, and Australian. Together, the Nearctic and Palaearctic form Neogaea (the New World), while other regions form Palaeogaea (The Old World). Wallace also recognized subregions, four per region (Table 1.1), which correspond largely to established plant regions.

Modern methods of numerical classification produce similar regions to the Sclater–Wallace scheme, but there are differences. Figure 1.3b shows faunal regions of the world based on mammal distributions (C. H. Smith 1983). There are four regions – Holarctic, Latin American, Afro-Tethyan, and Island – and ten subregions. Each subregion is as unique as it can be compared with all other subregions.

SUMMARY

The distribution of organisms is determined by ecology, history, and geography. Most distributions result from a combination of all three factors. Biological and geological evolution have acted together to produce the biogeographical regions and subregions seen today.

ESSAY QUESTIONS

1 Describe the characteristic animals in Wallace's biogeographical regions.

2 Describe the chief botanical regions of the world.

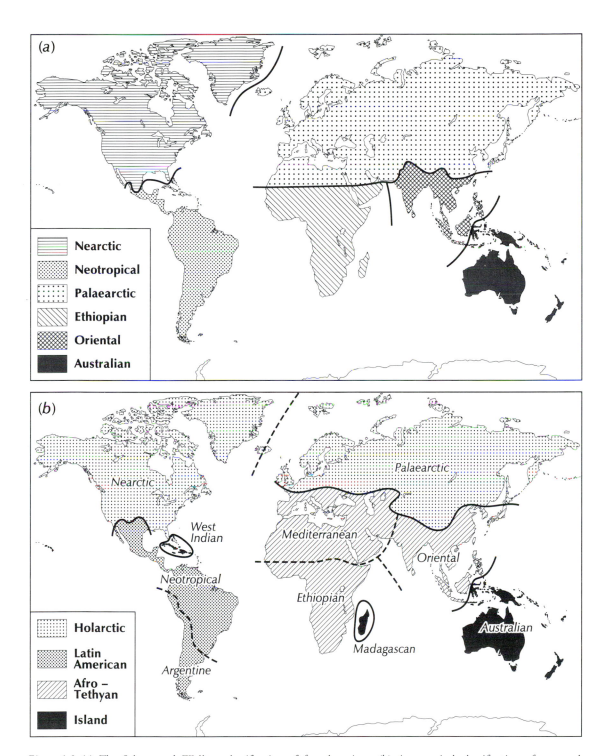

Figure 1.3 (a) The Sclater and Wallace classification of faunal regions. (b) A numerical classification of mammal distributions showing four main regions and ten subregions.
Source: After C. H. Smith (1983)

Table 1.1 Biogeographical regions and subregions, as defined by Alfred Russel Wallace

Region	Subregion
Palaeogaea (Old World)	
Palaearctic	North Europe
	Mediterranean
	Siberia
	Manchuria (or Japan)
Ethiopian	East Africa
	West Africa
	South Africa
	Madagascar
Oriental	Hindustan (or central India)
	Ceylon (Sri Lanka)
	Indo-China (or Himalayas)
	Indo-Malaya
Australian	Austro-Malaya
	Australia
	Polynesia
	New Zealand
Neogaea (New World)	
Neotropical	Chile
	Brazil
	Mexico
	Antilles
Nearctic	California
	Rocky Mountains
	Alleghenies
	Canada

Source: After Wallace (1876)

FURTHER READING

Cox, C. B. and Moore, P. D. (1993) *Biogeography: An Ecological and Evolutionary Approach*, 5th edn, Oxford: Blackwell.

George, W. (1962) *Animal Geography*, London: Heinemann.

Pears, N. (1985) *Basic Biogeography*, 2nd edn, Harlow: Longman.

Tivy, J. (1992) *Biogeography: A Study of Plants in the Ecosphere*, 3rd edn, Edinburgh: Oliver & Boyd.

2

LIFE AND THE ENVIRONMENT

Life is adapted to nearly all Earth surface environments. This chapter covers:

- places to live
- climatic constraints on living
- other physical constraints on living
- ways of living

LIVING SPACE: HABITATS AND ENVIRONMENTS

Individuals, species, and populations, both marine and terrestrial, tend to live in particular places. These places are called **habitats**. Each habitat is characterized by a specific set of environmental conditions – radiation and light, temperature, moisture, wind, fire frequency and intensity, gravity, salinity, currents, topography, soil, substrate, geomorphology, human disturbance, and so forth.

A place to live: habitats

Habitats come in all shapes and sizes, occupying the full sweep of geographical scales. They range from small (microhabitats), through medium (meso-habitats) and large (macrohabitats), to very large (mega-habitats). **Microhabitats** are a few square centimetres to a few square metres in area (Table 2.1). They include leaves, the soil, lake bottoms, sandy beaches, talus slopes, walls, river banks, and paths. **Meso-habitats** have areas up to about 10,000 km²; that is, a 100 × 100 kilometre square, which is about the size of Cheshire, England. Each main mesohabitat is influenced by the same regional climate, by similar features of geomorphology and soils, and by a similar

set of disturbance regimes. Deciduous woodland, caves, and streams are examples. **Macrohabitats** have areas up to about 1,000,000 km², which is about the size of Ireland. **Megahabitats** are regions more than 1,000,000 km² in extent. They include continents and the entire land surface of the Earth.

Landscape ecologists, who have an express interest in the geographical dimension of ecosystems, recognize three levels of 'habitat' – region, landscape, and landscape element. These correspond to large-scale, medium-scale, and small-scale habitats. Some landscape ecologists are relaxing their interpretation of a landscape to include smaller and larger scales – they have come to realize that a beetle's view or a bird's view of the landscape is very different from a human's view.

Landscape elements

Landscape elements are similar to microhabitats, but a little larger. They are fairly uniform pieces of land, no smaller than about 10 m, that form the building blocks of landscapes and regions. They are also called ecotopes, biotopes, geotopes, facies, sites, tesserae, landscape units, landscape cells, and landscape prisms. These terms are roughly equivalent to landscape element, but have their own special meanings (see Forman 1995; Huggett 1995).

Table 2.1 Habitat scales

Scale[a]	Approximate area (km²)	Terminology applied to landscape units at same scale[b]		
		Fenneman (1916)	Linton (1949)	Whittlesey (1954)
Microhabitat (small)	<1	–	Site	–
Mesohabitat (medium)	1–10	–	–	–
	10–100	–	Stow	Locality
	100–1,000	District	Tract	District
	1,000–10,000	Section	Section	–
Macrohabitat (large)	10,000–100,000	Province	Province	Province
	100,000–1,000,000	Major division	Major division	Realm
Megahabitat (very large)	>1,000,000	–	Continent	–

Note: [a] These divisions follow Delcourt and Delcourt (1988)
[b] The range of areas associated with these regional landscape units is meant as a rough-and-ready guide rather than precise limits

Landscape elements are made of individual trees, shrubs, herbs, and small buildings. There are three basic kinds of landscape element – patches, corridors, and background matrixes:

1 **Patches** are fairly uniform (homogeneous) areas that differ from their surroundings. Woods, fields, ponds, rock outcrops, and houses are all patches.
2 **Corridors** are strips of land that differ from the land to either side. They may interconnect to form networks. Roads, hedgerows, and rivers are corridors.
3 **Background matrixes** are the background ecosystems or land-use types in which patches and corridors are set. Examples are deciduous forest and areas of arable cultivation.

Landscape elements include the results of human toil – roads, railways, canals, houses, and so on. Such features dominate the landscape in many parts of the world and form a kind of 'designer mosaic'. Designed patches include urban areas, urban and suburban parks and gardens (greenspaces), fields, cleared land, and reservoirs. Designed corridors include hedgerows, roads and railways, canals, dikes, bridle paths, and footpaths. There is also a variety of undesigned patches – waste tips, derelict land, spoil heaps, and so on.

Landscapes

Landscape elements combine to form **landscapes**. A landscape is a mosaic, an assortment of patches and corridors set in a matrix, no bigger than about 10,000 km². It is 'a heterogeneous land area composed of a cluster of interacting ecosystems that is repeated in similar form throughout' (Forman and Godron 1986: 11). By way of example, the recurring cluster of interacting ecosystems that feature in the landscape around the author's home, in the foothills of the Pennines, includes woodland, field, hedgerow, pond, brook, canal, roadside, path, quarry, mine tip, disused mining incline, disused railway, farm building, and residential plot.

Regions

Landscapes combine to form **regions**, more than about 10,000 km^2 in area. They are collections of landscapes sharing the same macroclimate. All Mediterranean landscapes share a seasonal climate characterized by mild, wet winters and hot, droughty summers.

The bare necessities: habitat requirements

It is probably true to say that no two species have exactly the same living requirements. There are two extreme cases – fussy species or habitat specialists and unfussy species or habitat generalists – and all grades of 'fussiness' between.

Habitat specialists

Habitat specialists have very precise living requirements. In southern England, the red ant, *Myrmica sabuleti*, needs dry heathland with a warm south-facing aspect that contains more than 50 per cent grass species, and that has been disturbed within the previous five years (N. R. Webb and Thomas 1994). Other species are less pernickety and thrive over a wider range of environmental conditions. The three-toed woodpecker (*Picoides tridactylus*) lives in a broad swath of cool temperate forest encircling the Northern Hemisphere. Races of the common jay (*Garrulus glandarius*) occupy a belt of oak and mixed deciduous woodland stretching from Britain to Japan.

Habitat generalists

A few species manage to eke out a living in a great array of environments. The human species (*Homo sapiens*) is the champion **habitat generalist** – the planet Earth is the human habitat. In the plant kingdom, the broad-leaved plantain (*Plantago major*), typically a species of grassland habitats, is found almost everywhere except Antarctica and the dry parts of North Africa and the Middle East. In the British Isles, it seems indifferent to climate and soil conditions, being found in all grasslands on acid and alkaline soils alike. It also lives on paths, tracks, disturbed habitats (spoil heaps, demolition sites, arable land), pasture and meadows, road verges, river banks, mires, skeletal habitats, and as a weed in lawns and sports fields. In woodland, it is found only in relatively unshaded areas along rides. It is not found in aquatic habitats or tall herb communities.

Life's limits: ecological tolerance

Organisms live in virtually all environments, from the hottest to the coldest, the wettest to the driest, the most acidic to the most alkaline. Understandably, humans tend to think of their 'comfortable' environment as the norm. But moderate conditions are anathema to the micro-organisms that love conditions fatal to other creatures. These are the **extremophiles** (Madigan and Marrs 1997). An example is high-pressure-loving microbes (barophiles) that flourish in deep-sea environments and are adapted to life at high pressures (Bartlett 1992). Many other organisms are adapted to conditions that, by white western human standards, are harsh, though not so extreme as the conditions favoured by the extremophiles. Examples are hot deserts and Arctic and alpine regions.

Limiting factors

A limiting factor is an **environmental factor** that slows down population growth. The term was first suggested by Justus von Liebig (1840), a German agricultural chemist. Liebig noticed that the growth of a field crop is limited by whichever nutrient happens to be in short supply. A field of wheat may have ample phosphorus to yield well, but if another nutrient, say nitrogen, should be lacking, then the yield will be reduced. No matter how much extra phosphorus is applied in fertilizer, the lack of nitrogen will limit wheat yield. Only by making good the nitrogen shortage could yields be improved. These observations led to Liebig to establish a '**law of the minimum**': the productivity, growth, and reproduction of organisms will be constrained if one or more environmental factors lies below its limiting level.

Later, ecologists established a 'law of the maximum'. This law applies where population growth is curtailed by an environmental factor exceeding an upper limiting level. In a wheat field, too much phosphorus is as harmful as too little – there is an upper limit to nutrient levels that plants can tolerate.

Tolerance range

For every environmental factor (such as temperature and moisture) there is a lower limit below which a species cannot live, an optimum range in which it thrives, and an upper limit above which it cannot live (Figure 2.1). The upper and lower bounds define the **tolerance range** of a species for a particular environmental factor. The bounds vary from species to species. A species will prosper within its optimum range of tolerance; survive but show signs of physiological stress near its tolerance limits; and not survive outside its tolerance range (Shelford 1911). **Stress** is a widely used but troublesome idea in ecology. It may be defined as 'external constraints limiting the rates of resource acquisition, growth or reproduction of organisms' (Grime 1989).

Each **species** (or **race**) has a characteristic tolerance range (Figure 2.2). Stenoecious species have a wide tolerance; euryoecious species have a narrow tolerance. All species, regardless of their tolerance range, may be adapted to the low end (oligotypic), to the middle (mesotypic), or to the high end (polytypic) of an environmental gradient. Take the example of photosynthesis in plants. Plants adapted to cool temperatures (oligotherms) have photosynthetic optima at about 10°C and cease to photosynthesize above 25°C. Temperate-zone plants (mesotherms) have optima between 15°C and 30°C. Tropical plants (polytherms) may have optima as high as 40°C. Interestingly, these optima are not 'hard and fast'. Cold-adapted plants are able to shift their photosynthetic optima towards higher temperatures when they are grown under warmer conditions.

Ecological valency

Tolerance may be wide or narrow and the optimum may be at low, middle, or high positions along an environmental gradient. When combined, these contingencies produce six grades of **ecological valency** (Figure 2.2). The glacial flea (*Isotoma saltans*) has a narrow temperature tolerance and likes it cold. It is

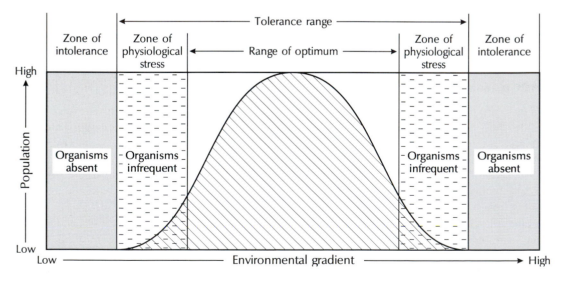

Figure 2.1 Tolerance range and limits.
Source: Developed from Shelford (1911)

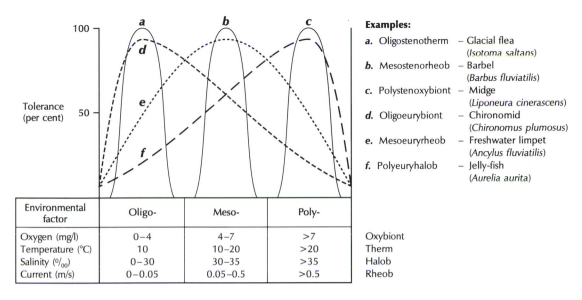

Examples:

a. Oligostenotherm – Glacial flea
(*Isotoma saltans*)

b. Mesostenorheob – Barbel
(*Barbus fluviatilis*)

c. Polystenoxybiont – Midge
(*Liponeura cinerascens*)

d. Oligoeurybiont – Chironomid
(*Chironomus plumosus*)

e. Mesoeuryrheob – Freshwater limpet
(*Ancylus fluviatilis*)

f. Polyeuryhalob – Jelly-fish
(*Aurelia aurita*)

Environmental factor	Oligo-	Meso-	Poly-	
Oxygen (mg/l)	0–4	4–7	>7	Oxybiont
Temperature (°C)	10	10–20	>20	Therm
Salinity (⁰/₀₀)	0–30	30–35	>35	Halob
Current (m/s)	0–0.05	0.05–0.5	>0.5	Rheob

Figure 2.2 Ecological valency, showing the amplitude and position of the optimum.
Source: After Illies (1974)

an oligostenotherm. The midge *Liponeura cinerascens* has a narrow oxygen-level tolerance at the high end of the oxygen-level gradient. It is a polystenoxybiont. Other examples are shown on Figure 2.2.

WARM AND WET: CLIMATIC FACTORS

Flower power: radiation and light

The Sun is the primary source of radiation for the Earth. It emits electromagnetic radiation across a broad spectrum, from very short wavelengths to long wavelengths (Box 2.1). The visible portion (sunlight) is the effective bit for photosynthesis. It is also significant in heating the environment. Long-wave (infrared) radiation emitted by the Earth is locally important around volcanoes, in geothermal springs, and in hydrothermal vents in the deep-sea floor. These internal sources of energy are tapped by unusual organisms, including the thermophiles and hyperthermophiles that like it very hot (p. 19).

Three aspects of **solar radiation** influence photo-synthesis – the intensity, the quality, and the photo-period or duration. The intensity of solar radiation is the amount that falls on a given area in a unit of time. Calories per square centimetre per minute (cal/cm²/ min) were once popular units, but Watts per square metre (W/m²) or kiloJoules per hectare (kJ/ha) are metric alternatives. The average annual solar radiation on a horizontal ground surface ranges from about 800 kJ/ha over subtropical deserts to less than 300 kJ/ha in polar regions. Equatorial regions receive less radiation than the subtropics because they are cloudier. A value of 700 kJ/ha is typical.

The quality of solar radiation is its wavelength composition. This varies from place to place depending on the composition of the atmosphere, different components of which filter out different parts of the electromagnetic spectrum. In the tropics, about twice as much ultraviolet light reaches the ground above 2,500 m than at sea level. Indeed, ultraviolet light is stronger in all mountains – hence incautious humans may unexpectedly suffer sunburn at ski resorts.

Photoperiod is seasonal variations in the length of day and night. This is immensely important ecologically because day-length, or more usually

---

Box 2.1

THE ELECTROMAGNETIC SPECTRUM EMITTED BY THE SUN

Electromagnetic radiation pours out of the Sun at the speed of light. Extreme ultraviolet radiation with wavelengths in the range 30 to 120 nanometres (nm) occupies the very short end of the spectrum. Ultraviolet light extends to wavelengths of 0.4 micrometres (μm). Visible light has wavelengths in the range 0.4 to 0.8 μm. This is the portion of the electromagnetic spectrum humans can see. Infrared radiation has wavelengths longer than 0.8 μm. It grades into radio frequencies with millimetre to metre wavelengths. The Sun emits most intensely near 5 μm, which is in the green band of the visible light. This fact might help to account for plants being green – they reflect the most intense band of sunlight.

---

night-length, stimulates the timing of daily and seasonal rhythms (breeding, migration, flowering, and so on) in many organisms. Short-day plants flower when day-length is below a critical level. The cocklebur (*Xanthium strumarium*), a widespread weed in many parts of the world, flowers in spring when, as days become longer, a critical night-length is reached (Ray and Alexander 1966). Long-day plants flower when day-length is above a critical level. The strawberry tree (*Arbutus unedo*) flowers in the autumn as the night-length increases. In its Mediterranean home, this means that its flowers are ready for pollination when such long-tongued insects as bees are plentiful. Day-neutral plants flower after a period of vegetative growth, irrespective of the photoperiod.

In the high Arctic, plant growth is telescoped into a brief few months of warmth and light. Positive **heliotropism** (growing towards the Sun) is one way that plants can cope with limited light. It is common in Arctic and alpine flowers. The flowers of the Arctic avens (*Dryas integrifolia*) and the Arctic poppy (*Papaver radicatum*) track the Sun, turning at about 15° of arc per hour (Kevan 1975; see also Corbett *et al*. 1992). Their corollas reflect radiation onto their reproductive parts. The flowers of the alpine snow buttercup (*Ranunculus adoneus*) track the Sun's movement from early morning until mid-afternoon (Stanton and Galen 1989). Buttercup flowers aligned parallel to the Sun's rays reach mean internal temperatures several degrees Celsius above ambient air temperature. Internal flower temperature is significantly reduced as a flower's angle of deviation from the Sun increases beyond 45°.

Arctic and alpine animals and plants also have to cope with limited solar energy. Herbivores gear their behaviour to making the most of the short summer. Belding ground squirrels (*Spermophilus beldingi*), which live at high elevations in the western United States, are active for four or five summer months, and they must eat enough during that time to survive the winter on stored fat (Morhardt and Gates 1974). To do this, their body temperature fluctuates by 3–4°C (to a high of 40°C) so that valuable energy is not wasted in keeping body temperature constant. Should they need to cool down, they go into a burrow or else adopt a posture that lessens exposure to sunlight. A constant breeze cools them during the hottest part of the day.

Some like it hot: temperature

Broadly speaking, average annual temperatures are highest at the equator and lowest at the poles. Temperatures also decrease with increasing elevation. The average annual temperature range is an important ecological factor. It is highest deep in high-latitude continental interiors and lowest over oceans, especially tropical oceans. In north-east Siberia, an annual temperature range of 60°C is not uncommon, whereas the range over equatorial oceans is less than about 3°C. Land lying adjacent to oceans, especially land on the western seaboard of continents, has an annual temperature range around the 11°C mark. These large differences in annual temperature range reflect differences in **continentality** (or **oceanicity**) – the winter temperatures of places near oceans will be less cold.

Many aspects of temperature affect organisms, including daily, monthly, and annual extreme and mean temperatures, and the level of temperature variability. Different aspects of temperature are relevant to different species and commonly vary with the time of year and the stage in an organism's life cycle. Temperature may be limiting at any stage of an organism's life cycle. It may affect survival, reproduction, and the development of seedlings and young animals. It may affect competition with other organisms and susceptibility to predation, parasitism, and disease when the limits of temperature tolerance are approached. Many flowering plants are especially sensitive to low temperatures between germination and seedling growth.

Microbes and temperature

Heat-loving microbes (**thermophiles**) reproduce or grow readily in temperatures over 45°C. **Hyperthermophiles**, such as *Sulfolobus acidocaldarius*, prefer temperatures above 80°C, and some thrive above 100°C. The most resistant hyperthermophile discovered to date is *Pyrolobus fumarii*. This microbe flourishes in the walls of 'smokers' in the deep-sea floor. It multiplies in temperatures up to 113°C.

Below 94°C it finds it too cold and stops growing! Only in small areas that are intensely heated by volcanic activity do high temperatures prevent life. Cold-loving microbes (**psychrophiles**) are common in Antarctic sea ice. These communities include photosynthetic algae and diatoms, and a variety of bacteria. *Polaromonas vacuolata*, a bacterium, grows best at about 4°C, and stops reproducing above 12°C. Lichens can photosynthesize at −30°C, providing that they are not covered with snow. The reddish-coloured snow alga, *Chlamydomonas nivalis*, lives on ice and snow fields in the polar and nival zones, giving the landscape a pink tinge during the summer months.

Animals and temperature

In most animals, temperature is a critical limiting factor. Vital metabolic processes are geared to work optimally within a narrow temperature band. Cold-blooded animals (**poikilotherms**) warm up and cool down with environmental temperature (Figure 2.3a). They can assist the warming process a little by taking advantage of sunny spots or warm rocks. Most warm-blooded animals (**homeotherms**) maintain a constant body temperature amidst varying ambient

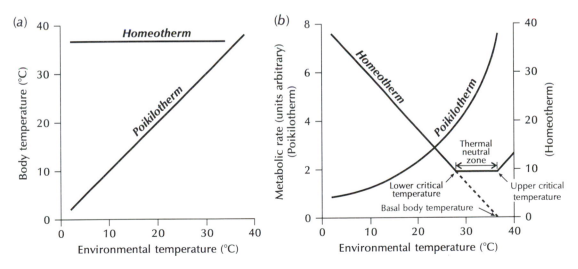

Figure 2.3 Temperature control in poikilotherms and homeotherms.
Source: After Bartholomew (1968)

conditions (Figure 2.3a). They simply regulate the production and dissipation of heat. The terms 'cold-blooded' and 'warm-blooded' are misleading because the body temperature of some 'cold-blooded' animals may rise above that of 'warm-blooded' animals.

Each homeothermic species has a characteristic **thermal neutral zone**, a band of temperature within which little energy is expended in heat regulation (Figure 2.3b) (Bartholomew 1968). Small adjustments are made by fluffing or compressing fur, by making local changes in the blood supply, or by changing position. The bottom end of the thermal neutral zone is bounded by a **lower critical temperature**. Below this temperature threshold, the body's central heating system comes on fully. The colder it gets, the more oxygen is needed to burn fuel for heat. Animals living in cold environments are well insulated – fur and blubber can reduce the lower

critical temperature considerably. An Arctic fox (*Alopex lagopus*) clothed in its winter fur rests comfortably at an ambient temperature of –50°C without increasing its resting rate of metabolism (Irving 1966). Below the lower critical temperature, the peripheral circulation shuts down to conserve energy. An Eskimo dog may have a deep body temperature of 38°C, the carpal area of the forelimb at 14°C, and foot pads at 0°C (Irving 1966) (Figure 2.4). Hollow hair is also useful for keeping warm. It is found in the American pronghorn (*Antilocapra americana*), an even-toed ungulate, and enables it to stay in open and windswept places at temperatures far below 0°C. The polar bear (*Ursus maritimus*) combines hollow hair, a layer of blubber up to 11 cm thick, and black skin to produce a superb insulating machine. Each hair acts like a fibre-optic cable, conducting warming ultra-violet light to the heat-absorbing black skin. This

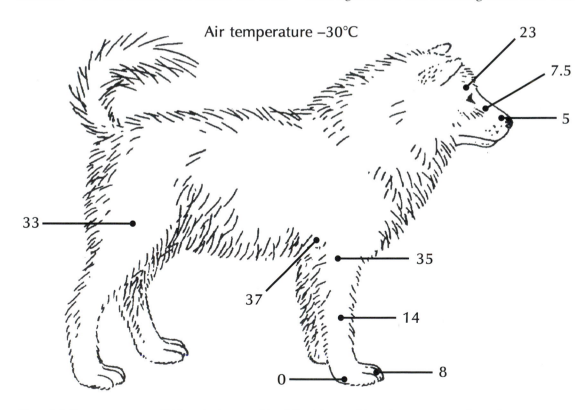

Figure 2.4 Temperatures at an Eskimo dog's extremities (°C) with an ambient air temperature of –30°C.
Source: After Irving (1966)

heating mechanism is so efficient that polar bears are more likely to overheat than to chill down, which partly explains their ponderousness. Many animals also have behavioural patterns designed to minimize heat loss. Some roll into a ball, some seek shelter. Herds of deer or elk seek ridge tops or south-facing slopes.

Above the **upper critical temperature**, animals must lose heat to prevent their overheating. Animals living in hot environments can lose much heat. Evaporation helps heat loss, but has an unwanted side-effect – precious water is lost. Small animals can burrow to avoid high temperatures at the ground surface. In the Arizona desert, United States, most rodents burrow to a depth where hot or cold heat stress is not met with. Large size is an advantage in preventing overheating because the surface area is relatively greater than the body volume. Many desert mammals are adapted to high temperatures. The African rock hyrax (*Heterohyrax brucei*) has an upper critical temperature of 41°C. Camels (*Camelus* spp.), oryx (*Oryx* spp.), common eland (*Taurotragus oryx*), and gazelle (*Gazella* spp.) let their body temperatures fluctuate considerably over a 24-hour period, falling to about 35°C toward dawn and rising to over 40°C during the late afternoon. In an ambient temperature of 45°C sustained for 12 hours under experimental conditions, an oryx's temperature rose above 45°C and stayed there for 8 hours without injuring the animal (Taylor 1969). It has a specialized circulatory system that helps it to survive such excessive overheating.

Many mammal species are adapted to a limited range of environmental temperatures. Even closely related groups display significant differences in their ability to endure temperature extremes. The lethal ambient temperatures for four populations of wood-rats (*Neotoma* spp.) in the western United States showed differences between species, and between populations of the same species living in different states (Figure 2.5).

Plants and temperature

Temperature affects many processes in plants, including photosynthesis, respiration, growth, reproduction, and

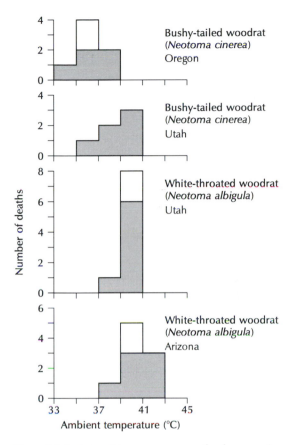

Figure 2.5 Lethal ambient temperatures for four populations of woodrats (*Neotoma*) living in the western United States. The numbers of deaths, shown by the shaded areas, are based on four-hour exposures.
Source: After Brown (1968)

transpiration. Plants vary enormously in their ability to tolerate either heat or cold. There are five broad categories of cold tolerance (Table 2.2). **Chilling-sensitive** plants, which are mostly tropical, are damaged by temperatures lower than 10°C. **Chilling-resistant (frost-sensitive)** plants can survive at temperatures below 10°C, but are damaged when ice forms within their tissues. **Frost-resistant** plants make physiological changes that enable them to survive temperatures as low as about −15°C. **Frost-tolerant** plants survive by withdrawing water from their cells, so preventing ice forming. The withdrawal of water also increases the concentration in sap and

Table 2.2 Temperature tolerance in plants

Temperature sensitivity	Minimum temperature (°C)	Life-form
Chilling-sensitive	>10	Broad-leaved evergreen
Chilling-resistant (frost sensitive)	0 to 10	Broad-leaved evergreen
Frost-resistant	−15 to 10	Broad-leaved evergreen
Frost-tolerant	−40 to −15	Broad-leaved deciduous
Cold-tolerant	<−40	Broad-leaved evergreen and deciduous; boreal needle-leaved

Source: After Woodward (1992)

protoplasm, which acts as a kind of antifreeze, and lowers freezing point. Temperatures down to about −40°C can be tolerated in this way. **Cold-tolerant** plants, which are mostly needle-leaved, can survive almost any subzero temperature.

Cold tolerance varies enormously at different seasons in some species. Willow twigs (*Salix* spp.) collected in winter can survive freezing temperatures below −150°C; the same twigs in summer are killed by a temperature of −5°C (Sakai 1970). Similarly, the red-osier dogwood (*Cornus stolonifera*), a hardy shrub from North America, could survive a laboratory test at −196°C by midwinter when grown in Minnesota (Weiser 1970). Nonetheless, dogwoods native to coastal regions with mild climates are often damaged by early autumn frosts. Plants growing on Mt Kurodake, Hokkaido Province, Japan, are killed by temperatures of −5°C to −7°C during the growing season. In winter, most of the same plants survive freezing to −30°C, and the willow ezo-mame-yanagi (*Salix pauciflora*), mosses, and lichens will withstand a temperature of −70°C (Sakai and Otsuka 1970). **Acclimatization** or cold hardening accounts for these differences. The coastal dogwoods did not acclimatize quickly enough. Timing is important in cold resistance, but absolute resistance can be altered. Many plants use the signal of short days in autumn as an early warning system. The short days trigger metabolic changes that stop the plant growing and produce resistance to cold. Many plant species, especially deciduous plants in temperate regions, need chilling during winter if they are to grow well the following summer. Chilling requirements are specific to species. They are often necessary for buds to break out of dormancy, a process called **vernalization**.

Many plants require a certain amount of 'warmth' during the year. The total 'warmth' depends on the growing season length and the growing season temperature. These two factors are combined as 'day-degree totals' (Woodward 1992). **Day-degree totals** are the product of the growing season length (the number of days for which the mean temperature is above standard temperature, such as freezing point or 5°C), and the mean temperature for that period. The Iceland purslane (*Koenigia islandica*), a tundra annual, needs only 700 day-degrees to develop from a germinating seed to a mature plant producing seeds of its own. The small-leaved lime (*Tilia cordata*), a deciduous tree, needs 2,000 day-degrees to complete its reproductive development (Pigott 1981). Trees in tropical forests may need up to 10,000 day-degrees to complete their reproductive development.

Excessive heat is as detrimental to plants as excessive cold. Plants have evolved resistance to **heat stress**, though the changes are not so marked as resistance to cold stress (see Gates 1980, 1993: 69–72). Different parts of plants acquire differing degrees of heat resistance, but the pattern varies between species. In some species the uppermost canopy leaves are often the most heat resistant; in other species, it is the middle canopy leaves, or the leaves at the base of the plant. Temperatures of about 44°C are usually injurious to evergreens and shrubs from cold-winter regions. Temperate-zone trees are damaged at

50–55°C, tropical trees at 45–55°C. Damage incurred below about 50°C can normally be repaired by the plant; damage incurred above that temperature is most often irreversible. Exposure time to excessive heat is a critical factor in plant survival, while exposure time to freezing temperatures is not.

Quenching thirst: moisture

Protoplasm, the living matter of animal and plant cells, is about 90 per cent water – without adequate moisture there can be no life. Water affects land animals and plants in many ways. Air humidity is important in controlling loss of water through the skin, lungs, and leaves. All animals need some form of water in their food or as drink to run their excretory systems. Vascular plants have an internal plumbing system – parallel tubes of dead tissue called xylem – that transfers water from root tips to leaves. The entire system is full of water under stress (capillary pressure). If the **water stress** should fall too low, disaster may ensue – germination may fail, seedlings may not become established, and, should the fall occur during flowering, seed yields may be severely cut. An overlong drop in water stress kills plants, as anybody who has tried to grow bedding plants during a drought and hose-pipe ban will know.

Bioclimates

On land, precipitation supplies water to ecosystems. Plants cannot use all the precipitation that falls. A substantial portion of the precipitation evaporates and returns to the atmosphere. For this reason, **available moisture** (roughly the precipitation less the evaporation) is a better guide than precipitation to the usable water in a terrestrial ecosystem. This point is readily understood with an example. A mean annual rainfall of 400 mm might support a forest in Canada, where evaporation is low, but might support a dry savannah in Tanzania, where evaporation is high.

Available moisture largely determines soil water levels, which in turn greatly influence plant growth. For a plant to use energy for growth, water must be available. Without water, the energy will merely heat and stress the plant. Similarly, for a plant to use water for growth, energy must be obtainable. Without an energy source, the water will run into the soil or run off unused. For these reasons, temperature (as a measure of energy) and moisture are **master limiting factors** that act in tandem. In tropical areas, temperatures are always high enough for plant growth and precipitation is the limiting factor. In cold environments, water is usually available for plant growth for most of the year – low temperatures are the limiting factor. This is true, too, of limiting factors on mountains where lower altitudinal limits are set by heat or water or both, and upper altitudinal limits are set by a lack of heat.

So important are precipitation and temperate that several researchers use them to characterize bioclimates. **Bioclimates** are the aspects of climate that seem most significant to living things. The most widely used bioclimatic classification is the 'climate diagram' devised by Heinrich Walter. This is the system of summarizing ecophysiological conditions that makes David Bellamy 'feel like a plant' (Bellamy 1976: 141)! **Climate diagrams** portray climate as a whole, including the seasonal round of precipitation and temperature (Figure 2.6). They show at a glance the annual pattern of rainfall; the wet and dry seasons characteristic of an area, as well as their intensity, since the evaporation rate is directly related to temperature; the occurrence or non-occurrence of a cold season, and the months in which early and late frost have been recorded. Additionally, they provide information on such factors as mean annual temperature, mean annual precipitation, the mean daily minimum temperature during the coldest month, the absolute minimum recorded temperature, the altitude of the station, and the number of years of record.

Wet environments

Plants are very sensitive to water levels. Hydrophytes are water plants and root in standing water. Helophytes are marsh plants. Mesophytes are plants that live in normally moist but not wet conditions. Xerophytes are plants that live in dry conditions.

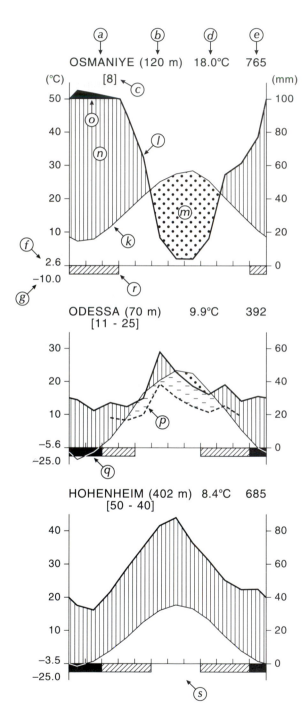

Wetlands support **hydrophytes** and **helophytes**. The common water crowfoot (*Ranunculus aquatalis*) and the bog pondweed (*Potamogeton polygonifolius*) are hydrophytes; the greater bird's-foot trefoil (*Lotus uliginosus*) is a helophyte. These plants manage to survive by developing a system of air spaces in their roots, stems, or leaves. The air spaces provide buoyancy and improve internal ventilation. **Mesophytes** vary greatly in their ability to tolerate flooding. In the southern United States, bottomland hardwood forests occupy swamps and river floodplains. They contain a set of tree species that can survive in a flooded habitat. The water tupelo (*Nyssa aquatica*), which is found in bottomland forest in the southeastern United States, is well adapted to such wet conditions.

Flooding or high soil-moisture levels may cause seasonal changes in mammal distributions. The mole

Figure 2.6 Examples and explanation of climate diagram. The letters denote the following. *a* Weather station. *b* Altitude (m above mean sea-level). *c* Number of years of observation. Where there are two figures, the first refers to temperature measurements and the second the precipitation measurements. *d* Mean annual temperature (°C). *e* Mean annual precipitation (mm). *f* Mean daily maximum temperature during the coldest month (°C). *g* Absolute minimum (lowest recorded) temperature (°C). *k* Curve of mean monthly temperature (1 scale graduation = 10°C). *l* Curve of mean monthly precipitation (1 scale graduation = 20 mm). *m* Relatively arid period or dry season (dotted). *n* Relatively humid period or wet season (vertical bars). *o* Mean monthly rainfall above 100 mm with the scale reduced by a factor of 0.1 (the black area in Osmaniye). *p* Curve for precipitation on a smaller scale (1 scale graduation = 30 mm). Above it, horizontal broken lines indicate the relatively dry period or dry season (shown for Odessa). *q* Months with a mean daily minimum temperature below 0°C (black boxes below zero line). *r* Months with an absolute minimum temperature below 0°C (diagonal lines). *s* Average duration of period with daily mean temperature above 0°C (shown as the number of days in standard type); alternatively, the average duration of the frost-free period (shown as the number of days in italic type, as for Honenheim). Mean daily maximum temperature during the warmest month (*h*), absolute maximum (highest recorded) temperature (*i*), and mean daily temperature fluctuation (*j*) are given only for tropical stations with a diurnal climate, and are not shown in the examples. *Source*: After Walter and Lieth (1960–7)

rats (*Cryptomys hottentotus*) in Zimbabwe focus their activity around the bases of termite mounds during the rainy seasons as they rise a metre or so above the surrounding grassland and produce relatively dry islands in a sea of waterlogged terrain (Genelly 1965).

Many organisms are fully adapted to watery environments and always have been – the colonization of dry land is a geological recent event. Some vertebrates have returned to an aquatic existence. Those returning to the water include crocodiles, turtles, extinct plesiosaurs and ichthyosaurs, seals, and whales. Some, including the otter, have adopted a semi-aquatic way of life.

Dry environments

Plants are very sensitive to drought and aridity poses a problem of survival. Nonetheless, species of algae grow in the exceedingly dry Gobi desert. Higher plants survive in arid conditions by xerophytic adaptations – drylands support **xerophytes**. One means of survival is simply to avoid the drought as seeds (pluviotherophytes) or as below-ground storage organs (bulbs, tubers, or rhizomes). Other xerophytic adaptations enable plants to retain enough water to keep their protoplasts wet, so avoiding desiccation. Water is retained by several mechanisms. A very effective mechanism is water storage. Succulents are plants that store water in leaves, stems, or roots. The saguaro

---

Box 2.2

REPTILES IN DESERTS

Lizards are abundant in deserts in the daytime whereas mammals are not. The reason for this is not a reduction of evaporative water loss through the skin. Cutaneous water loss is about the same in mammals and reptiles. However, reptiles from dry habitats do have a lower skin permeability. Therefore, they lose less water through the skin than do reptiles from moist environments. The tropical, tree-living green iguana (*Iguana iguana*) loses about 4.8 mg/cm^2/day through the skin and 3.4 mg/cm^2/day through respiration; the desert-dwelling chuckwalla (*Sauromalus obesus*), which is active in daytime, loses about 1.3 mg/cm^2/day through the skin and 1.1 mg/cm^2/day through respiration. The difference between mammals and reptiles lies in three reptilian characteristics that predispose them for water conservation in arid environments – low metabolic rates; nitrogenous waste excretion as uric acid and its salts; and, in many taxa, the presence of nasal salt glands (an alternative pathway of salt excretion to the kidneys). Low metabolic rates mean less frequent breathing, which means that less water is lost from the lungs. Uric acid is only slightly soluble in water and precipitates in urine to form a whitish, semi-solid mass. Water is left behind and may be reabsorbed into the blood and used to produce more urine. This recycling of water is useful to reptiles because their kidneys are unable to make urine with a higher osmotic pressure than that of their blood plasma. The potassium and sodium ions that do not precipitate are reabsorbed in the bladder. This costs energy, so why do it? The answer lies in a third water-conserving mechanism in reptiles – extra-renal salt excretion. In at least three reptilian groups – lizards, snakes, and turtles – there are some species that have salt glands. These glands make possible the selective transport of ions out of the body. They are most common in lizards, where they have been found in five families. In these families, a lateral nasal gland excretes salt. The secretions of the glands are emptied into the nasal passages and are expelled by sneezing and shaking of the head. Salt glands are very efficient at excreting. The total osmotic pressure of salt glands may be more than six times that of urine produced by the kidney. This explains the paradox of salt uptake from urine in bladders. As ions are actively reabsorbed, water follows passively and the animal recovers both water and ions from the urine. The ions can then be excreted through the salt gland at much higher concentrations. There is thus a proportional reduction in the amount of water needed to dispose of the salt.

Box 2.3

MAMMALS IN DESERTS

Rodents are the dominant small mammals in arid environments. Population densities may be higher in deserts than in temperate regions. As with reptiles, several features of rodent biology predispose them to desert living. Many rodents are nocturnal and live in burrows. Although night-time activity might be thought to avoid the heat stress of the day, heat stress can also occur at night when deserts can be cold. In rodents (and birds and lizards), there is a countercurrent water-recycler in the nasal passages that is important in the energy balance of these organisms. While breathing in, air passes over the large surface area of the nasal passages and is warmed and moistened. The surface of the nasal passages cool by evaporation in the process. When warm, saturated air from the lungs is breathed out, it condenses in the cool nasal passages. Overall, this process saves water and energy. Indeed, the energetic savings are so great that it is unlikely that a homeotherm could survive without this system – ethmoturbinal bones in the fossil *Cynognathus* is persuasive evidence that mammal-like reptiles (therapsids) were homeotherms. However, this countercurrent exchange of heat and moisture is not an adaptation to desert life; it is an inevitable outcome of the anatomy and physiology of the nasal passages.

Water is also lost in faeces and urine. Rodents generally can produce fairly dry faeces and concentrated urine. Kangaroo rats, sand rats, and jerboas can produce urine concentrations double to quadruple the urine concentration in humans. The spinifex hopping mouse or dargawarra (*Notomys alexis*), which lives throughout most of the central and western Australian arid zone, is a hot contender for 'world champion urine concentrator'; its urine concentration is six times higher than in humans (Plate 2.1). Low evaporative water loss through nocturnal habits, concentrated urine, and fairly dry faeces mean that many desert rodents are independent of water – they can get all the water they need

from air-dried seeds. Part of this water comes from the seeds and part comes from the oxidation of food (metabolic water). Interestingly, the water content of some desert plants varies with the relative humidity of the air. In parts of semi-arid Africa, *Diasperma* leaves have a water content of 1 per cent by day, but at night, when the relative humidity increases, their water content rises to 30 per cent. The leaves are forage for the oryx (*Oryx gazella*). By feeding at night, the oryx takes in 5 litres of water, on which it survives through several water-conserving mechanisms (Taylor 1969). The banner-tailed kangaroo rat (*Dipodomys spectabilis*) from Arizona, which is not as its name implies a pouched mammal, stores several kilograms of plant material in its burrow where it is exposed to a relatively high relative humidity of the burrow atmosphere (Schmidt-Nielsen and Schmidt-Nielsen 1953). Hoarding food in a burrow provides not only a hedge against food shortages, but also an enhanced source of water. In Texas, burrows of the plains pocket gopher (*Geomys bursarius*) have relative humidities of 86–95 per cent (Kennerly 1964). In sealed burrows, the humidities can be up to 95 per cent while the soil of the burrow floor contains only 1 per cent water. Nonetheless, although high temperatures and low humidities are avoided in burrows, other stresses do occur – carbon dioxide concentrations may be between ten and sixty times greater than in the normal air.

Plate 2.1 Spinifex hopping mouse (*Notomys alexis*) – world champion urine concentrator
Photograph by Pat Morris

cactus (*Carnegiea gigantea*) and the barrel cactus (*Ferocactus wislizeni*) are examples. The barrel cactus stores so much water that it has been used as an emergency water supply by Indians and other desert inhabitants. Many succulents have a crassulacean acid metabolism (CAM) pathway for carbon dioxide assimilation. This kind of photosynthesis involves carbon dioxide being taken in at night with stomata wide open, and then being used during the day with stomata closed to protect against transpiration losses. Other xerophytes (sometimes regarded as true xerophytes) do not store water, but have evolved very effective ways of reducing water loss – leaves with thick cuticles, sunken and smaller stomata, leaves shed during dry periods, improved water uptake through wide spreading or deeply penetrating root systems, and improved water conduction. The big sagebrush (*Artemisia tridentata*) is an example of a true xerophyte. The creosote bush (*Larrea divaricata*) simply endures a drought by ceasing to grow when water is not available.

Plants vary enormously in their ability to withstand water shortages. Young plants suffer the worst. Drought resistance is measured by the specific survival time (the time between the point when the roots can no longer take up water and the onset of dessicative injury). The specific survival time is 1,000 hours for prickly-pear cactus, 50 hours for Scots pine, 16 hours for oak, 2 hours for beech, and 1 hour for forget-me-not (Larcher 1975: 172).

Desert-dwelling animals face a problem of water shortage, as well as high daytime temperatures. They have overcome these problems in several remarkable ways (Boxes 2.2 and 2.3).

Snow

This is a significant ecological factor in polar and some boreal environments. A snow cover that persists through the winter is a severe hardship to large mammals. To most North American artiodactyls, including deer, elk, bighorn sheep, and moose, even moderate snow imposes a burden by covering some food and making it difficult to find. In mountainous areas, deer and elk avoid deep snow by abandoning summer ranges and moving to lower elevations. South-facing slopes and windswept ridges, where snow is shallower or on occasions absent, are preferred at these times. In area of relatively level terrain, deer and moose respond to deep snow by restricting their activities to a small area called 'yards' where they establish trails through the snow. Prolonged winters and deep snow take a severe toll on deer and elk populations.

For small mammals, snow is a blessing. It forms an insulating blanket, a sort of crystalline duvet, under which is a ground-surface micro-environment where activity, including breeding in some species, continues throughout the winter. To these small mammals, which include shrews (*Sorex*), pocket gophers (*Thomomys*), voles (*Microtus, Clethrionomys, Phenacomys*), and lemmings (*Lemmus, Dicrostonys*), the most stressful times are autumn, when intense cold descends but snowfall has not yet moderated temperatures at the ground surface, and in the spring, when rapid melting of a deep snowpack often results in local flooding. Another advantage of a deep snowpack is that green vegetation may be available beneath it, and several species make tunnels to gain access to food.

Even in summer, snow may be important to some mammals. Alpine or northern snowfields commonly last through much of the summer on north-facing slopes and provide a cool microclimate unfavoured by insects. Caribou (*Rangifer tarandus*) and bighorn sheep (*Ovis* spp.) sometimes congregate at these places to seek relief from pesky warble flies.

The big picture: climatic zones

Terrestrial ecozones

On land, characteristic animal and plant communities are associated with nine basic climatic types, variously called zonobiomes (Walter 1985), ecozones (Schultz 1995), and ecoregions (R. G. Bailey 1995, 1996) (Figure 2.7):

1 **Polar and subpolar zone.** This zone includes the Arctic and Antarctic regions. It is associated with

tundra vegetation. The Arctic tundra regions have low rainfall evenly distributed throughout the year. Summers are short, wet, and cool. Winters are long and cold. Antarctica is an icy desert, although summer warming around the fringes is causing it to bloom.

2 **Boreal zone.** This is the cold-temperate belt that supports coniferous forest (taiga). It usually has cool, wet summers and very cold winters lasting at least six months. It is found only in the Northern Hemisphere where it forms a broad swath around the pole – it is a circumpolar zone.

3 **Humid mid-latitude zone.** This zone is the temperate or nemoral zone. In continental interiors it has a short, cold winter and a warm, or even hot, summer. Oceanic regions, such as the British Isles, have warmer winters and cooler, wetter summers. This zone supports broad-leaved deciduous forests.

4 **Arid mid-latitude zone.** This is the cold-temperate (continental) belt. The difference between summer and winter temperatures is marked and rainfall is low. Regions with a dry summer but only a slight drought support temperate grasslands. Regions with a clearly defined drought period and a short wet season support cold desert and semi-desert vegetation.

5 **Tropical and subtropical arid zone.** This is a hot desert climate that supports thorn and scrub savannahs and hot deserts and semi-deserts.

6 **Mediterranean subtropical zone.** This is a belt lying between roughly 35° and 45° latitude in both hemispheres with winter rains and summer drought. It supports sclerophyllous (thick-leaved), woody vegetation adapted to drought and sensitive to prolonged frost.

7 **Seasonal tropical zone.** This zone extends from roughly 25° to 30° North and South. There is a marked seasonal temperature difference. Heavy rain in the warmer summer period alternates with extreme drought in the cooler winter period. The annual rainfall and the drought period increase with distance from the equator. The vegetation is tropical grassland or savannah.

8 **Humid subtropical zone.** This zone has almost no cold winter season, and short wet summers. It is the warm temperate climate in Walter's zonobiome classification. Vegetation is subtropical broad-leaved evergreen forest.

9 **Humid tropical zone.** This torrid zone has rain all year and supports evergreen tropical rain forest. The climate is said to be diurnal because it varies more by day and night than it does through the seasons.

Marine ecozones

The **marine biosphere** also consists of 'climatic' zones, which are also called ecozones. The main surface-water marine ecozones are the polar zone, the temperate zone, and the tropical zone (R. G. Bailey 1996: 161):

1 **Polar zone.** Ice covers the polar seas in winter. Polar seas are greenish, cold, and have a low salinity.

2 **Temperate zone.** Temperate seas are very mixed in character. They include regions of high salinity in the subtropics.

3 **Tropical zone.** Tropical seas are generally blue, warm, and have a high salinity.

Biomes

Each ecozone supports several characteristic communities of animals and plants known as **biomes** (Clements and Shelford 1939). The deciduous forest biome in temperate western Europe is an example. It consists largely of woodland with areas of heath and moorland. A plant community at the biome scale – all the plants associated with the deciduous woodland biome, for example – is a **plant formation**. An equivalent animal community has no special name; it is simply an animal community. Smaller communities within biomes are usually based on plant distribution. They are called **plant associations**. In England, associations within the deciduous forest biome include beech forest, lowland oak forest, and ash forest. Between biomes are transitional belts where the climate changes from one type to the next. These are called **ecotones**.

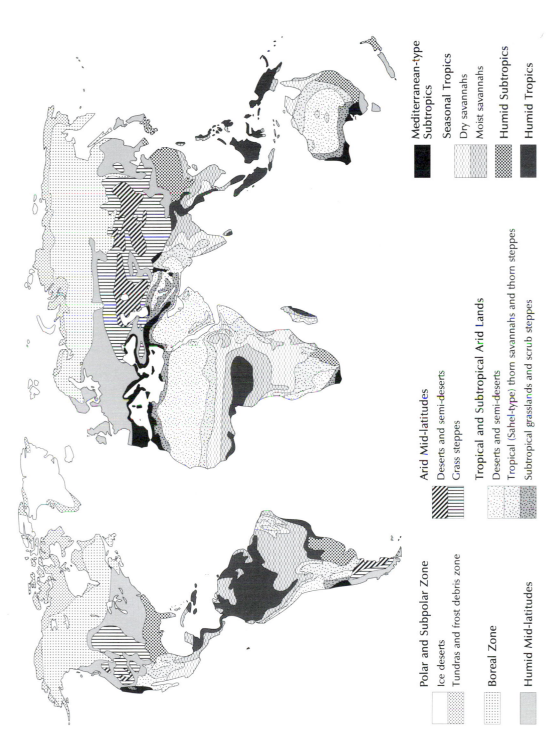

Polar and Subpolar Zone

Ice deserts

Tundras and frost debris zone

Boreal Zone

Humid Mid-latitudes

Mediterranean-type Subtropics

Seasonal Tropics

Dry savannahs

Moist savannahs

Humid Subtropics

Humid Tropics

Arid Mid-latitudes

Deserts and semi-deserts

Grass steppes

Tropical and Subtropical Arid Lands

Deserts and semi-deserts

Tropical (Sahel-type) thorn savannahs and thorn steppes

Subtropical grasslands and scrub steppes

Figure 2.7 Ecozones of the world.
Source: After Schultz (1995)

Zonobiomes

All the biomes around the world found in a particular ecozone constitute a **zonobiome**. A plant community at the same large scale is a **formation-type** or **zonal plant formation**. The broad-leaved temperate forests of western Europe, North America, eastern Asia, southern Chile, south-east Australia and Tasmania, and most of New Zealand comprise the humid temperate zonobiome. Between the zonobiomes are transitional belts where the climate changes from one type to the next. These are called **zonoecotones**.

Freshwater communities (lakes, rivers, marshes, and swamps) are part of continental zonobiomes. They may be subdivided in various ways. Lakes, for instance, may be well mixed (**polymictic** or **oligomictic**) or permanently layered (**meromictic**). They may be wanting in nutrients and biota (**oligotrophic**) or rich in nutrients and algae (**eutrophic**). A **thermocline** (where the temperature profile changes most rapidly) separates a surface-water layer mixed by wind (**epilimnon**) from a more sluggish, deep-water layer (**hypolimnon**). And, as depositional environments, lakes are divided into a **littoral** (near-shore) zone, and a **profundal** (basinal) zone.

Marine ecozones, and the deep-water regions, consist of biomes (equivalent to terrestrial zonobiomes). The chief **marine biomes** are the intertidal (estuarine, littoral marine, algal bed, coral reef) biome, the open sea (pelagic) biome, the upwelling zone biome, the benthic biome, and the hydrothermal vent biome.

Orobiomes

Mountain areas possess their own biomes called **orobiomes**. The basic environmental zones seen on ascending a mountain are submontane (colline, lowland), montane, subalpine, alpine, and nival. On south-facing slopes in the Swiss Alps around Cortina, the **submontane belt** lies below about 1,000 m. It consists of oak forests and fields. The **montane belt** ranges from about 1,000 m to 3,000 m. The bulk of it is Norway spruce (*Picea abies*) forest, with scat-

tered beech (*Fagus sylvatica*) trees at lower elevations. Mountain pines (*Pinus montana*) with scattered Swiss stone pines (*Pinus cembra*) grow near the tree line. The **subalpine belt** lies between about 3,000 m and 3,500 m. It contains diminutive forests of tiny willow (*Salix* spp.) trees, only a few centimetres tall when mature, within an alpine grassland. The **alpine belt**, which extends up to about 4,000 m, is a meadow of patchy grass and a profusion of alpine flowers – poppies, gentians, saxifrages, and many more. The mountain tops above about 4,000 m lie within the **nival zone** and are covered with permanent snow and ice.

SOILS, SLOPES, AND DISTURBING AGENCIES: PEDOGEOMORPHIC FACTORS AND DISTURBANCE

Soil and substrate

Soil and substrate influence animals and plants, both at the level of individual species and at the level of communities.

Microbes and substrate

Acid-loving microbes (**acidophiles**) prosper in environments with a pH below 5. *Sulfolobus acidocaldarius*, as well as liking it hot, also likes it acid. Alkali-loving microbes (**alkaliphiles**) prefer an environment with a pH above 9. *Natronobacterium gregoryi* lives in soda lakes.

Salt-loving microbes live in intensely saline environments. They survive by producing large amounts of internal solutes that prevents rapid dehydration in a salty medium. An example is *Halobacterium salinarium*.

Plants and substrate

Plants seem capable of adapting to the harshest of substrates. **Saxicolous** vegetation grows on cliffs, rocks, and screes, some species preferring rock crevices (**chasmophytes**), others favouring small ledges where detritus and humus have collected (**chomophytes**). In the Peak District of Derbyshire, England, maiden-hair spleenwort (*Asplenium trichomanes*) is a common chasmophyte and the wallflower (*Cheiranthus cheiri*) is a common and colourful chomophyte (Anderson and Shimwell 1981: 142). Perhaps the most extreme adaptation to a harsh environment is seen in the mesquite trees (*Prosopis tamarugo* and *Prosopis alba*) that grow in the Pampa del Tamagural, a closed basin, or salar, in the rainless region of the Atacama Desert, Chile. These plants manage to survive on concrete-like carbonate surfaces (Ehleringer *et al.* 1992). Their leaves abscise (are shed) and accumulate to depths of 45 cm. Because there is virtually no surface water, the leaves do not decompose and nitrogen is not incor-porated back into the soil for recycling by plants. The thick, crystalline pan of carbonate salts prevents roots from growing into the litter. To survive, the trees have roots that fix nitrogen in moist subsurface layers, and extract moisture and nutrients from groundwater at depths of 6–8 m or more through a tap root and a mesh of fine roots lying between 50 and 200 cm below the salt crust. A unique feature of this ecosystem is the lack of nitrogen cycling.

Calcicoles (or **calciphiles**) are plants that favour such calcium-rich rocks as chalk and limestone. Calcicolous species often grow only on soil formed in chalk or limestone. An example from England, Wales, and Scotland is the meadow oat-grass (*Helicto-trichon pratense*), the distribution of which picks out the areas of chalk and limestone and the calcium-rich schists of the Scottish Highlands. Other examples are traveller's joy (*Clematis vitalba*), the spindle tree (*Euonymus europaeus*), and the common rock-rose (*Helianthemum nummularium*). **Calcifuges** (or **calci-phobes**) avoid calcium-rich soils, preferring instead acidic soils developed on rocks deficient in calcium. An example is the wavy hair-grass (*Deschampsia flexuosa*). However, many calcifuges are seldom entirely restricted to exposures of acidic rocks. In the limestone Pennine dales, the wavy hair-grass can be found growing alongside meadow oat-grass. **Neutro-philes** are acidity 'middle-of-the-roaders'. They tend to grow in the range pH 5–7. In the Pennine dales, strongly growing, highly competitive grasses

that make heavy demands on water and nutrient stores are the most common neutrophiles.

Animals and substrate

Some animals are affected by soil and substrate. For instance, the type and texture of soil or substrate is critical to two kinds of mammals: those that seek diurnal refuge in burrows, and those that have modes of locomotion suited to relatively rough surfaces. **Burrowing species**, which tend to be small, may be confined to a particular kind of soil. For instance, many desert rodents display marked preferences for certain substrates. In most deserts, no single species of rodent is found on all substrates; and some species occupy only one substrate. Four species of pocket mice (*Perognathus*) live in Nevada, United States (Hall 1946). Their preferences for soil types are largely complementary: one lives on fairly firm soils of slightly sloping valley margins; the second is restricted to slopes where stones and cobbles are scattered and partly embedded in the ground; the third is associated with the fine, silty soil of the bottomland; and the fourth, a substrate generalist, can survive on a variety of soil types.

Saxicolous species grow in, or live among, rocks. Some woodrats (*Neotoma*) build their homes exclu-sively in cliffs or steep rocky outcrops. The dwarf shrew (*Sorex nanus*) seems confined to rocky areas in alpine and subalpine environments. Even some salt-atorial species are adapted to life on rocks. The Australian rock wallabies (*Petrogale* and *Petrodorcas*) leap adroitly among rocks (Colour plate 1). They are aided in this by traction-increasing granular patterns on the soles of their hind feet. Rocky Mountain pikas (*Ochotona princeps*) in the southern Rocky Mountains, United States, normally live on talus or extensive piles of gravel (Hafner 1994). Those living near Bodie, a ghost town in the Sierra Nevada, utilize tail-ings of abandoned gold mines (A. T. Smith 1974, 1980). The yellow-bellied marmot (*Marmota flaviven-tris*) is another saxicolous species, and commonly occurs with the Rocky Mountain pika. The entire life style of African rock hyraxes (*Heterohyrax*, *Procavia*) is built around their occupancy of rock piles and cliffs.

Most of their food consists of plants growing among, or very close by, rocks. Their social system is bonded by the scent of urine and faeces on the rocks. The rocks provide useful vantage points to keep an eye out for predators, hiding places, and an economical means of conserving energy.

Pedobiomes

Within zonobiomes, there are areas of **intrazonal** and **azonal soils** that, in some cases, support a distinctive vegetation. These non-zonal vegetation communities are **pedobiomes** (Walter and Breckle 1985). Several different pedobiomes are distinguished on the basis of soil type: lithobiomes on stony soil, psammobiomes on sandy soil, halobiomes on salty soil, helobiomes in marshes, hydrobiomes on waterlogged soil, peinobiomes on nutrient-poor soils, and amphibiomes on soils that are flooded only part of the time (e.g. river banks and mangroves). Pedobiomes commonly form a mosaic of small areas and are found in all zonobiomes. There are instances where pedobiomes are extensive: the Sudd marshes on the White Nile, which cover 150,000 km^2; fluvio-glacial sandy plains; and the nutrient poor soils of the Campos Cerrados in Brazil.

A striking example of a **lithobiome** is found on serpentine. The rock serpentine and its relatives, the serpentinites, are deficient in aluminium. This leads to slow rates of clay formation, which explains the characteristic features of soils formed on serpentinites: they are high erodible, shallow, and stock few nutrients. These peculiar features have an eye-catching influence on vegetation (Brooks 1987; Baker *et al*. 1992). Outcrops of serpentine support small islands of brush and bare ground in a sea of forest and grassland. These islands are populated by native floras with many endemic species (Whittaker 1954).

The ups and downs of living: topography

Many topographic factors influence ecosystems. The most influential factors are altitude, aspect, inclination, and insularity.

Altitude

Altitude exerts a strong influence on animals and plants. The plant communities girdling the Earth as broad zonal belts are paralleled in the plant communities encircling mountains – orobiomes (p. 30). Individual animals and plant species often occur within a particular elevational band, largely owing to climatic limits of tolerance. However, altitudinal ranges are influenced by a host of environmental factors, and not just climatic ones. For instance, tree-lines are not always purely the result of climatic constraints on tree growth, but involve pedological and biotic interactions as well.

Aspect

Aspect (compass direction) strongly affects the climate just above the ground and within the upper soil layers. For this reason, virtually all landscapes display significant differences in soil and vegetation on adjacent north-facing (distal) and south-facing (proximal) slopes, and on the windward and leeward sides of hills and mountains. In the Northern Hemisphere, south-facing slopes tend to be warmer, and so more prone to drought, than north-facing slopes. The difference may be greater than imagined. In a Derbyshire dale, England, the summer mean temperature was 3°C higher on a south-facing slope than on a north-facing slope (Rorison *et al*. 1986), a difference equivalent to a latitudinal shift of hundreds of kilometres! These differences affect plant growth. The Lägern is an east–west trending mountain range near Baden in the Swiss Jura. Cool north-facing slopes are separated from warm south-facing slopes by a half-metre-wide mountain ridge. The south-facing slope supports warmth-loving vegetation – downy oak (*Quercus pubescens*) woodland with pale-flowered orchid (*Orchis pallens*), wonder violet (*Viola mirabilis*), and bastard balm (*Melittis melissophyllum*). The north-facing slope supports a beech (*Fagus sylvatica*) forest with such subalpine plants as alpine penny-cress (*Thlaspi alpestre*) and green spleenwort (*Asplenium viride*) (Stoutjesdijk and Barkman 1992: 78). The change over the narrow mountain ridge is equivalent botanically to about 1,000 km of latitude.

The geographical distributions of some animal species are influenced by aspect. On the south-facing slopes of the Kullaberg Peninsula, south-west Sweden, several arthropod species are found far to the north of their main range. The species include the moth *Idaea dilutaria*, the beetle *Danacea pallipes*, and the spider *Theridion conigerum* (Ryrholm 1988). Also in Europe, Alpine marmots (*Marmota marmota*) prefer to burrow on south-facing slopes, although summer temperatures on these slopes are high enough to limit their feeding time above ground and north-facing slopes would seem a wiser choice. They may dig burrows in south-facing slopes because conditions for hibernating are better (Türk and Arnold 1988).

Differences between the climate of windward and leeward slopes may also be consequential. Large mountain ranges cast a **rain shadow** in their lee that is sufficient to alter the vegetation. In north-west North America, the wet coastal mountain ranges support luxuriant Sitka spruce (*Picea sitchensis*), Douglas fir (*Pseudotsuga menziesii*), and hemlock (*Tsuga* spp.) forests. Further inland, in the rain shadow of the Cascade Mountains, the valley of the Columbia River is a desert supporting sagebrush and tufts of drought-resistant bunch grasses.

Inclination

Slope is the inclination of the ground from the horizontal. It affects vegetation through its influence on soil moisture and on substrate stability. Slope commonly varies in a regular sequence over undulating terrain – from summit, down hillside, to valley bottom – to form a geomorphic catena or toposequence. The geomorphic toposequence is mirrored in **soil** and **vegetation toposequences**. Figure 2.8 shows a soil and vegetation toposequence on Hodnet Heath, Shropshire, England. Humus–iron podzols form on dry summit slopes and support either dry heather (*Calluna vulgaris*) heath or silver birch (*Betula pendula*) wood with bracken (*Pteridium aquilinium*) and wavy hair-grass (*Deschampsia flexuosa*). Gley podzols form on the wetter mid-slopes and support damp cross-leaved heath (*Erica tetralix*) and heather, with purple moor-grass (*Molinia caerulea*) and common

cotton-grass (*Eriophorum angustifolium*) becoming more important towards the bottom end. Peaty gley soils and peat soils form in the wet lower slopes and support purple moor-grass, common cotton-grass, and bog mosses (*Sphagnum* spp.), with tussocks of hare's-tail cotton-grass (*Eriophorum vaginatum*) in bogs and rushes (*Juncus* spp.) in acid pools of water.

Topographic factors normally work in tandem with other factors. This is seen at small and large scales. On the small scale, aspect and topographical position are good predictors of funnel ant (*Aphaenogaster longiceps*) nests in a 20-ha eucalypt-forest site at Wog Wog, south-eastern New South Wales (Nicholls and McKenzie 1994). On a large scale, elevation and altitudinal range partly explain the distribution of sixty-one willow (*Salix*) species in Europe, though latitude and July mean temperature have somewhat more explanatory power (Myklestad and Birks 1993).

Insularity

Insularity is the relative isolation of an ecosystem. It has a significant influence on individuals and communities. True islands have abnormally high rates of extinction, house many endemic and relict species, favour the evolution of dwarf, giant, and flightless forms, and are often inhabited by good dispersers. Habitat islands – woods within fields, remnant patches of heathland, mountain tops, parks in cities, and so on – are becoming increasingly common as the land cover becomes more and more fragmented. The survival of populations in fragmenting habitat islands is currently being assessed using metapopulation models (Chapter 4).

Disturbance

Disturbance is any event that disrupts an ecosystem. Disturbing agencies may be physical or biological. Grazing, for example, disturbs vegetation communities, although it originates within the same ecosystem as the plants that it disturbs. It is useful to think of disturbance as a continuum between purely external agencies and purely internal agencies. Classical

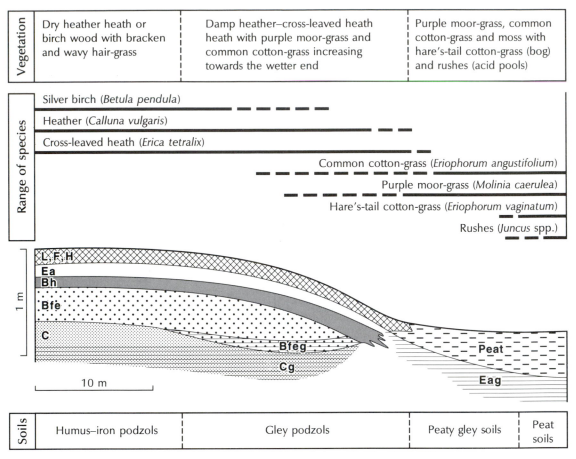

Figure 2.8 Soil and vegetation toposequences on Hodnet Heath, Shropshire, England. The letters L, F, H, etc. are soil horizons.
Source: After Burnham and Mackney (1968)

'disturbance' lies at the external agency end of the continuum. It is the outcome of physical processes acting at a particular time, creating sharp patch boundaries, and increasing resource availability. Internal community cycles, such as the **hummock– hollow cycle** in peatland, lie at the internal end of the disturbance continuum and are driven by internal community dynamics.

Ecosystems may be physically disturbed by strong winds (which uproot trees), fire, flood, landslides, and lightning; and biotically disturbed by pests, pathogens, and the activities of animals and plants, including the human species. The effects of these disturbing agencies on ecosystems can be dramatic. **Pathogens**, for example, are forceful disrupters of ecosystems at small and medium scales, witness the efficacy of Dutch elm disease in England and the chestnut blight in the Appalachian region eastern United States (p. 80).

Some disturbances act essentially randomly within a landscape to produce disturbance patches. Strong winds commonly behave in this way. The patches produced by random disturbance can be extensive: the biomantles formed by burrowing animals are a case in point (D. L. Johnson 1989, 1990), as are the patches of eroded soil created by grizzly bears (*Ursus arctos horribilis*) excavating dens, digging for food, and trampling well-established trails (Butler 1992; see also Butler 1995). Second, some disturbances, such as fire, pests, and pathogens, tend to start at a point within a landscape and then spread to other parts. In both cases, the disturbances operate in a heterogeneous manner because some sites within landscapes will be more susceptible to disturbance agencies than others.

Wind

Tree-throw by strong winds, and to lesser extent other factors, may have a considerable impact on communities. In many forests, disturbance by up-rooting is the primary means by which species rich-ness is maintained (Schaetzl *et al.* 1989a, b). A fallen tree creates a gap in the forest that seems vital to community and vegetation dynamics and succes-sional pathways: it provides niches with much sunlight for pioneer species, encourages the release of suppressed, shade-tolerant saplings, and aids the recruitment of new individuals. Even tropical forests, once thought immune to physical disturbance, have been shown to contain seemingly haphazard patterns of tree forms, age classes, and species. The tropical rain forests of the Far East may be analysed as 'gap phases' (Whitmore 1975). Gaps are openings made in the forests by various disturbances, and the phases are the stages of tree growth in the gaps, from seedling to maturity and death. Small gaps, about 0.04 ha, are opened up by the fall of individual large trees, with crowns 15–18 m in diameter. Lightning strikes open up gaps with an area of about 0.6 ha. Local storms cause larger gaps (up to 80 ha), while typhoons and tornadoes destroy even larger areas. These gaps are an integral part of the tropical forest system.

Fire

Fire is relatively common in many terrestrial environments. It influences the structure and func-tion of some plant species and many communities (e.g. Whelan 1995). The effects of fire are beneficial and detrimental. There are three main detrimental effects. First, many organisms are lost from the com-munity. Second, after a severe fire, the ground is left vulnerable to soil erosion. Third, minerals are lost in smoke and through volatilization, which releases large quantities of nitrogen and sulphur. Beneficial effects are that dead litter is burnt to ash, which boosts mineral recycling, and that nitrogen-fixing legumes often appear in the aftermath of a moderate surface (as opposed to crown) fire. In areas where fires are frequent, many plants are tolerant of fire, and some require a fire to prompt seed release and germination. In the Swartboskloof catchment, near Stellenbosch, Cape Province, South Africa, mountain fynbos (Mediterranean-type) plants possess a wide range of regeneration strategies and fire-survival mechanisms (van Wilgen *et al.* 1992). Most species can sprout again after a fire, and are resilient to a range of fire regimes. Few species are reliant solely on seeds for regeneration. Those that are obligate reseeders regenerate from seeds stored in the soil, and just a few, such as *Protea neriifolia*, maintain seed stores in the canopy. Most of the fynbos plants flower within in a year of a fire. Some species, including *Watsonia borbonica* and *Cyrtanthus ventricosus*, are stimulated into flowering by fires, though the trigger appears to act indirectly through changes in the environment (such as altered soil temperature regimes), and not directly through heat damage to leaves or apical buds.

ADAPTING TO CIRCUMSTANCES: NICHES AND LIFE-FORMS

Ways of living

Organisms have evolved to survive in the varied conditions found at the Earth's surface. They have come to occupy nearly all habitats and to have filled multifarious roles within food chains.

Ecological niche

An organism's **ecological niche** (or simply **niche**) is

its 'address' and 'profession'. Its address or home is the habitat in which it lives, and is sometimes called the **habitat niche**. Its profession or occupation is its position in a food chain, and is sometimes called the **functional niche**. A skylark's (*Alauda arvensis*) address is open moorland (and, recently, arable farmland); its profession is insect-cum-seed-eater. A merlin's (*Falco columbarius*) address is open country, especially moorland; its profession is a bird-eater, with skylark and meadow pipit (*Anthus pratensis*) being its main prey. A grey squirrel's (*Sciurus carolinensis*) habitat niche is a deciduous woodland; its profession is a nut-eater (small herbivore). A grey wolf's (*Canis lupus*) habitat niche is cool temperate coniferous forest, and its profession is large-mammal-eater.

A distinction is drawn between the fundamental niche and the realized niche. The **fundamental** (or **virtual**) **niche** circumscribes where an organism would live under optimal physical conditions and with no competitors or predators. The **realized** (or **actual**) **niche** is always smaller, and defines the 'real-world' niche occupied by an organism constrained by biotic and abiotic limiting factors.

A niche reflects how an individual, species, or population interacts with and exploits its environment. It involves adaptation to environmental conditions. The **competitive exclusion principle** (Chapter 5) precludes two species occupying identical niches. However, a group of species, or **guild**, may exploit the same class of environmental resources in a similar way (R. B. Root 1967; Simberloff and Dayan 1991). In an oak woodland, one guild of birds forages for arthropods from the foliage of oak trees; another catches insects in the air; another eats seeds. The foliage-gleaning guild in a California oak woodland includes members of four families: the plain titmouse (*Parus inornatus*, Paridae), the blue-gray gnatcatcher (*Polioptila caerulea*, Sylviidae), the warbling vireo and Hutton's vireo (*Vireo gilvus* and *Vireo huttoni*, Vireonidae), and the orange-crowned warbler (*Vermivora celata*, Parulidae) (R. B. Root 1967).

Ecological equivalents

Although each niche is occupied by only one species, different species may occupy the same or similar niches in different geographical regions. These species are **ecological equivalents** or **vicars**. A grassland ecosystem contains a niche for large herbivores living in herds. Bison and the pronghorn antelope occupy this niche in North America; antelopes, gazelles, zebra, and eland in Africa; wild horses and asses in Europe; the pampas deer and guanaco in South America; and kangaroos and wallabies in Australia. As this example shows, quite distinct species may become ecological equivalents through historical and geographical accidents.

Many bird guilds have ecological equivalents on different continents. The nectar-eating (nectivore) guild has representatives in North America, South America, and Africa. In Chile and California the representatives are the hummingbirds (Trochilidae) and the African representatives are the sunbirds (Nectariniidae). One remarkable convergent feature between hummingbirds and sunbirds is the iridescent plumage.

Plant species of very different stock growing in different areas, when subjected to the same environmental pressures, have evolved the same life-form to fill the same ecological niche. The American cactus and the South African euphorbia, both living in arid regions, have adapted by evolving fleshy, succulent stems and by evolving spines instead of leaves to conserve precious moisture.

Life-forms

The structure and physiology of plants and, to a lesser extent, animals are often adapted for life in a particular habitat. These structural and physiological adaptations are reflected in life-form and often connected with particular ecozones.

The **life-form** of an organism is its shape or appearance, its structure, its habits, and its kind of life history. It includes overall form (such as herb, shrub, or tree in the case of plants) and the form of individual features (such as leaves). Importantly, the dominant types of plant in each ecozone tend to have a life-form finely tuned for survival under in that climate.

Plant life-forms

A widely used classification of **plant life-forms**, based on the position of the shoot-apices (the tips of branches) where new buds appear, was designed by Christen Raunkiaer in 1903 (see Raunkiaer 1934). It distinguishes five main groups: therophytes, cryptophytes, hemicryptophytes, chamaephytes, and phanerophytes (Box 2.4).

A **biological spectrum** is the percentages of the different life-forms in a given region. The 'normal spectrum' is a kind of reference point; it is the percentages of different life-forms in the world flora. Each ecozone possesses a characteristic biological spectrum that differs from the 'normal spectrum'. Tropical forests contain a wide spectrum of life-forms, whereas in extreme climates, with either cold or dry seasons, the spectrum is smaller (Figure 2.10). As a rule of thumb, very predictable, stable climates, such as humid tropical climates, support a wider variety of plant life-forms than do regions with inconstant climates, such as arid, Mediterranean, and alpine climates. Alpine regions, for instance, lack trees, the dominant life-form being dwarf shrubs (chamaephytes). In the Grampian Mountains, Scotland, 27 per cent of the species are chamaephytes, a figure three times greater than the percentage of chamaephytes in the world flora (Tansley 1939). Some life-forms appear to be constrained by climatic factors. Megaphanerophytes (where the regenerating parts stand over 30 m from the ground) are found only where the mean annual temperature of the warmest month is 10°C or more. Trees are confined to places where the mean summer temperature exceeds 10°C, both altitudinally and latitudinally. This uniform behaviour is somewhat surprising as different taxa are involved in different countries. Intriguingly, dwarf shrubs, whose life cycles are very similar to those of trees, always extend to higher altitudes and latitudes than do trees (Grace 1987).

Individual parts of plants also display remarkable adaptations to life in different ecozones. This is very true of leaves. In humid tropical lowlands, forest trees have evergreen leaves with no lobes. In regions of Mediterranean climate, plants have small, sclerophyllous evergreen leaves. In arid regions, stem succulents without leaves, such as cacti, and plants with entire leaf margins (especially among evergreens) have evolved. In cold wet climates, plants commonly possess notched or lobed leaf margins.

Animal life-forms

Animal life-forms, unlike those of plants, tend to match taxonomic categories rather than ecozones. Most mammals are adapted to basic habitats and may be classified accordingly. They may be adapted for life in water (aquatic or swimming mammals), underground (**fossorial** or burrowing mammals), on the ground (**cursorial** or running, and **saltatorial** or leaping mammals), in trees (**arboreal** or climbing mammals), and in the air (**aerial** or flying mammals) (Osburn *et al.* 1903). None of these habitats is strongly related to climate. That is not to say that animal species are not adapted to climate: there are many well-known cases of adaptation to marginal environments, including deserts (pp. 25–6) (see Cloudsley-Thompson 1975b).

Autoecological accounts

Detailed habitat requirements of individual species require careful and intensive study. A ground-breaking study was the **autoecological accounts** prepared for plants around Sheffield, England (Grime *et al.* 1988). About 3,000 km^2 were studied in three separate surveys by the Natural Environment Research Council's Unit of Comparative Plant Ecology (formerly the Nature Conservancy Grassland Research Unit). The region comprises two roughly equal portions: an 'upland' region, mainly above 200 m and with mean annual precipitation more than 850 mm, underlain by Carboniferous Limestone, Millstone Grit, and Lower Coal Measures; and a drier, 'lowland' region overlying Magnesian Limestone, Bunter Sandstone, and Keuper Marl.

Box 2.4

PLANT LIFE FORMS

Phanerophytes

Phanerophytes (from the Greek *phaneros*, meaning visible) are trees and large shrubs. They bear their buds on shoots that project into the air and are destined to last many years. The buds are exposed to the extremes of climate. The primary shoots, and in many cases the lateral shoots as well, are negatively geotropic (they stick up into the air). Weeping trees are an exception. Raunkiaer (1934) divided phanerophytes into twelve subtypes according to their bud covering (with bud-covering or without it), habit (deciduous or evergreen), and size (mega, meso, micro, and nano); and three other subtypes – herbaceous phanerophytes, epiphytes, and stem succulents. A herbaceous example is the scaevola, *Scaevola koenigii*. Phanerophytes are divided into four size classes: megaphanerophytes (> 30 m), mesophanerophytes (8–30 m), microphanerophytes (2–8 m) and nanophanerophytes (< 2 m).

Chamaephytes

Chamaephytes (from the Greek *khamai*, meaning on the ground) are small shrubs, creeping woody plants, and herbs. They bud from shoot-apices very close to the ground. The flowering shoots project freely into the air but live only during the favourable season. The persistent shoots bearing buds lie along the soil, rising no more than 20–30 cm above it.

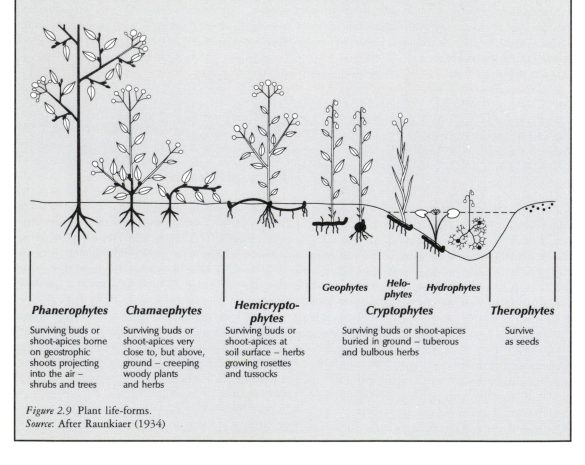

		Geophytes	Helo-phytes	Hydrophytes

Phanerophytes	*Chamaephytes*	*Hemicrypto-phytes*	*Cryptophytes*	*Therophytes*
Surviving buds or shoot-apices borne on geostrophic shoots projecting into the air – shrubs and trees	Surviving buds or shoot-apices very close to, but above, ground – creeping woody plants and herbs	Surviving buds or shoot-apices at soil surface – herbs growing rosettes and tussocks	Surviving buds or shoot-apices buried in ground – tuberous and bulbous herbs	Survive as seeds

Figure 2.9 Plant life-forms.
Source: After Raunkiaer (1934)

Suffructicose chamaephytes have erect aerial shoots that die back to the ground when the unfavourable season starts. They include species of the Labiatae, Caryophyllaceae, and Leguminosae. Passive chamaephytes have procumbent persistent shoots – they are long, slender, comparatively flaccid, and heavy and so lie along the ground. Examples are the greater stitchwort (*Stellaria holostea*) and the prostrate speedwell (*Veronica prostrata*). Active chamaephytes have procumbent persistent shoots that lie along the ground because they are transversely geotropic in light (take up a horizontal position in response to gravity). Examples are the heath speedwell (*Veronica officinalis*), the crowberry (*Empetrum nigrum*), and the twinflower (*Linnaea borealis*). Cushion plants are transitional to hemicryptophytes. They have very low shoots, very closely packed together. Examples are the hairy rock-cress (*Arabis hirusa*) and the houseleek (*Sempervivum tectorum*).

Hemicryptophytes

Hemicryptophytes (from the Greek *kryptos*, meaning hidden) are herbs growing rosettes or tussocks. They bud from shoot-apices located in the soil surface. They include protohemicryptophytes (from the base upwards, the aerial shoots have elongated internodes and bear foliage leaves) such as the vervain (*Verbena officinalis*), partial rosette plants such as the bugle (*Ajuga reptans*), and rosette plants such as the daisy (*Bellis perennis*).

Cryptophytes

Cryptophytes are tuberous and bulbous herbs. They are even more 'hidden' than hemicryptophytes – their buds are completely buried beneath the soil, thus affording them extra protection from freezing and drying. They include geophytes (with rhizomes, bulbs, stem tubers, and root tuber varieties) such as the purple crocus (*Crocus vernus*), helophytes or marsh plants such as the arrowhead (*Sagittaria sagittifolia*), and hydrophytes or water plants such as the rooted shining pondweed (*Potamogeton lucens*) and the free-swimming frogbit (*Hydrocharis morsus-ranae*).

Therophytes

Therophytes (from the Greek *theros*, meaning summer) or annuals are plants of the summer or favourable season and survive the adverse season as seeds. Examples are the cleavers (*Galium aparine*), the cornflower (*Centaurea cyanus*), and the wall hawk's-beard (*Crepis tectorum*).

An example of the 'autoecological accounts' for the bluebell (*Hyacinthoides non-scripta*) is shown in Figure 2.11. The bluebell is a polycarpic perennial, rosette-forming geophyte, with a deeply buried bulb. It appears above ground in the spring, when it exploits the light phase before the development of a full summer canopy. It is restricted to sites where the light intensity does not fall below 10 per cent of the daylight between April and mid-June, in which period the flowers are produced. Shoots expand during the late winter and early spring. The seeds are gradually shed, mainly in July and August. The leaves are normally dead by July. There is then a period of aestivation (dormancy during the dry season). This ends in the autumn when a new set of roots forms. The plant cannot replace damaged leaves and is very vulnerable to grazing, cutting, or trampling. Its foliage contains toxic glycosides and, though sheep and cattle will eat it, rabbits will not. Its reproductive strategy is intermediate between a stress-tolerant ruderal and a competitor–stress-tolerator–ruderal (p. 103). It extends to 340 m around Sheffield, but is known to grow up to 660 m in the British Isles. It is largely absent from skeletal habitats and steep slopes. The bluebell commonly occurs in woodland. In the Sheffield survey, it was recorded most frequently in broad-leaved plantations. It was also common in scrub and woodland overlying either acidic or limestone beds, but less frequent in coniferous plantations. It occurs in upland areas on waste ground and heaths, and occasionally in unproductive pastures, on spoil heaps, and on cliffs. In woodland habitats, it

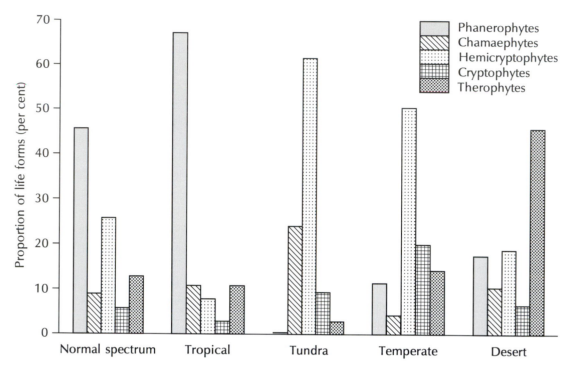

Figure 2.10 The proportion of plant life-forms in various ecozones. The 'normal' spectrum was constructed by selecting a thousand species at random.
Source: From data in Raunkiaer (1934) and Dansereau (1957)

grows more frequently and is significantly more abundant on south-facing slopes. However, in unshaded habitats, it prefers north-facing slopes. It is not found in wetlands. It can grow on a wide range of soils, but it most frequent and more abundant in the pH range 3.5–7.5. It is most frequent and abundant in habitats with much tree litter and little exposed soil, though it is widely distributed across all bare-soil classes.

SUMMARY

All living things live in particular places – habitats. Habitats range in size from a few cubic centimetres to the entire ecosphere. Species differ in their habitat requirements, the span going from habitat generalists, who live virtually anywhere, to habitat specialists, who are very choosy about their domicile. Species are constrained by limiting factors in their environment. Limiting factors include moisture, heat, and nutrient levels. Each species has a characteristic tolerance range and ecological valency. Limiting climatic factors are radiation and light, various measures of temperature (e.g. annual mean, annual range, occurrence of frost), various measures of the water balance (e.g. annual precipitation, effective precipitation, drought period, snow cover), windiness, humidity, and many others. Of these climatic factors, temperature and water are master limiting factors. They are summarized in climate diagrams that characterize bioclimates. Ecozones are large climatic regions sharing the same kind of climate. The Mediterranean ecozone is an example. Ecozones are equivalent to zonobiomes, which in turn are composed of similar biomes. Soil and substrate, topography, and disturbance also influence life. Certain soils produce distinctive vegetation types called pedobiomes, such as that associated with soils

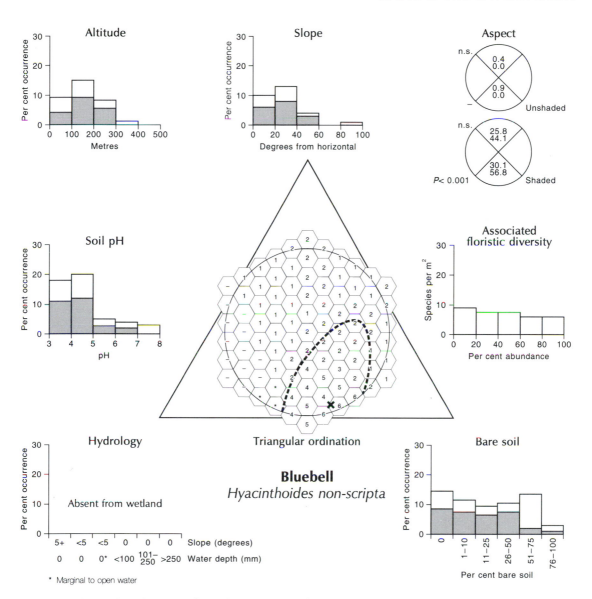

Figure 2.11 Autoecological accounts for the bluebell (*Hyacinthoides non-scripta*) in the Sheffield region, England. For the subject species, bluebell in the example, the blank bars show the percentage of simple occurrences over each of the environmental classes, and the shaded bars show the percentage of samples in which 20 per cent or more of the quadrat subsections contained rooted shoots. The 'triangular ordination' is explained in Figure 4.8. On the 'aspect' diagram, n.s. stands for 'not significant'.
Source: After Grime *et al.* (1988)

formed on serpentine. Several topographic features influence life – altitude (which produces orobiomes), aspect, inclination (which along a hillslope produces vegetation catenae), and insularity (the relative isolation of a community or ecosystem). Disturbance by biotic and abiotic factors is an important environmental factor. Examples are disturbance by wind and fire. Life has to adapt to environmental conditions in the ecosphere. There are several 'ways of living', each of which corresponds to an ecological niche. Ecological equivalents (or vicars) are species from different stock, and living in different parts of the world, that have adapted to the same environmental constraints. Adaptation to environmental conditions is also seen in life-forms. The overall habitat preferences of an individual species require detailed and intense study. They may be summarized as autoecological accounts.

ESSAY QUESTIONS

1 How useful are bioclimatic classifications to ecological biogeographers?

2 How important is disturbance in shaping communities?

3 Why are ecological equivalents so common?

FURTHER READING

Cox, C. B. and Moore, P. D. (1993) *Biogeography: An Ecological and Evolutionary Approach*, 5th edn, Oxford: Blackwell.

Dansereau, P. (1957) *Biogeography: An Ecological Perspective*, New York: Ronald Press.

Forman, R. T. T. (1995) *Land Mosaics: The Ecology of Landscapes and Regions*, Cambridge: Cambridge University Press.

Huggett, R. J. (1995) *Geoecology: An Evolutionary Approach*, London: Routledge.

Kormondy, E. J. (1996) *Concepts of Ecology*, 4th edn, Englewood Cliffs, NJ: Prentice Hall.

Larcher, W. (1995) *Physiological Plant Ecology: Ecophysiology and Stress Physiology of Functional Groups*, 3rd edn, Berlin: Springer.

Stoutjesdijk, P. and Barkman, J. J. (1992) *Microclimate, Vegetation and Fauna*, Uppsala, Sweden: Opulus Press.

Viles, H. A. (ed.) (1988) *Biogeomorphology*, Oxford: Basil Blackwell.

Whelan, R. J. (1995) *The Ecology of Fire*, Cambridge: Cambridge University Press.

3

THE DISTRIBUTION OF ORGANISMS

All species and other groups of organism have a particular geographical range or distribution. This chapter covers:

- kinds of distribution
- how organisms spread
- how species distributions are broken apart
- human impacts on species distributions

COSMOPOLITAN AND PAROCHIAL: PATTERNS OF DISTRIBUTION

All species, genera, families, and so forth have a geographical range or distribution. Distributions range in size from a few square metres to almost the entire terrestrial globe. Their boundaries are determined by the physical environment, by the living environment, and by history. They tend to follow a few basic patterns – large or small, widespread or restricted, continuous or broken.

Large and small, widespread and restricted

An **endemic** species lives in only one place, no matter how large or small that place should be. A species can be endemic to Australia or endemic to a few square metres in a Romanian cave (when it is a micro-endemic). A **pandemic** species lives in all places. The puma or cougar (*Felis concolor*), for example, is a pandemic species because it occupies nearly all the western New World, from Canada to Tierra del Fuego (Colour plate 2). It is also an endemic species of this region because it is found nowhere else. **Cosmopolitan** species inhabit the whole world, though not necessarily in all places. It is possible for

a cosmopolitan species to occur in numerous small localities in all continents. As a rule, pandemic or cosmopolitan species have widespread distributions, whereas endemic species have restricted distributions. The Romanian hamster (*Mesocricetus newtoni*) is restricted and endemic (Figure 3.1). The capybara (*Hydrochoerus hydrochaeris*), the largest living rodent, is endemic to South America. It is also a pandemic, ranging over half the continent. The distinction between widespread and restricted often rests on the occurrence or non-occurrence of species within continents or biogeographical regions (Table 3.1).

Micro-endemic species

Some species have an extremely restricted (**micro-endemic**) distribution, living as a single population in a small area. The Devil's Hole pupfish (*Cyprinodon diabolis*) is restricted to one thermal spring issuing from a mountainside in south-west Nevada (Moyle and Williams 1990). The Amargosa vole (*Microtus californicus scirpensis*) lives in a single watershed in California, United States (Murphy and Freas 1988). The rare black hairstreak butterfly (*Strymonidia pruni*) is confined to a few sites in central England and in central and eastern Europe.

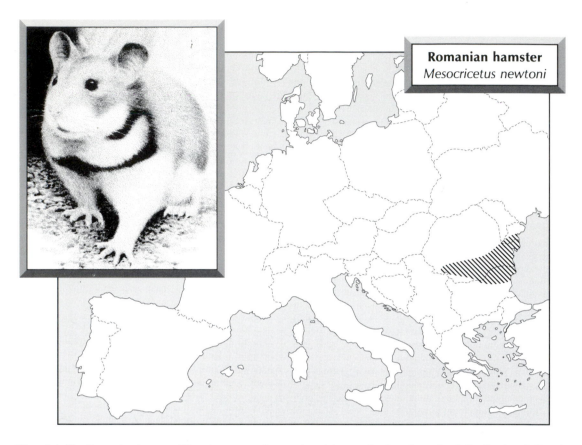

Figure 3.1 The Romanian hamster (*Mesocricetus newtoni*), a species with an endemic and restricted distribution. It lives in a narrow strip along the Black Sea coast in Bulgaria, Romania, and adjacent parts of the Ukraine.
Source: After Bjärvall and Ullström (1986)

Endemic plant families

Two restricted and endemic flowering-plant families are the Degeneriaceae and the Leitneriaceae (Figure 3.2). The Degeneriaceae consists of a single tree species, *Degeneria vitiensis*, that grows on the island of Fiji. The Leitneriaceae also consists of a single species – the Florida corkwood (*Leitneria floridana*). This deciduous shrub is native to swampy areas in the south-eastern United States where it is used as floats for fishing nets.

Cosmopolitan plant families

Two widespread flowering-plant families are the sunflower family (Compositae or Asteraceae) and the grass family (Graminae) (Figure 3.3). The Compositae is one of the largest families of flowering plants. It contains around 1,000 genera and 25,000 species. It is found everywhere except the Antarctic mainland, though it is poorly represented in tropical rain forests. The grass family comprises about 650 genera and 9,000 species, including all the world's cereal crops (including rice). Its distribution is world-wide, ranging from inside the polar circle to the equator. It is the chief component in about one-fifth of the world's vegetation. Few plant formations lack grasses; some (steppe, prairie, and savannah) are dominated by them.

Table 3.1   Species classed according to range size and cosmopolitanism

Range size		Degree of cosmopolitanism		
	Endemic (peculiar)	Characteristic (shared between two biogeographical regions)	Semi-cosmopolitan (shared between three or four biogeographical regions)	Cosmopolitan (shared between five or more biogeographical regions)
Microscale	Black hairstreak butterfly (*Strymonidia pruni*) Amargosa vole (*Microtus californicus scirpensis*) Devil's Hole pupfish (*Cyprinidon diabolis*)	*Friesea oligorhopala* – a collembolan species found in Tripoli (Libya), Malta, Bahía Blanca (Argentina), and Santiago (Chile)	Skua (*Stercorarius skua*) – a coastal bird of Antarctica, southern South America, Iceland, and the Faeroes	Stenotypic ornamental plants
Mesoscale	California vole (*Microtus californicus*) Romanian hamster (*Mesocricetus newtoni*)	Rose pelican (*Pelecanus onocrotalus*) – a bird shared by central Asia and southern Africa	Olivaceus warbler (*Hippolais pallida*) – a passerine bird shared by southern Europe, the Near East, and scattered places in the Ethiopian region	Cormorant (*Phalacrocorax carbo*) – a bird from the Palaearctic, Oriental, Ethiopian, Australian, and Nearctic regions
Macro- and megascale	Capybara (*Hydrochoerus hydrochaeris*) – a pandemic	Puma or cougar (*Felis concolor*) – a pandemic of the New World	Black heron (*Nycticorax nycticorax*) – a pandemic bird of South America, North America, Africa, and Eurasia	Human (*Homo sapiens*) – a pandemic cosmopolite

Source: After Rapoport (1982)

Zonal climatic distributions

Some animal and plant distributions follow climatic zones (Figure 3.4). Five relatively common zonal patterns are **pantropical** (throughout the tropics), **amphitropical** (either side of the tropics), **boreal** (northern), **austral** (southern), and **temperate** (middle latitude). The sweetsop and soursop family (Annonaceae), consisting of about 2,000 trees and shrub species, is pantropical, though centred in the Old World tropics. The sugarbeet, beetroot, and spinach family (Chenopodiaceae) consists of about 1,500 species, largely of perennial herbs, widely distributed either side of the tropics in saline habitats. The arrowgrass family (Scheuchzeriaceae) comprises a single genus (*Scheuchzeria*) of marsh plants that are restricted to a cold north temperate belt, and are especially common in cold *Sphagnum* bogs. The

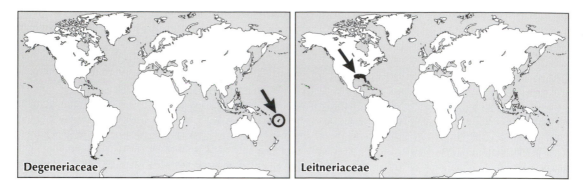

Figure 3.2 Two restricted and endemic plant families: the Degeneriaceae and Leitneriaceae.
Source: After Heywood (1978)

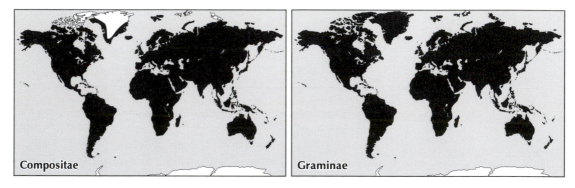

Figure 3.3 Two widespread plant families: the sunflower family (Compositae or Asteraceae) and the grass family (Graminae).
Source: After Heywood (1978)

poppy family (Papaveraceae) has some 250 species of mainly herbaceous annuals or perennials that are confined largely to the north temperate zone.

Continuous and broken

Plant and animal distributions have a third basic pattern – they tend to be either **continuous** or else **broken (disjunct)**. Several factors cause broken distributions, including geological change, climatic change, evolution, and jump dispersal by natural and human agencies.

Evolutionary disjunctions

Evolutionary disjunctions occur under the following circumstances. A pair of sister species evolves on either side of an area occupied by a common ancestor. The common ancestor then becomes extinct. The extinction leaves a disjunct species pair. This mechanism may account for some amphitropical disjunct species, including the woody genera *Ficus* and *Acacia* in Sonoran and Chihuahuan Deserts (North America) and the Chilean and Peruvian Deserts (South America) (Raven 1963).

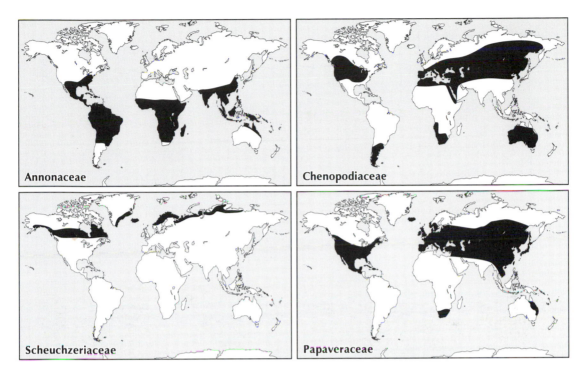

Figure 3.4 Zonal climatic distributions of four plant families: the pantropical sweetsop and soursop family (Annonaceae), the amphitropical sugarbeet, beetroot, and spinach family (Chenopodiaceae), the cold temperate arrowgrass family (Scheuchzeriaceae), and the north temperate poppy family (Papaveraceae).
Source: After Heywood (1978)

Jump dispersal disjunctions

Jump dispersal is the rapid passage of individual organisms across large distances, often across inhospitable terrain. The jump takes less time than the individuals' life-spans. Plants and animals that survive a long-distance jump and found a new colony lead to 'jump' disjunctions. The process is probably common, especially in plants. There are about 160 temperate or cool temperate plant species or species groups that have amphitropical distributions in the Americas (Raven 1963). Most of these arose from jump dispersal. Exceptions are members of the woodland genera *Osmorhiza* and *Sanicula* (Raven 1963). Tropical montane species of these genera almost bridge the gap in the disjunction. This suggests that the disjunctions are evolutionary in origin. The

groups spread slowly along mountain chains. Later, members occupying the centre of the distribution became extinct, leaving the surviving members on either side of the tropics.

Humans carry species to all corners of the globe. In doing so, they create disjunct distributions. A prime example is the mammals introduced to New Zealand. There are fifty-four such introduced species (C. M. King 1990). Twenty came directly or indirectly from Britain and Europe, fourteen from Australia, ten from the Americas, six from Asia, two from Polynesia, and two from Africa. The package contained domestic animals for farming and household pets, and feral animals for sport or fur production. Farm animals included sheep, cattle, and horses. Domestic animals included cats and dogs. Sporting animals included pheasant, deer, wallabies, and rabbits. The Australian possum was introduced to start a fur industry. Wild boar and goats were liberated on New Zealand by Captain James Cook. Many other species were introduced – European blackbirds, thrushes, sparrows, rooks, yellowhammers, chaffinches, budgerigars, hedgehogs, hares, weasels, stoats, ferrets, rats, mice. Several species failed to establish themselves. These failed antipodean settlers include bandicoots, kangaroos, racoons, squirrels, bharals, gnus, camels, and zebras.

Geological disjunctions

Geological disjunctions are common in the southern continents, which formed a single land mass (Gondwana) during the Triassic period but have subsequently fragmented and drifted apart. Ancestral populations living on Gondwana were thus split and evolved independently. Their Gondwanan origin is reflected in present day distributions. This is seen in many flowering plant families. The protea, banksia, and grevillea family (Proteaceae) is one of the most prominent families in the Southern Hemisphere (Figure 3.5). It provides numerous examples of past connections between South American, South African, and Australian floras. The genus *Gevuina*, for instance, has three species, of which one is native to Chile and the other two to Queensland and New Guinea. The breakup of Pangaea is also responsible

for the disjunct distribution of the flightless running birds, or ratites (p. 71).

Climatic disjunctions

Climatic disjunctions result from a once widespread distribution being reduced and fragmented by climatic change. An example is the magnolia and tulip tree family (Magnoliaceae) (Figure 3.5). This family consists of about twelve genera and about 220 tree and shrub species native to Asia and America. The three American genera (*Magnolia*, *Talauma*, and *Liriodendron*) also occur in Asia. Fossil forms show that the family was once much more widely distributed in the Northern Hemisphere, extending into Greenland and Europe. Indeed, the Magnoliaceae were formerly part of an extensive Arcto-Tertiary deciduous forest that covered much Northern Hemisphere land until the end of the Tertiary period, when the decline into the Ice Age created cold climates.

Relict groups

These are produced by environmental changes, particularly climatic changes, and evolutionary processes that lead to a shrinking distribution.

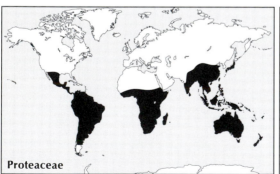
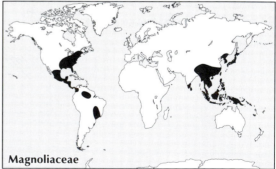

Figure 3.5 The protea, banksia, and grevillea family (Proteaceae), and the magnolia and tulip tree family (Magnoliaceae). The Proteaceae originated on Gondwana and have survived on the all southern continents. The Magnoliaceae once formed part of extensive Northern Hemisphere deciduous forests that retreated from high latitudes during the Quaternary ice age.
Source: After Heywood (1978)

Climatic relics

Climatic relics are survivors of organisms that formerly had larger distributions. The alpine marmot (*Marmota marmota*), a large ground squirrel, lives on alpine meadows and steep rocky slopes in the Alps (Figure 3.6). It has also been introduced into the Pyrenees, the Carpathians, and the Black Forest. During the Ice Age it lived on the plains area of central Europe. With Holocene warming, it became restricted to higher elevations and its present distribution, which lies between 1,000 m and 2,500 m, is a relict of a once much wider species distribution.

The Norwegian mugwort (*Artemisia norvegica*) is a small alpine plant. During the last ice age, it was widespread in northern and central Europe. Climatic warming during the Holocene epoch has left it stranded in Norway, the Ural Mountains, and two isolated sites in Scotland.

Evolutionary relics

Evolutionary relics are survivors of ancient groups of organism. The tuatara (*Sphenodon punctatus*) is the only native New Zealand reptile (Plate 3.1). It is the sole surviving member the reptilian order Rhyncocephalia, and is a 'living fossil'. It lives nowhere else in the world. Why has the tuatara survived on New Zealand but other reptiles (and marsupials

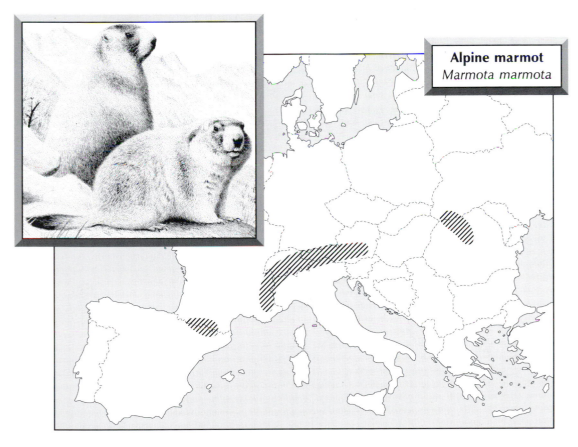

Alpine marmot
Marmota marmota

Figure 3.6 The alpine marmot (*Marmota marmota*) – a climatic relict.
Source: After Bjärvall and Ullström (1986)

Plate 3.1 Tuatara (*Sphenodon punctatus*)
Photograph by Pat Morris

and monotremes) have not? New Zealand in the Cretaceous period lay at latitude 60° to 70° S. The climate would have been colder, and the winter nights longer. Most reptiles and mammals could not have tolerated this climate. The tuatara has a low metabolic rate, remaining active at a temperature of 11°C, which is too cold to allow activity in any other reptile. It may have survived because it can endure cold conditions. Its long isolation on an island, free from competition with other reptiles and mammals, may also have helped.

Cycads belong to the family Cycadaceae, which comprises nine genera and about a hundred species, all of which are very rare and have highly restricted distributions confined to the tropics and subtropics (Colour plate 3). Early members of the cycads were widely distributed during the Mesozoic era – specimens have been unearthed in Oregon, Greenland, Siberia, and Australia. They may have been popular items on dinosaurian menus. The reduction of cycad distribution since the Mesozoic may have partly resulted from competition with flowering plants (which reproduce more efficiently and grow faster). Climatic change may also have played a role. Over the last 65 million years, tropical climates slowly pulled back to the equatorial regions as the world climates underwent a cooling. Cycads are therefore in part climatic and in part evolutionary relics. They commonly maintain a foothold in isolated regions – their seeds survive prolonged immersion in sea water and the group has colonized many Pacific islands.

Range size and shape

Interesting relationships

The areas occupied by species vary enormously. In Central and North America, the average area occupied by bear species (Ursidae) is 11.406 million km²; for cat species (Felidae) it is 5.772 million km²; for squirrel, chipmunk, marmot, and prairie dog species (Sciuridae) it is 0.972 million km²; and for pocket gopher species (Geomyidae) it is 0.284 million km² (Rapoport 1982: 7). The range occupied by individual species spans 100 km² for such rodents as the pocket mouse *Heteromys desmarestianus* to 20.59 million km² for the wolf (*Canis lupus*).

Species range in Central and North America is related to feeding habits (Table 3.2). Large carnivores tend to occupy the largest ranges. Large herbivores come next, followed by smaller mammals. The larger range of carnivores occurs in African species, too. The mean species range for carnivores (Canidae, Felidae, and Hyaenidae) is 8.851 million km²; the mean species range for herbivores (Bovidae, Equidae, Rhinocerotidae, and Elephantidae) is 3.734 million km².

Geographical regularities in ranges

The size and shape of ranges are related. Figure 3.7a is a scattergraph of the greatest north–south and the east–west range dimensions for North American snake species (Brown 1995). In the graph, ranges equidistant in north–south and east–west directions will lie on the line of equality that slopes at 45 degrees upwards from the origin (Figure 3.7b). Ranges stretched out in a north–south direction will lie below the line of equality; ranges that are stretched in an east–west direction lie above the line of equality. The pattern for North American snakes is fairly plain (Figure 3.7a). Small ranges lie mainly above the line (these are stretched in a north–south direction); large ranges tend to lie below the line (these are stretched in an east–west direction).

There is a plausible reason for these patterns, which are also found in lizards, birds, and mammals (Brown 1995: 110–11). Species with small geographical ranges are limited by local or regional environmental conditions. The major mountain ranges, river valleys, and coastlines in North America run roughly north to south. The soils, climates, and vegetation associated with these north–south physical geographical features may determine the boundaries of small-range species. Large-range species are distributed over much of the continent. Their ranges cannot therefore be influenced by local and regional environmental factors. Instead, they may be limited by large-scale climatic and vegetational patterns, which display a zonary arrangement, running east–west in wide latitudinal belts.

Table 3.2 Mean geographical ranges of Central and North American mammal species grouped by order

Order	Mean range area (millions of km²)
Carnivora (carnivores)	6.174
Artiodactyla (even-toed ungulates)	5.072
Lagomorpha (rabbits, hares, and pikas)	1.926
Chiroptera (bats)	1.487
Marsupialia (marsupials)	1.130
Insectivora (insectivores)	1.117
Xenarthra or Edentata (anteaters, sloths, and armadillos)	0.889
Rodentia (rodents)	0.764
Primates (primates)	0.249

Source: After Rapoport (1982)

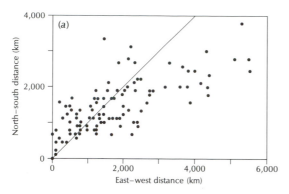
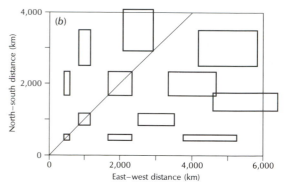

Figure 3.7 (a) A scattergraph of the greatest north–south versus the greatest east–west dimension of North American snake species ranges. The line represents ranges whose two dimensions are equal. (b) Some example shapes and sizes of ranges. For convenience, the ranges are shown as squares and rectangles, but they could be circles, ellipses, or any other shape. The diagonal line represents ranges with two equal dimensions. Above the line, ranges are stretched in a north–south direction; below the line, they are stretched in and east–west direction. *Source*: After Brown (1995)

Range size tends to increase with increasing latitude. In other words, on average, geographical ranges are smaller in the tropics than they are near the poles. Moreover, species whose ranges are centred at increasing higher latitudes (nearer the poles) tend to be distributed over an increasingly wider range of latitudes. This relationship is called **Rapoport's rule** (Stevens 1989), after its discoverer (see Rapoport 1982). The same pattern holds for altitudinal distributions: within the same latitude, the altitudinal range of a species increases with the midpoint elevation of the range (Stevens 1992). These biogeographical patterns are basic, holding for organisms from land mammals to coniferous trees. Five hypotheses may explain them – continental geometry, congenial environments, constraints on dispersal, climatic change, and tropical competition (Brown 1995: 112–14) (Box 3.1).

Rapoport's rule does not apply everywhere. The range size of Australian mammals does not increase from the tropics towards the poles (F. D. M. Smith *et al.* 1994). Latitudinal and longitudinal variations in range size correlate with continental width. Moreover, the arid centre of Australia contains the fewest mammal species with the largest ranges, while the moist and mountainous east coast, and the monsoonal north, contain large numbers of species with small ranges.

Distribution within a geographical range

Home range

Individuals of a species, especially vertebrate species, may have a **home range**. A home range is the area traversed by an individual (or by a pair, or by a family group, or by a social group) in its normal activities of gathering food, mating, and caring for young (Burt 1943). Home ranges may have irregular shapes and may partly overlap, although individuals of the same species often occupy separate home ranges. Land iguanas (*Conolophus pallidus*) on the island of Sante Fe, in the Galápagos Islands, have partly overlapping home ranges near cliffs and along plateaux edges (Figure 3.8a) (Christian *et al.* 1983). Red foxes (*Vulpes vulpes*) in the University of Wisconsin arboretum have overlapping ranges, too (Figure 3.8b) (Ables 1969).

The size of home ranges varies enormously (see Vaughan 1978: 306). In mammals, small home ranges are held by the female prairie vole (*Microtus ochrogaster*), mountain beaver (*Aplodontia rufa*), and the common shrew (*Sorex araneus*), at 0.014 ha, 0.75 ha, and 1.74 ha, respectively. The American badger (*Taxidea taxus*) has a medium-size home range of 8.5 km². A female grizzly bear (*Ursus arctos*) with three

---

Box 3.1

ACCOUNTING FOR REGULARITIES IN RANGE SIZE

Continental geometry

Rapoport's rule and the tendency of small ranges to occur at low latitudes follow from the geometry of North America. The continent tapers from north to south so species become more tightly packed in lower latitudes. Against this idea, species ranges become smaller before the tapering becomes marked. Moreover, it does not account for the elevational version of Rapoport's rule.

Congenial environments

The abiotic environments in low latitudes are more favourable and less variable than those in high latitudes. Small-range species living in low latitudes can avoid extinction by surviving in 'sink' habitats, and recolonize favourable 'source' habitats and re-establish populations after local extinctions.

Constraints on dispersal

Barriers to dispersal are more severe in low latitudes, which may contribute to the evolution and persistence of narrowly endemic species. The logic of this idea is that a species living at high latitudes could pass over mountains in summer without experiencing harsher conditions than it experiences normally. In low latitudes, however, species used to living in tropical lowlands and trying to cross a mountain pass at an equivalent elevation would experience conditions never experienced before, no matter what the season. The same would be true of a tropical montane species trying to cross tropical lowlands in the winter.

Climatic change

Large-range species have adapted to the Pleistocene climatic swings at high latitudes.

Tropical competition

Range size increases and species diversity decreases in moving from the equator towards the poles. Interaction between species may be a stronger limiting factor in lower latitudes.

---

yearlings has a large home range of 203 km², and a pack of eight timber wolves (*Canis lupus*) has a very large home range of 1,400 km².

Territory

Some animals actively defend a part of their home range against members of the same species. This core area, which does not normally include the peripheral parts of the home range, is called the **territory**. Species that divide geographical space in this way are territorial species. Territories may be occupied permanently or temporarily. Breeding pairs of the tawny owl (*Strix aluco*) stay within their own territory during adulthood. Great tits (*Parus major*) establish territories only during the breeding season. The 'i'iwi (*Vestiaria coccinea*), a Hawaiian honeycreeper, exhibits territorial behaviour when food is scarce.

Habitat selection

As the conditions near the margins of ecological tolerance create stress, it follows that the geographical range of a species is strongly influenced by its ecological tolerances. It is generally true that species with wide ecological tolerances are the most widely distributed. A species will occupy a habitat that meets its tolerance requirements, for it simply could not survive elsewhere. Nonetheless, even where a population is large and healthy, not all favourable habitat inside its geographical range will necessarily be occupied, and there may be areas outside its geographical range

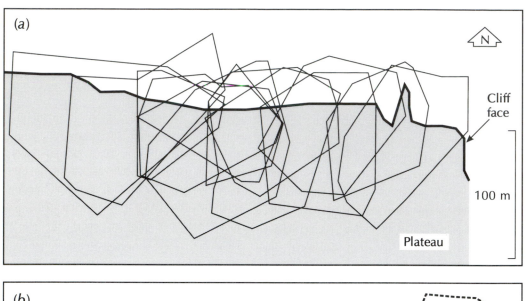

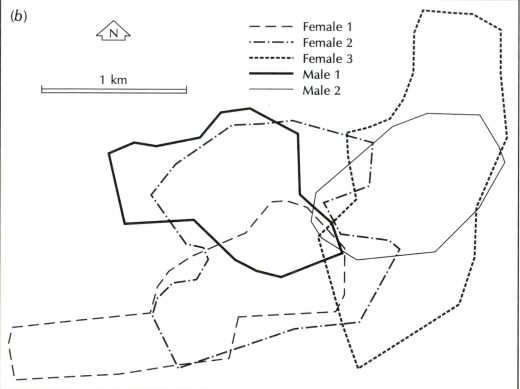

Figure 3.8 Home ranges. (a) Land iguanas (*Conolophus pallidus*) on the island of Sante Fe, in the Galápagos Islands. (b) Red fox (*Vulpes vulpes*) home ranges in the University of Wisconsin arboretum.
Sources: (a) After Christian *et al*. 1983; (b) After Ables (1969)

where it could live. In many cases, individuals 'choose' not to live in particular habitats, a process called **habitat selection**. Not a great deal is known about habitat selection, but it does occur and limits the distribution of some species.

Habitat selection is prevalent in birds. An early study was carried out in the Breckland of East Anglia, England (Lack 1933). The wheatear (*Oenanthe oenanthe*) lives on open heathland in Britain. It nests in old rabbit burrows. It does not occur in newly forested heathland lacking rabbits. Nesting-site selection thus excludes it from otherwise suitable habitat. The tree pipit (*Anthus trivialis*) and the meadow pipit (*Anthus pratensis*) are both ground nesters and feed on the same variety of organisms, but the tree pipit breeds only in areas with one or more tall trees. In consequence, in many treeless areas in Britain, the tree pipit is not found alongside the meadow pipit. David Lack found some tree pipits breeding in one treeless area close to a telegraph pole. The pipits used the pole merely as a perch on which to land at the end of their aerial song. Meadow pipits sing a similar song but land on the ground. This finding suggests that the tree pipit does not colonize heathland simply because it likes a perch from which to sing. The conclusion of the Breckland study was that the heathland and pine planation birds had a smaller distribution than they otherwise might because they selected habitat to live in. In short, they were choosy about where they lived.

Distributional limits

Plants

Many distributional boundaries of plant species seem to result from extreme climatic events causing the failure of one stage of the life cycle (Grace 1987). The climatic events in question may occur rarely, say once or twice a century, so the chances of observing a failure are slim. Nonetheless, edges of plant distributions often coincide with **isolines** of climatic variables. The northern limit of madder (*Rubia peregrina*) in northern Europe sits on the 40°F (4.4°C) mean January isotherm (Salisbury 1926). Holly (*Ilex aquifolium*) is confined to areas where the mean annual temperature of the coldest month exceeds –0.5°C, and, like madder, seems unable to withstand low temperatures (Iversen 1944). Several frost-sensitive plant species, including the Irish heath (*Erica erigena*), St Dabeoc's heath (*Daboecia cantabrica*), large-flowered butterwort (*Pinguicula grandifola*), and sharp rush (*Juncus acutus*), occur only in the extreme west of the British Isles where winter temperatures are highest. Other species, such as the twinflower (*Linnaea borealis*) and chickweed-wintergreen (*Trientalis europaea*), have a northern or north-eastern distribution, possibly because they have a winter chilling requirement for germination that southerly latitudes cannot provide (Perring and Walters 1962). Low summer temperatures seem to restrict the distribution of such species as the stemless thistle (*Cirsium acaule*). Near to its northern limit, this plant is found mainly on south-facing slopes, for on north-facing slopes it fails to set seed (Pigott 1974). The distribution of grey hair-grass (*Corynephorus canescens*) is limited by the 15°C mean isotherm for July. This may be because its short life-span (2 to 6 years) means that, to maintain a population, seed production and germination must continue unhampered (J. K. Marshall 1978). At the northern limit of grey hair-grass, summer temperatures are low, which delays flowering, and, by the time seeds are produced, shade temperatures are low enough to retard germination.

The small-leaved lime tree (*Tilia cordata*) ranges across much of Europe. Its northern limit in England and Scandinavia is marked by the mean July 19°C isotherm (Pigott 1981; Pigott and Huntley 1981). The tree requires 2,000 growing day-degrees to produce seeds by sexual reproduction. But for the lime tree to reproduce, the flowers must develop and then the pollen must germinate and be transferred through a pollen tube to the ovary for fertilization. The pollen fails to germinate at temperatures at or below 15°C, germinates best in the range 17°C to 22°C, and germinates, but less successfully up to about 35°C. A complicating factor is that the growth of the pollen tube depends on temperature. The growth rate is maximal around 20–25°C, diminishing fast at higher and lower temperatures. Indeed, the extension of the pollen tube becomes rapid above 19°C, which suggest why the northern limit is marked by the 19°C mean July isotherm.

Several models use known climatic constraints on plant physiology to predict plant species distribution. One study investigated the climatic response of boreal tree species in North America (Lenihan 1993). Several climatic predictor variables were used in a regression model. The variables were annual snowfall, day-degrees, absolute minimum temperature, annual soil-moisture deficit, and actual evapotranspiration summed over summer months. Predicted patterns of species' dominance probability closely matched observed patterns (Figure 3.9). The results suggested that the boreal tree species respond individually to different combinations of climatic constraints. Another study used a climatic model to predict the distribution of woody plant species in Florida, United States (Box *et al*. 1993). The State of Florida is small enough for variations in substrate to play a major role in determining what grows where. Nonetheless, the model predicted that climatic factors, particularly winter temperatures, exert a powerful influence, and in some cases a direct control, on species' distri-butions. Predicted distributions and observed distrib-ution of the longleaf pine (*Pinus palustris*) and the Florida poison tree (*Metopium toxiferum*) are shown in Figure 3.10. The predictions for the longleaf pine are very good, except for a narrow strip near the cen-tral Atlantic coast. The match between predicted and observed distributions is not so good for the Florida poison tree. The poison tree is a subtropical species and the model was less good at predicting the distribution of subtropical plants.

Animals

Many bird species distributions are constrained by such environmental factors as food abundance, climate, habitat, and competition. Distributions of 148 species of North American land birds wintering in the conterminous Unites States and Canada, when compared with environmental factors, reveal a consis-tent pattern (T. L. Root 1988a, b). Six environmental factors were used: average minimum January temper-

ature; mean length of the frost-free period; potential vegetation; mean annual precipitation; average gen-eral humidity; and elevation. Isolines for average minimum January temperature, mean length of the frost-free period, and potential vegetation correlated with the northern range limits of about 60 per cent, 50 per cent, and 64 per cent, respectively, of the wintering bird species. Figure 3.11 shows the winter distribution and abundance of the eastern phoebe (*Sayornis phoebe*). The northern boundary is constrained by the −4°C isotherm of January minimum temper-ature. Just two environmental factors − potential vegetation and mean annual precipitation − coincided with eastern range boundaries, for about 63 per cent and 40 per cent of the species, respectively. On the western front, mean annual precipitation distribution coincided for 36 per cent of the species, potential vegetation 46 per cent, and elevation 40 per cent.

Why should the northern boundary of so many wintering bird species coincide with the average minimum January temperature? The answer to this poser appears to lie in metabolic rates (T. L. Root 1988b). At their northern boundary, the calculated mean metabolic rate in a sample of fourteen out of fifty-one passerine (song birds and their allies) species was 2.49 times greater than the basal metabolic rate (which would occur in the thermal neutral zone, see p. 20). This figure implies that the winter ranges of these fourteen bird species are restricted to areas where the energy needed to compensate for a colder environment is not greater than around 2.5 times the basal metabolic rate. The estimated mean metabolic rate for thirty-six of the remaining thirty-seven passerine species averaged about 2.5. This '2.5 rule' applies to birds whose body weight ranges from 5 g in wrens to 448 g in crows, whose diets range from seeds to insects, and whose northern limits range from Florida to Canada − a remarkable finding.

SPREADING: DISPERSAL AND RANGE CHANGE

Some organisms − information is too scanty to say how many − have an actual geographical range that

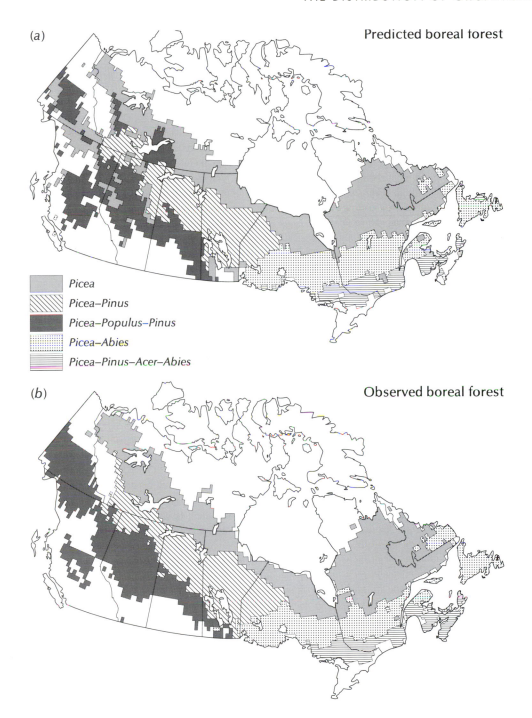

Figure 3.9 Boreal forest types in Canada. (a) Predicted forest types using a regression model. (b) Observed forest types. *Source*: After Lenihan (1993)

Figure 3.10 Predicted and observed longleaf pine (*Pinus palustris*) and the Florida poison tree (*Metopium toxiferum*) distributions in Florida, United States.
Source: After Box *et al.* (1993)

is smaller than their potential geographical range. In other words, some species do not occupy all places that their ecological tolerances would allow. Often, a species has simply failed to reach the 'missing bits'. Dispersal is the process whereby organisms colonize new areas. It is a vast subject that has long occupied the minds of ecologists and biogeographers.

Life on the move

Organisms disperse. They do so in at least three different ways (Pielou 1979: 243):

1 **Jump dispersal** is the rapid transit of individual organisms across large distances, often across inhospitable terrain. The jump takes less time than the life-span of the individual involved. Insects carried over sea by the wind is an example (see p. 60).

2 **Diffusion** is the relatively gradual spread of populations across hospitable terrain. It takes place over many generations. Species that expand their ranges little by little are said to be diffusing. Examples include the American muskrat (*Ondatra zibethicus*), spreading in central Europe after five

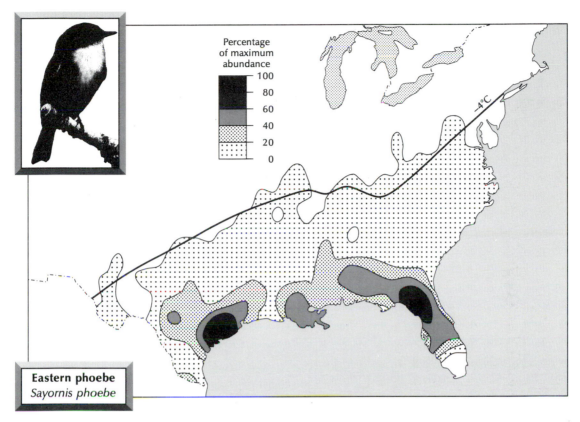

Figure 3.11 Winter distribution and abundance of the eastern phoebe (*Sayornis phoebe*). The northern boundary is constrained by the −4°C average minimum January isotherm.
Source: Map after T. L. Root (1988a, b); picture from Stokes and Stokes (1996)

individuals were introduced by a Bohemian landowner in 1905 (Elton 1958). It now inhabits Europe in many millions.

3 **Secular migration** is the spread or shift of a species that takes place very slowly, so slowly that the species undergoes evolutionary change while it is taking place. By the time population arrives in a new region it will differ from the ancestral population in the source area. South American members of the family Camelidae (the camel family) – the llama (*Lama glama*), vicuña (*Lama vicugna*), guanaco (*Lama guanicoe*), and alpaca (*Lama pacos*) – are examples. They are all descended from now extinct North American ancestors that underwent a secular migration during Pliocene times over the then newly created Isthmus of Panama.

Agents of dispersal

Species may disperse by **active** movement (digging, flying, walking, or swimming), or by **passive** carriage. Passive dispersal is achieved with the aid of physical agencies (wind, water) or biological agencies (other organisms, including humans). These various modes of transport are given technical names – **anemochore** for wind dispersal, **thalassochore** for sea dispersal, **hydrochore** for water dispersal, **anemohydrochore** for a mixture of wind and water dispersal, and **biochore** for hitching a ride on other organisms. Figure 3.12 shows the means by which colonists were carried to Rakata, which lies in the Krakatau Island group, after the volcanic explosion of 1883. Notice that sea-dispersed (thalassochore) species, most of which live

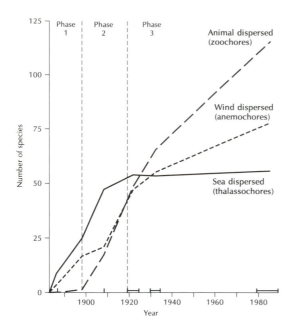

Figure 3.12 The means by which the spermatophyte flora reached Rakata, in the Krakatau Island group, from 1883 to 1989. The collation periods are indicated on the horizontal axis. Human introduced species are excluded. *Source*: After Bush and Whittaker (1991)

along strand-lines, are rapid colonizers. The anemochores comprise three ecological groups. The very early colonists are mostly ferns, grasses, and composites (members of the Compositae), which are common in early pioneer habitats. A second group is dominated by forest ferns, orchids, and Asclepiadaceae (milkweed, butterfly flower, and wax plant family). These second-phase colonists require more humid conditions. Numerically, most of them are epiphytes. The third group consists of seven primarily wind-dispersed trees. Animal-dispersed (zoochore) organisms are the slowest to colonize. They are mainly carried by birds and bats.

The efficacy of organisms as agents of dispersal is surprising. A study carried out in the Schwäbische Alb, south-west Germany, shows just how effective sheep are at spreading populations of wild plants by dispersing their seeds (Fischer *et al.* 1996). A sheep was specially tamed to stand still while it was groomed for seeds. Sixteen searches, each covering half of the

fleece (it was difficult to search all parts of the animal), produced 8,511 seeds from 85 species. The seeds were a mixture of hooked, bristled and smooth forms. They included sweet vernal-grass (*Anthoxanthum odoratum*), large thyme (*Thymus pulegioides*), common rock-rose (*Helianthemum nummularium*), lady's bedstraw (*Galium verum*), and salad burnet (*Sanuisorba minor*).

Good and bad dispersers

Dispersal abilities vary enormously. This is evident in records of the widest ocean gap known to have been crossed by various land animals, either by flying, swimming, or on rafts of soil and vegetation (Figure 3.13). The 'premier league' of transoceanic dispersers is occupied by bats and land birds, insects and spiders, and land molluscs. Lizards, tortoises, and rodents come next, followed by small carnivores. The poorest dispersers are large mammals and freshwater fish.

Not all large mammals are necessarily inept at crossing water. It pays to check their swimming proficiency before drawing too many biogeographical conclusions from their distributions. Fossil elephants, mostly pygmy forms, are found on many islands: San Miguel, Santa Rosa, and Santa Cruz, all off the Californian coast; Miyako and Okinawa, both off China; Sardinia, Sicily, Malta, Delos, Naxos, Serifos, Tilos, Rhodes, Crete, and Cyprus, all in the Mediterranean Sea; and Wrangel Island, off Siberia. Before reports on the proficiency of elephants as swimmers (D. L. Johnson 1980), it was widely assumed that elephants must have walked to these islands from mainland areas, taking advantage of former land bridges (though vicariance events are also a possibility). Now it is known that elephants could, apparently, have swum to the islands, new explanations for the colonization of the islands are required.

Supertramps are ace dispersers. They move with ease across ocean water and reproduce very rapidly. They were first recognized on the island of Long, off New Guinea (Diamond 1974). Long was devegetated and defaunated about two centuries ago by a volcanic explosion. The diversity of bird species is now far

**Plate 1 Yellow-footed rock wallaby
(Petrogale xanthropus)**
Photograph by T & P Gardner/Nature Focus

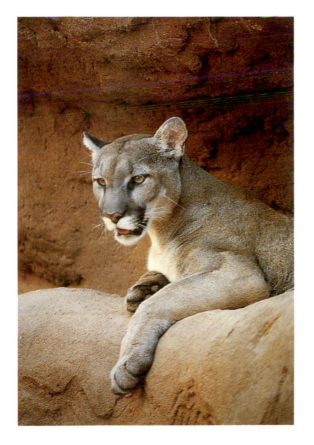

Plate 2 Puma or mountain lion (Felis concolor)
Photograph by Pat Morris

Plate 3 A cycad *(Zamia lindenii)* in Ecuador
Photograph by Pat Morris

**Plate 4 Common cassowary
*(Casuarius casuarius)***
Photograph by Pat Morris

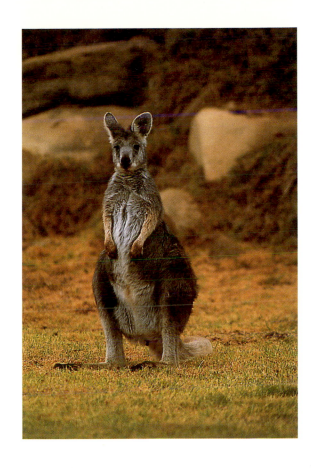

Plate 5 Common wallaroo or euro
(Macropus robustus)
Photograph by T & P Gardner/Nature Focus

Plate 6 Leadbeater's possum
(Gymnobelideus leadbeateri)
Photograph by Peter R. Brown/Nature Focus

Plate 7 Honey possum or noolbender *(Tarsipes rostratus)*
Photograph by Babs & Bert Weiss/Nature Focus

Plate 8 Cattle egrets *(Bubulcus ibis)*
Photograph by Pat Morris

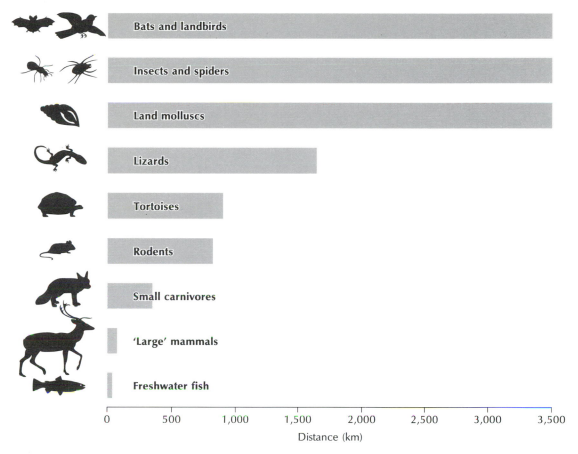

Figure 3.13 The widest ocean gaps crossed by terrestrial animals. The distances are extremes and probably not typical of the groups.
Source: After Gorman (1979)

higher than would be expected (Figure 3.14). Of the forty-three species present on the island, nine were responsible for the high density. These were the supertramps. They specialize in occupying islands too small to maintain stable, long-lasting populations, or islands devastated by catastrophic disturbance – volcanic eruptions, tsunamis, or hurricanes. They are eventually ousted from these islands by competitors that can exploit resources more efficiently and that can survive at lower resource abundances.

Dispersal routes

The ease and rate at which organisms disperse depend on two things: the topography and climate of the terrain over which they are moving and the wanderlust of a particular species. Topography and climate may impose constraints upon dispersing organisms. Obviously, dispersal will be more easily accomplished over hospitable terrain than over inhospitable terrain. Obstacles or **barriers** to dispersal may be classified according to the 'level of difficulty' in crossing them. George Gaylord Simpson (1940) suggested three types:

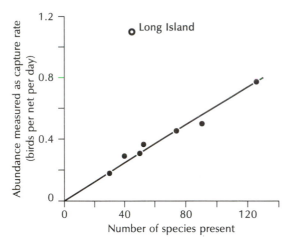

Figure 3.14 Bird abundance on various islands between New Guinea and New Britain. Abundance was measured by the daily capture rate in nets. Netting yields, which reflect total population densities, increase with the local number of species. Long Island is exceptional because population density is abnormally high for an island of its size. This high density is due largely to the presence of supertramp species.
Source: After Diamond (1974)

1 'Level 1' barriers are **corridors** – routes through hospitable terrain that allow the unhindered passage of animals or plants in both directions.

2 'Level 2' barriers are **filter routes**. An example is a land bridge combined with a climatic barrier that bars the passage of some migrants. The Panamanian Isthmus is such a route since it filters out species that cannot tolerate tropical conditions.

3 'Level 3' barriers are **sweepstakes routes**. They reflect the fact that, as in gambling, there are always a small number of winners compared with losers. In biology, the winners are those few lucky individuals that manage to survive a chance journey by water or by air and succeed in colonizing places far from their homeland.

Simpson's terms apply primarily to connections between continents. A more recent scheme, though not dissimilar to Simpson's, is more applicable to connections between islands and continents and island with other islands (E. E. Williams 1989). It recognizes five types of connections:

1 **Stable land bridges**. These are 'filter routes' in Simpson's sense. Faunas are free to move in both directions.

2 **Periodically interrupted land bridges**. These are akin to stable land bridges, but there are two differences. First, access in both directions is periodically interrupted by water gaps. Second, faunas on one or both sides may suffer extinction, owing to the loss of area during times of separation.

3 **'Noah's arks'**. These are fragments of lithospheric plates carrying entire faunas with them from one source area to another (see p. 76). Ordinarily, transport on Noah's arks is one-way.

4 **Stepping stone islands**. These are a fairly permanent or temporary series of islands separated by moderate to small water gaps. Traffic could be two-way, but often goes from islands of greater diversity to islands of lesser diversity.

5 **Oceanic islands**. These are situated a long way from the mainland and they receive 'waifs'. They are similar to stepping stone islands, but there will be fewer arrivals that arrive at much longer time intervals. They are not true sweepstakes routes because the chances of arriving depends on species characteristics – some species have a much better chance of arriving than others.

Of course, for terrestrial animals, crossing land is not so difficult as crossing water, and many large mammals have dispersed between biogeographical regions. There is probably no barrier that cannot be crossed given enough time:

One morning [in Glacier Park], dark streaks were observed extending downward at various angles from saddles or gaps in the mountains to the east of us. Later in the day, these streaks appeared to be much longer and at the lower end of each there could be discerned a dark speck. Through binoculars these spots were seen to be animals floundering downward in the deep, soft snow. As they reached lower levels not so far distant, they proved to be porcupines. From every little gap there poured forth a dozen or twenty, or in one case actually fifty five, of these animals, wallowing down to the timberline on the west side. Hundreds of porcupines were crossing the main range of the Rockies.

(W. T. Cox 1936: 219)

Dispersal today

Dispersal undoubtedly occurs at present but it is normally difficult to observe. The chief problem in detecting dispersal in action is that detailed species distributions are scarce. Most instances of organisms moving to new areas probably pass unnoticed – is a new sighting an individual that has moved in from elsewhere, or is it an individual that was born in the area but has not been seen before? In addition, it is sometimes difficult to account for a range change. Species ranges alter through dispersal and local extinctions. Acting in tandem, dispersal and extinction may lead to **range expansion** (through all or any of the dispersal processes), to **range contraction** (from local extinction), or to **range 'creep'** (through a mixture of spread and local extinction). It is far from easy to establish the processes involved in actual cases.

Despite the problems of studying range changes, there are several amazing cases of present-day dispersal resulting from human introductions. **Introduced species** are taken to new areas, accidentally or purposely, by people. Not all introductions survive; some gain a foothold but progress little further; others go rampant and swiftly colonize large tracts of what is to them uncharted territory. Just four species of amphibians and reptiles live in Ireland, compared with twelve on the British mainland. The species are the natterjack toad (*Bufo calamita*), the common newt (*Triturus vulgaris*), the common or viviparous lizard (*Lacerta vivipara*), and the common frog (*Rana temporaria*). The common frog was introduced into ditches in University Park, Trinity College Dublin, in 1696. It still flourishes there today, and has spread to the rest of Ireland. So, why have only three species of amphibians and reptiles colonized Ireland? One explanation is that other newts and toads did establish bridgeheads, but they died out because they were unable to sustain large enough colonies for successful invasion. This view is lent some support by the present distribution of the natterjack toad in Ireland – it is restricted to a small part of Kerry and shows no signs of spreading.

A fine example of a **rampant disperser** is the European starling (*Sturnus vulgaris*). This bird has successfully colonized North America, South Africa, Australia, and New Zealand. Its spread in North America was an indubitable ecological explosion – within sixty years it had colonized the entire United States and much of Canada. There were several 'false starts' or failed introductions during the nineteenth century when attempts to introduce the bird in the United States failed. For example, in 1899, twenty pairs of starlings were released in Portland, Oregon, but they vanished. Then in April 1890, eighty birds were released in Central Park, New York, and in March the following year a further eighty birds were released. Within ten years the European starling was firmly established in the New York City area. From that staging post, it expanded its range very rapidly, colonizing some 7,000,000 km^2 in fifty years (Figure 3.15). The speed of dispersal was due to the irregular migrations and wanderings of non-breeding 1- and 2-year-old starlings. Adult birds normally use the same breeding ground year after year and do not colonize new areas. The roaming young birds frequented faraway places. Only after five to twenty years of migration between the established breeding grounds and the new sites, did the birds take up permanent residence and set up new breeding colonies. For example, the European starling was first reported in California in 1942. It first nested there in 1949. Large-scale nesting did not occur until after 1958.

Dispersal in the past

Centres-of-origin and dispersal

Historical biogeography began as a scholarly debate about the restocking of the Earth from Mount Ararat after Noah's Flood. The debate evolved into the classical theory of centres-of-origin and dispersal first proposed by Charles Robert Darwin and Alfred Russel Wallace. Darwin argued that all species have a centre of origin from which they disperse; and that, because barriers impede their movement, species spread slowly enough for natural selection to cause them to change while dispersing. He saw dispersal as

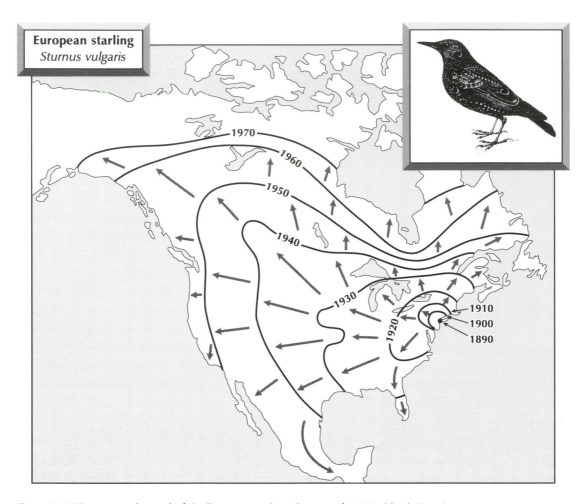

Figure 3.15 The westward spread of the European starling (*Sturnus vulgaris*) in North America.
Source: Map after Kessel (1953) and Perrins (1990); picture from Saunders (1889)

a phenomenon of overarching importance, and his **centre-of-origin–dispersal model** became the ruling theory in historical biogeography. It was developed persuasively in the twentieth century by Ernst Mayr, George Gaylord Simpson, and Philip J. Darlington.

South American invasions

A classic example of a dispersal model is the mammalian invasions of South America (Simpson 1980). Although South America is now connected to North America, it was an island-continent for most of the last 65 million or so. On two occasions during that time, a land connection with North America, probably through a chain of islands, developed for a few million years and was then lost. During these times, and in recent times, mammals invaded South America from the north. Four phases of invasion are recognized (L. G. Marshall 1981a) (Figure 3.16):

1 **Phase I**, which occurred in the Late Cretaceous period and earliest Palaeocene epoch, was an invasion by 'old timers' or 'ancient immigrants'. Three orders of mammal invaded – only two of

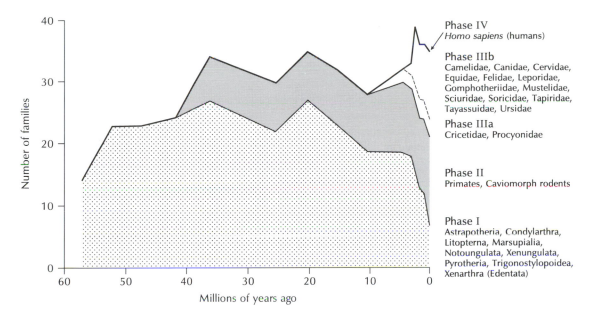

40

30

Number of families

20

10

0

60   50   40   30   20   10   0

Millions of years ago

Phase IV
Homo sapiens (humans)

Phase IIIb
Camelidae, Canidae, Cervidae,
Equidae, Felidae, Leporidae,
Gomphotheriidae, Mustelidae,
Sciuridae, Soricidae, Tapiridae,
Tayassuidae, Ursidae

Phase IIIa
Cricetidae, Procyonidae

Phase II
Primates, Caviomorph rodents

Phase I
Astrapotheria, Condylarthra,
Litopterna, Marsupialia,
Notoungulata, Xenungulata,
Pyrotheria, Trigonostylopoidea,
Xenarthra (Edentata)

Figure 3.16 The history of South American mammals.
Source: After L. G. Marshall (1981a)

which are represented in South America today – the marsupials, the xenarthrans (edentates in some classifications), and the condylarthrans. Once isolated again, these ancestral stocks evolved independently of mammals elsewhere and some unique forms arose. The first marsupial invaders were didelphoids and included ancestors of the modern opossums (Box 3.2). The borhyaenids were dog-like marsupial carnivores. They bore a strong resemblance to the Australian thylacine or Tasmanian wolf, the likeness being the result of convergent evolution. *Thylacosmilus*, a marsupial equivalent of the sabre-toothed tigers, was also a splendid example of convergent evolution (Figure 3.18). The xenarthrans ('strange-jointed mammals') were ancestors of living armadillos, sloths, and anteaters, and extinct glyptodonts (Figure 3.19). The condylarthrans produced a spectacular array of endemic orders of hoofed mammals, all of which are now extinct – the condylarthra, litopterns, notoungulates, astrapotheres, trigonstylopoidea, pyrotheres, and xenungulates (Figure 3.20).

2 **Phase II** involved a brief 'window of dispersal opportunity' arising in the Late Eocene and Early Oligocene epochs, from about 40 million to 36 million years ago. During this time, a series of islands linked the two American continents. Two groups of mammal invaded South America – primates and ancestors of the caviomorph rodents. These were the 'ancient island hoppers'. After having arrived in South America, both groups underwent an impressive adaptive radiation to produce the great variety of rodents found in South America today (as well as some interesting extinct forms – the Late Pleistocene *Telicomys gigantissimus* was nearly as large as a rhinoceros) and the New World monkeys.

3 **Phase III** began in the Late Miocene epoch, some 6 million years ago. Then, the Bolivar Trough connected the Caribbean Sea with the Pacific Ocean and deterred the passage of animals (Figure 3.21). Phase IIIa saw the first mammals rafting across the seaway on clumps of soil and vegetation. These 'ancient mariners' were members of two families – the 'field mouse' family (Cricetidae)

Box 3.2

A CURIOUS SOUTH AMERICAN MARSUPIAL

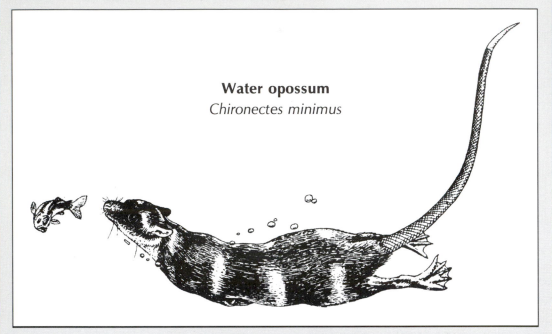

Water opossum

Chironectes minimus

Figure 3.17 The yapok or water opossum (*Chironectes minimus*).
Source: From Rodríguez de la Fuente (1975)

The yapok or water opossum (*Chironectes minimus*) is a curious living didelphid marsupial. It is an otter-like aquatic carnivore with webbed hind feet, living along river banks and eating mostly crayfish, fish, and frogs (Figure 3.17). It is the only aquatic marsupial in the world. The female's pouch is closed by a sphincter muscle when diving to prevent her babies from drowning.

and the racoon and its allies family (Procyonidae). By 3 million years ago, there was a complete land connection – the Panamanian land bridge – that furnished a gateway for faunal interchange between North and South America. Phase IIIb began as a flood of mammals simply walked into South America. Members of many families were involved – camels (Camelidae), dogs (Canidae), deer (Cervidae), horses (Equidae), cats (Felidae), rabbits (Leporidae), mastodons (Gomphotheriidae), weasels (Mustelidae), squirrels (Sciuridae), shrews (Soricidae), mice (Muridae), tapirs (Tapir-

idae), peccaries (Tayassuidae), and bears (Ursidae). The traffic was two-way – many South American species travelled northwards and entered North America.

4 **Phase IV** started about 20,000 years ago (the exact date is debatable) as humans spread into South America.

The outcome of these waves of invasion is distinct **faunal strata** in the present mammals of South America. Anteaters, sloths, armadillos, and opossums are survivors of the first invasion. The caviomorph

Figure 3.18 Convergent evolution of the marsupial sabre-tooth *Thylacosmilus* and the placental sabre-tooth *Smilodon*.
Source: After L. G. Marshall (1981a)

rodents and New World monkeys are survivors of the second invasion. All other South American mammals (save recent introductions) are survivors the Great American Interchange.

SPLITTING: VICARIANCE AND RANGE CHANGE

What is vicariance?

Vicariance biogeography arose as an antidote to the hegemony of dispersal biogeography. Anti-dispersalist grumblings were heard early in the twentieth century. The first major critical onslaught was directed by Léon Croizat, a Franco-Italian scholar (Croizat 1958, 1964).

Croizat tested the Darwinian centre-of-origin–dispersal model by mapping the distributions of hundreds of plant and animal species. He found that species with quite different dispersal propensities and colonizing abilities had the same pattern of geographical distribution. He termed these shared geographical distributions generalized or standard tracks. Croizat argued that standard tracks do not represent lines of migration. Rather, they are the present distributions of a set of ancestral distributions, or a biota of which individual components are relict fragments. His reasoning was that widespread ances-

tral taxa had been fragmented by tectonic, sea-level, and climatic changes. He termed the fragmentation process 'vicariism' or 'vicariant form-making in immobilism'. Fortunately, it has become known as **vicariance**. Of course, to become widespread in the first place, a species must disperse. But vicariance biogeographers claim that ancestral taxa achieve widespread distribution through a mobile phase *in the absence of barriers*. They allow that some dispersal across barriers does occur, but feel that it is a relatively insignificant biogeographical process.

Continental fission

Continents break up, drift, and collide. In doing so, they make and break dispersal routes and alter species distributions. Overall, continental drift, and processes associated with it (such as mountain building), help to explain several features in animal and plant distributions for both living and fossil forms.

The tearing asunder of previously adjoining land masses causes the separation of ancestral populations of animals and plants. Once parted, the populations evolve independently and diverge. Eventually, they may become quite distinct from the ancestral population that existed before the land masses broke apart. There are several good examples of this process.

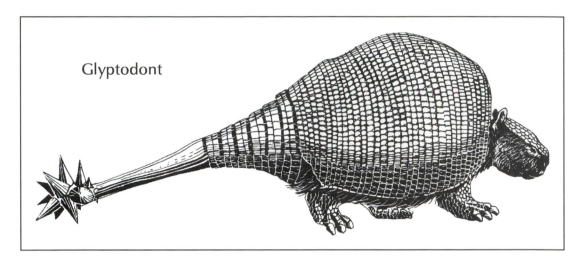

Figure 3.19 A glyptodont. This tanklike Pleistocene herbivore was descended from old xenarthran invaders.
Source: After L. G. Marshall (1981a)

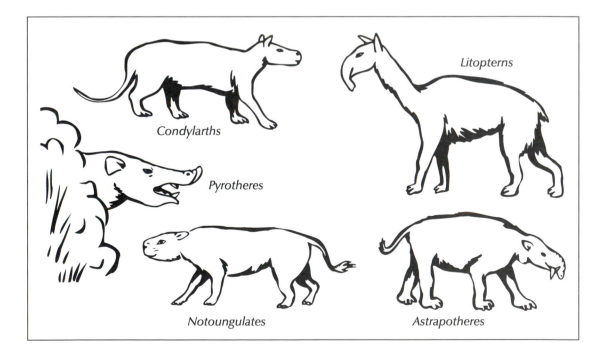

Figure 3.20 Unique South American mammals that evolved from Late Cretaceous condylarthran invaders.
Source: After Simpson (1980)

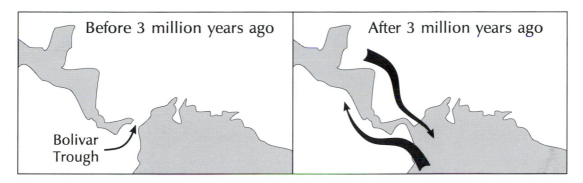

Figure 3.21 The Bolivar Trough and Panamanian Isthmus. Before 3 million years ago, a marine barrier separated Central America and South America. When the barrier dried up, the Panamanian land bridge was formed. Animals ventured northwards and southwards in the Great American Interchange.
Source: After L. G. Marshall (1981b)

Reptiles and mammal-like reptiles on Pangaea

The breakup of Pangaea, and in particular the southern part of it known as **Gondwana**, split several animal and plant populations into separate groups. In consequence, members of these ancient populations, and their living descendants, have disjunct distributions. One of the first pieces of fossil evidence used in support of the idea that Africa and South America were formerly joined was the distribution of a small reptile called *Mesosaurus* that lived about 270 million years ago, in the Permian period (Figure 3.22). *Mesosaurus* was about a metre long, slimly built with slender limbs and paddle-like feet. At the front end it had extended, slender jaws carrying very long and sharp teeth, probably used to catch fish or small crustaceans. The rear end was a long and deep tail, admirably suited to swimming. It surely spent much time in the water. The sediments in which it has been found suggest that it inhabited freshwater lakes and ponds. Its remains have been found only in southern Brazil and southern Africa. If, as the evidence suggests, it were a good swimmer, then it would have had a wider range than it apparently did have. The most parsimonious explanation is that Brazil and Africa abutted in the Late Palaeozoic era. As Alfred Sherwood Romer (1966: 117) said, 'although *Mesosaurus* was obviously a competent swimmer in

fresh waters, it is difficult to imagine it breasting the South Atlantic waves for 3,000 miles'.

Another disjunct distribution is displayed by *Lystrosaurus*, a squat, powerful, mammal-like reptile. *Lystrosaurus* lived in the Early Triassic period, about 245 million years ago. It was about a metre long, with a pair of downwards-pointing tusks. The position of its eyes, high on its head, suggest that it spent much time submerged with all but the top of its head below water. Specimens of *Lystrosaurus* have been found at localities in India, Antarctica, South Africa (all formerly part of Gondwana), and China (Figure 3.23). The distribution of *Lystrosaurus* on Gondwana accords with modern ranges of terrestrial reptiles, such as the snapping turtle, *Chelydra serpentina* (Colbert 1971). Its presence in China, in a place that is normally assumed to have been part of eastern Laurasia in Triassic times, presents a problem. One explanation, though an unlikely one, is that some individuals migrated from Gondwana, all the way round the head of the Tethys embayment, to eastern Laurasia. Another explanation is that Gondwana was bigger than is normally supposed, and extended beyond the northern edge of the Indian plate to include a large segment of what is now China (Crawford 1974). If this were the case, all the *Lystrosaurus* faunas would have lived on Gondwana, and the Chinese members of the population would not be anomalous. Another possibility, suggested by

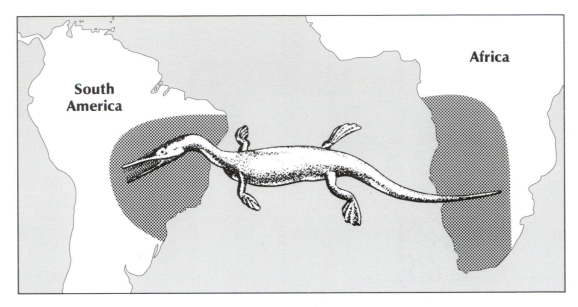

Figure 3.22 The distribution of *Mesosaurus* in the Permian period

work on plate tectonics in the South-east Asian region, is that in Late Permian times, or even before, large fragments of Australia (terranes) appear to have broken off and drifted northwards, colliding with Asia. These terranes could have carried, in the manner of Noah's arks, *Lystrosaurus* with them, or could have acted as a series of stepping stones making dispersal possible (C. B. Cox 1990: 125).

Cenozoic land mammals

During Cenozoic times, land-animal evolution was greatly affected by continental drift (Kurtén 1969). Fossil Cenozoic faunas, and particularly the mammal faunas, are fairly different on each continent. The most distinctive faunas are those of South America, Africa, and Australia – the southern continents. There are some thirty orders of mammal, almost two-thirds of which are alive today. All the mammalian orders probably had a common origin – an ancestor that lived in the Mesozoic era. The Cenozoic divergence of mammals may be due to the Late Mesozoic fragmentation of Pangaea. Interestingly, the land-dwelling animals (mostly reptiles) that lived in the Cenozoic era displayed far less divergence than the

mammals. By the end of the Cretaceous period, after about 75 million years of evolution, some seven to thirteen orders of reptiles had appeared, far fewer than the thirty or so produced by mammalian evolution over 65 million years. A possible explanation for this difference lies in the palaeogeography of the Mesozoic continents. For much of the Age of Reptiles, the continents were not greatly fragmented. There were two supercontinents: Laurasia lay to the north and Gondwana to the south. Rifts between the continents existed as early as the Triassic period, but they were not large enough to act as barriers to dispersal until well into the Cretaceous period. In Late Cretaceous and Early Tertiary times, when the mammals began to diversify, the rapid breakup of the former Pangaea, coupled with high sea-levels, led to the separation of several land masses and genetic isolation of animal populations. The result was divergence of early mammalian populations.

Pangaea and plant distributions

The breakup of Pangaea accounts for many disjunct plant distributions. Some plants have seeds unsuitable for jump dispersal. An example is the genus

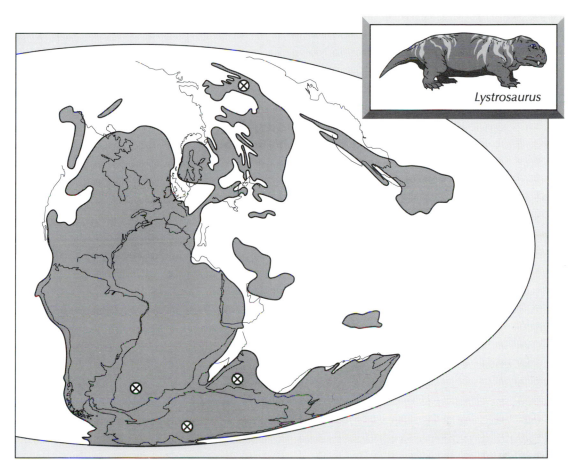

Figure 3.23 The distribution of *Lystrosaurus* in the Early Triassic period.
Source: Base map after A. G. Smith *et al.* (1994)

Nothofagus (the southern beeches) that consists of about sixty species of evergreen and deciduous trees and shrubs. Its present distribution is disjunct, being found on remnants of Gondwana – South America, New Zealand, Australia, New Caledonia, and New Guinea, but not Africa. Fossil *Nothofagus* pollen of Oligocene age has been found on Antarctica. The conclusion normally drawn is that the modern disjunct range of the genus has resulted from the breakup of Gondwana. The same explanation applies to plants of the protea, banksia, and grevillea family – the Proteaceae (p. 48).

Big, flightless birds

Continental breakup helps to explain the enigma of the large flightless birds. This group forms an avian superorder – the **ratites**. There are five families of living ratites and two extinct ones, all of which are thought to have arisen early in bird evolution from a common ancestor and thus are a monophyletic group. The moas (Dinornithidae) lived on New Zealand until a few hundred years ago; the kiwis (Apterigidae) still live on New Zealand (Plate 3.2). Emus (Dromaiidae) and cassowaries (Casuariidae)

Plate 3.2 Brown or spotted kiwi (*Apterix australis*)
Photograph by Pat Morris

both live in Australia, and the cassowaries are also found on New Guinea (Colour plate 4). The extinct dromorthinids are also Australian ratites. Ostriches (Struthionidae) live in Africa and Europe. The elephant birds (Aepyornithidae) lived on Madagascar and possibly Africa (e.g. Senut *et al.* 1995), but went extinct around the mid-seventeenth century. Rheas (Rheidae) live in South America.

With the exception of the chicken-sized kiwis, most of the ratites are well-built cursorial birds with huge hind limbs for fast and sustained running. The ostrich is the world's largest living bird, standing about 2.5 m high and weighing in at about 140 kg. The largest of the moas, *Diornis maximus*, stood about 3.5 m high and weighed around 240 kg; it was the tallest bird known to have lived. The largest elephant-bird, *Aepyornis maximus*, was a ponderous giant, with elephantine-style legs; it stood 3 m tall and weighed about 450 kg. How could such large birds have spread through the southern continents if they could not fly, or even if they could? A plausible hypothesis involves the early evolution of flightlessness and continental drift (Cracraft 1973, 1974). A flying ancestor of all the ratites probably lived in west Gondwana, or what is now South America. The flightless tinamous, partridge-like birds that have their own superorder (the Tinamae), may be descendants of this extinct bird that stayed at home in South America. Flightlessness may have evolved during the Cretaceous period, before continental separation was far advanced. The flightless birds could then have dispersed through the southern continents by walking to other parts of Gondwana. The ancestors of the New Zealand moas and kiwis must have walked through west Antarctica before New Zealand drifted away from the main land mass. Interestingly, a fossil ratite has recently been discovered in the Palaeogene La Meseta Formation, Seymour Island, Antarctica (Tambussi *et al*. 1994). The recently extinct elephant birds of Madagascar could have walked directly from South America. The emus and cassowaries may have walked through east Antarctica. The ostriches of Africa and Europe may have walked directly from South America, and the South American rheas presumably evolved from less adventurous South American flightless relatives.

The Gondwanan origin of the ratites is challenged by more recent work (e.g. Houde 1988). Forms ancestral to modern ratites appear to have evolved in North America and Europe from the Late Palaeocene to the Middle Eocene epochs. This would mean that living ratites are Southern Hemisphere relics of a widespread group that probably originated in the Early Tertiary period.

Greater Antillean insectivores

Vicariance events in the geological history of the Caribbean region may explain the distribution of **relict insectivores** in the Greater Antilles. There are two such insectivores – *Nesophontes* and *Solenodon*. *Nesophontes* lived on Cuba, Hispaniola, Puerto Rico, the Cayman Islands, and smaller surrounding islands. *Solenodon* still lives on Cuba and Hispaniola (Plate 3.3). They were, and in *Solenodon*'s case are, shrew-like animals but larger than a normal shrew. *Solenodon* is about 15–16 cm and weighs 40–46 g. It is nocturnal and lives in caves, burrows, and rotten trees.

One explanation of the relict insectivore distribution runs as follows (MacFadden 1980). In the Late Cretaceous, North American and South American land masses were joined by a land mass that was to become the Antilles. At this time, ancestral insectivores lived on North America and the proto-Antillean block. Early in the Cenozoic era, the proto-Antilles moved eastwards, relative to the rest of the Americas, carrying a cargo of insectivores with it. Later in the Cenozoic era, the proto-Antilles had reached their present position, and the gap between North and South America

North and South America was filled by the lower central American land mass. However, the mainland relatives of *Nesophontes* and *Solenodon* on North America were by now extinct, leaving their Antillean relatives as relicts of a once widespread distribution.

The history of the Greater Antillean insectivores may be interpreted in other ways. The traditional view is that they colonized the islands by over-water dispersal from the Americas – a sweepstakes route. Support for this interpretation came from a study of Caribbean tectonics, the composition of the fauna, and the fossil record in the Americas (Pregill 1981). The evidence suggested that the Antilles started life a volcanic archipelago in the Late Cretaceous epoch. Modern terrestrial vertebrates probably started arriving by over-water dispersal during the Oligocene epoch, when most of the living genera in the West Indies first appeared, and continued to do so through the Quaternary. The various living genera and species are not evenly distributed throughout the islands and the vertebrates are represented by remarkably few orders, families, and genera. This composition is consistent with an island biota built up though dispersal, rather than vicariance.

A reconstruction of geological history in the Caribbean region may offer a compromise between the dispersal and vicariance explanations of the Antillean insectivores and other animals and plants (Perfit and Williams 1989). About 130 million years ago, the proto-Antilles were a chain of volcanic islands lying along a subduction zone at the Pacific Ocean rim (Figure 3.24). They stretched 2,000 km between the west coast of Mexico and the coast of Ecuador. Some 100 million years ago, the North American and South American plates started to move westwards over the proto-Caribbean sea floor, and the islands drifted relatively eastwards, at the leading edge of the Caribbean plate. By 76 million years ago, Cuba struck the Yucatán region of Mexico. The remaining islands suffered uplift and deformation as they squeezed through the gap between the Yucatán and Colombia. A land bridge between the Americas (the proto-Costa Rica–Panama land bridge) had begun to grow along a new subduction line formed where the Pacific Ocean dived underneath the

Plate 3.3 A relict Antillean insectivore (*Solenodon*)
Photograph by Pat Morris

Caribbean plate. On occasions, the proto-Antillean islands may have formed a complete dry land connection between the Americas around this time. The proto-Antilles stayed close to the North American continent for the next 20 million years, and indeed were often connected to it. About 55 million years ago, Cuba hit the Bahamas bank, a large limestone platform, and became wedged there. The remaining islands kept moving eastwards, causing considerable shear against the beached Cuba. The shear stress caused the islands to break up and adopt their modern configuration. Puerto Rico broke away from Hispaniola about 35 million years ago; Hispaniola separated from Cuba about 23 million years ago.

These tectonic changes affected the biogeography of the area. Frogs of the genus *Eleutherodactylus*, which are all small (one from eastern Cuba is at 0.9 cm the smallest four-legged beast in the world), originally came from South America (Hedges 1989). They appear to have moved along the proto-Antillean island chain around 70 million to 80 million years ago. They disembarked when the island reached the Yucatán, and established a foothold on the North American continent. Today, there are 68 species in Mexico. The frogs that chose not to disembark have produced the 139 species found on the Greater Antillean islands at the present time. The traffic was two-way. While the *Eleutherodactylus* frogs went ashore onto North America, several North American species – including pines, butterflies, and todies (a kind of bird) – boarded the islands while they were docked. The presence of pines on the

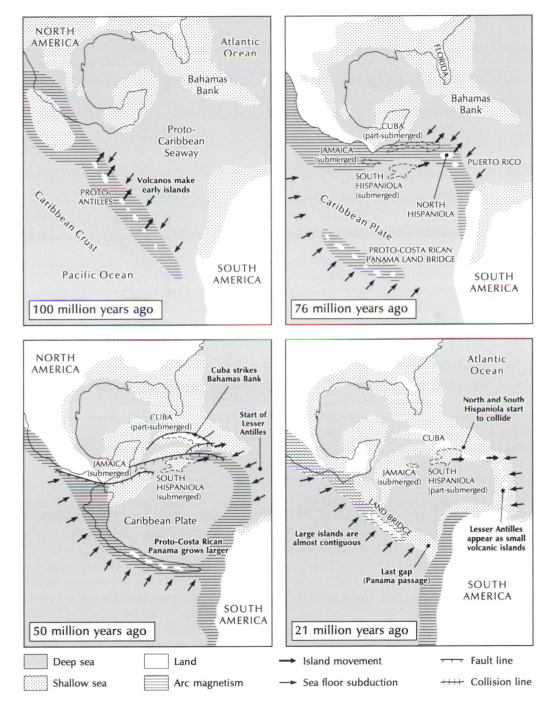

Figure 3.24 The geological history of the Caribbean region over the last 100 million years.
Source: After Perfit and Williams (1989) and Reddish (1996)

Caribbean islands is difficult to explain by dispersal because their heavy seeds are not designed to cross hundreds of kilometres of salt water. Similarly, many of the butterflies found on the Greater Antilles, including the fragile glasswings (family Ithomiidae) are not good dispersers. Todies are a family of birds endemic to the Caribbean. They are distantly related to the kingfishers. They are not powerful fliers. A 30-million-year-old tody from Wyoming at least allows the possibility that ancestors of the present birds could have boarded the islands early in their eastward travels, perhaps flying the short distance across to Cuba. At same time, ancestors of the primitive insectivores could have crossed over, either by walking or swimming the short distance. Most mammals had not evolved when the islands were docked against North America, but ancestors of these insectivores are known from fossils in North America to be at least 55 million to 60 million years old.

Continental fusion

Drifting continents eventually collide and fuse. When they do so, they unite to form a new single land mass. Uplift or a fall of sea-level may also forge a land bridge between two formerly separated land masses. A moving plate may carry species (indeed, entire faunas and floras) over vast distances, acting like a gigantic raft or **Noah's ark**; it may also carry a cargo of fossil forms, and is like a **Viking funeral ship** (McKenna 1973). The **biota** on these more-or-less isolated land masses should attain a steady state in which origination and extinction (turnover) will be roughly in balance. These biota should be saturated – all available niches should be filled.

Faunal mixing

When two continents collide, or a land bridge is formed between them, the fauna and flora of the two continents are free to mix. The mixing process has four possible outcomes (L. G. Marshall 1981a):

1 **Active competition** occurs between species or genera that occupy the same or very similar niches

on the different land masses. If the native form should be more efficient at exploiting resources, the invader will be expelled or kept out. Of course, such unsuccessful invasion attempts are seldom, if ever, recorded in the fossil record. The alternative outcome is that an invader is successful and ousts the native vicar.

2 **Passive replacement** may occur when, by chance, a native species sharing a niche with an invading species happens to go extinct. There is no competition involved – the surviving species was merely lucky to be in the right place at the right time.

3 Intruders may come across a biota with unexploited ecological niches and are able to insinuate themselves into the pre-existing community without having an overt affect on it. The **insinuators** are ecologically unique and have no discernible counterparts in the native biota. Insinuating species will boost the biotic diversity.

4 An invader may have a similar niche to a native species and character displacement (p. 118) takes place so that both species can live together. These invaders are **competitors-cum-insinuators**.

As the faunas and floras mix, so they come to resemble one another, though they may still retain distinctive features. The Great American Interchange is perhaps the most spectacular and best documented faunal mixing event. It resulted from the fusion of the North and South American continents and demonstrates many of the possible outcomes of faunal mixing.

The Great American Interchange

The faunas of North and South America began to mix at the close of the Miocene epoch, about 6 million years ago (p. 65). The **Great American Interchange** began as a trickle. It became a flood around 3 million years ago, when the Panamanian Isthmus formed, peaked around 2.5 million years ago, and is still going on at the present time.

It was once widely believed that the invasion of South America by North American species caused the extinction of many native taxa. The placental carnivores appeared and outcompeted marsupial carnivores, which became extinct. Likewise, South American 'ungulates' suffered heavy losses in competition with the northern invaders. Recent interpretations of the evidence paint a more complicated picture of events. Two cases illustrate this point (L .G. Marshall 1981a).

Before the invasions, South American 'ungulates' included litopterns (prototheres and macrauchenids) and notoungulates (toxodons, mesotheres, and hegetotheres). To these should be added five families of xenarthrans and even two of rodents that occupied the large herbivore niche. Once the Bolivar Trough was closed, many families of northern ungulates travelled south (mastodons, horses, tapirs, peccaries, camels, and deer). There is some evidence that the disappearance of native 'ungulates', especially the notoungulates, began before the appearance of northern rivals. The factors causing this decline were thus not related to the interchange. This conclusion is supported by the fact that no invading northern species were vicars (ecological equivalents) of southern species – they occupied slightly different niches. The most likely vicars were invading 'camels' and the camel-like macrauchenids. However, these two groups coexisted for 2.5 million years.

It was originally claimed that the South American dog-like borhyaenids (marsupials) declined to the point of extinction when in competition with invading placental carnivores from the north. However, it now seems clear that the borhyaenids were extinct before dogs, cats, and mustelids moved south. Some large omnivorous borhyaenids declined and fell when the waif members of the Procyonidae arrived in Phase IIIa (p. 65). For instance, *Stylocynus*, a large, omnivorous, bear-like borhyaenid, had a vicar in *Chapalmalania*, a large, bear-like procyonid. In turn, *Chapalmalania* vanished when members of the bear family (Ursidae) arrived. It is also possible that the placental sabre-tooths (felids) replaced the native marsupial sabre-tooths (thylacosmilids).

When the Panamanian Isthmus triggered the Great American Interchange, a large majority of land-mammal families crossed reciprocally between North and South America around 2.5 million years ago, in Late Pliocene times. Initially, there was an approximate balance between northward traffic and southward traffic. During the Quaternary, the interchange became decidedly unbalanced (S. D. Webb 1991). Groups of North American origin continued to diversify at exponential rates. In North America, extinctions more severely affected South American immigrants – six South American families were lost in North America, while two North American families were lost in South America.

RANGE CHANGE: HUMANS AND DISPERSAL

To an extent, the actual range of a species is a dynamic, statistical phenomenon that is constrained by the environment: in an unchanging habitat, the geographical range of a species can shift owing to the changing balance between local extinction and local invasion. And, it may also enlarge or contract owing to historical factors, as so plainly shown by the spread of many **introduced species** and chance colonizers in new, but environmentally friendly, regions. Humans have advertently and inadvertently aided and abetted the spread of many species. An example of a deliberate introduction is the coypu (*Myocastor coypus*), which, in the 1930s, was brought from South America to Britain for its fur (nutria). Numerous escapes occurred and it established itself in two areas: at a sewage farm near Slough, where a colony lived from 1940 to 1954, and in East Anglia, with a centre in the Norfolk Broads (Lever 1979). A concerted trapping programme is believed to have eradicated the coypu from Britain (Gosling and Baker 1989). An accidental introduction was the establishment of the ladybird (*Chilocorus nigritus*) in several Pacific islands, north-east Brazil, West Africa, and Oman after shipment from other areas (Samways 1989).

The success of many **introductions** is beyond question. Indeed, it is ironical that humans are brilliantly successful in the unintentional extermination of some species, mainly through habitat alteration and fragmentation, but hopelessly unsuccessful in the purposeful eradication introduced species that have become pests. Invasion success depends on the inter-action between the invader and the community it is invading. Predicting the fate and impact of a specific introduced species is very difficult (Lodge 1993). For instance, domestic and wild-type European rabbits have been liberated on islands all over the world (Flux and Fullagar 1992). The outcome of these introductions ranges from utter failure to rabbit den-sities so high that the island is stripped of vegetation and soil. Some rabbit populations have survived remarkably adverse conditions for up to a hundred years and then become extinct.

There are places where alien animals and plants are of major economic and conservation significance. A prime example is New Zealand (Atkinson and Cameron 1993). Plant introductions have averaged eleven species per year since European settlement in 1840, and distinctive landscapes are being increas-ingly altered by weeds. Many introduced animals act as disease vectors or threaten native biota. Recent studies of introduced wasps show adverse effects on honey-eating and insectivorous birds. Introduced possums prey on eggs and nestlings of native birds, damage native forests, and transmit bovine tuber-culosis.

A few examples drawn from the animal and fungal kingdoms will serve to illustrate the rates of spread and the impact that invaders may have on native communities.

Animal introductions

American mink and muntjac deer in Britain

The American mink (*Mustela vison*) is a medium-sized mustelid carnivore with a long body. It was introduced to British fur farms from North America in the 1930s. Some individuals escaped and soon became established in the wild. Mink is now found in many parts of the Britain and will continue to spread (Figure 3.25a). As a carnivore, it has had a rather different impact on native wildlife than other introduced mammals. A crucial question is whether the mink occupies a previously vacant niche with

an anticipated mild overall ecological impact, or whether it is a species that is endangering com-petitors such as the otter (*Lutra lutra*) and prey species (including fish stocks). A survey helped to resolve this issue, and drew five conclusions (Birks 1990). First, the mink is little threat to the otter, although it exacerbates otter decline in areas where the otter is already endangered. Second, the mink probably has, locally, aided the decline of water vole (*Arvicola terrestris*). Third, if the mink should have had any effects on waterfowl, then these have not been translated into widespread population declines. Fourth, there are grave potential risks of introducing mink to offshore islands. Fifth, the mink is not having a serious overall impact on fish stocks, at least in England and Wales.

The muntjac deer (*Muntiacus reevesi*) is small, standing about 50 cm at the shoulder (Plate 3.4). Its small size allows it to live in copses, thickets, neglected gardens, and even hedgerows (N. Chapman *et al*. 1994). Following the first releases from Woburn, Bedfordshire, in 1901, the numbers of free-living Reeves' muntjac in Britain remained low until the 1920s, when populations were largely confined to the woods around Woburn, and possibly also around Tring in Hertfordshire. However, in the 1930s and 1940s there were further deliberate introductions in selected areas some distance from Woburn. As a consequence, the subsequent spread of Reeves' munt-jac was from several foci (Figure 3.25b). The spread in the second half of this century has been aided by further deliberate and accidental releases, and by these means new populations continue to established themselves outside the main range. Thus, the natural spread has been much less impressive than previously assumed; even in areas with established populations it takes a long time for muntjac deer to colonize all the available habitat. The natural rate of spread is probably about 1 km a year, comparable to that of other deer species in Britain. Arable land classes are predominantly selected for, and marginal upland land classes tend to be avoided. However, long-estab-lished populations in areas such as Betws-y-Coed in Wales show that muntjac deer may persist in low numbers in atypical habitats.

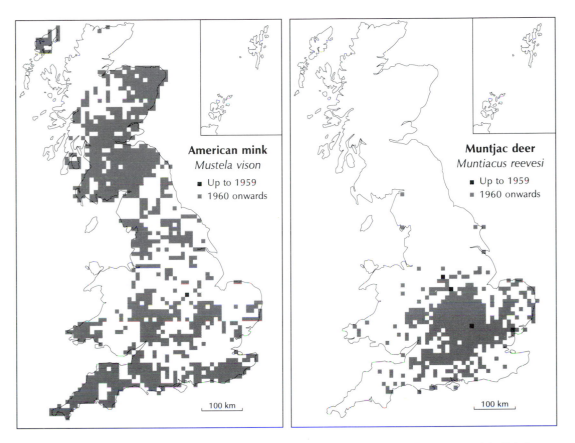

Figure 3.25 The distribution of the American mink (*Mustela vison*) and muntjac deer (*Muntiacus reevesi*) in Britain. *Source*: After H. R. Arnold (1993)

Introduced predators on islands

The Indian mongoose (*Herpestes auropunctatus*) is one of the most potent predators on diurnal ground-foraging lizards. It has been introduced to various islands world-wide in the hope of controlling rats and other vertebrate pests. Its success in doing so has been mixed. Its success in reducing and causing the extinction of native bird and reptile populations is spectacular (Case and Bolger 1991). This is demonstrated by a diurnal lizard census on Pacific islands with and without mongooses (Figure 3.26). Islands without mongooses have nearly a hundred times greater diurnal lizard abundance than islands with mongooses.

Other potent introduced island predators are rats and domestic cats and dogs. Cats and the tree rat (*Rattus rattus*) are capable of climbing trees and affect species that the mongoose is less likely to capture. New Zealand lizards became far less common after the mid-nineteenth century, owing to cover reduction through forest clearance and predation by cats. Present-day islands without mammalian predators around New Zealand house extraordinarily high numbers of lizards, with one lizard per 3 m^2 being recorded. The same pattern is found elsewhere. High numbers of diurnal lizards are found on rat-free Cousin Island in the Seychelles and on San Pedro Martir in the Sea of Cortez, Mexico. Some introduced reptiles have also caused extinctions. The introduced

Plate 3.4 Muntjac deer (*Muntiacus reevesi*)
Photograph by Pat Morris

brown tree snake (*Boiga irregularis*) has eliminated ten bird species and severely affected the lizard population on Guam, the largest of the Mariana Islands.

Introduced predators produce a distinctive biogeographical pattern, namely a **reciprocal co-occurrence pattern**. This means that many native amphibians and reptiles occur on islands without predators, but are absent from islands with predators. An example is the tuatara (*Sphenodon punctatus*) and its predator, the Polynesian rat (*Rattus exulans*), in New Zealand. On islands where the rat is present, the tuatara is either absent or not breeding. The largest surviving frog, the Hamilton frog (*Leiopelma hamiltoni*), occurs only on rat-free islands. On Viti Levu and Vanua Levu, the two largest islands of Fiji, the combination of cats and mongooses has proved devastating. Two

ground-foraging emos skinks (*Emoia nigra* and *Emoia trossula*) are locally extinct, and the two largest skinks in Fiji have not been seen on these islands for over a century, although they do survive on mongoose-free islands.

A fungal introduction

Around 1900, the American chestnut (*Castanea dentata*) comprised 25 per cent of the native eastern hardwood forest in the United States. It was a valuable forest species – its hardwood was used for furniture, its tall straight timbers for telegraph poles, its tannin for leather tanning, and its nuts for food. By 1950, most trees were dead or dying as a result of **chestnut blight**, a parasitic disease caused by a

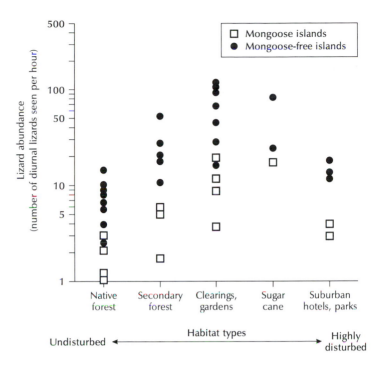

Figure 3.26 The effect of the Indian mongoose (*Herpestes auropunctatus*) on Pacific island lizards. The censuses were conducted during sunny days from 1984 to 1988. Each point represents the average of 2–4 censuses.
Source: After Case and Bolger (1991)

fungal parasite – the sac fungus, *Cryphonectria* (*Endothia*) *parasitica*. The fungus enters the host tree through a bark wound made by woodpeckers or bark-boring insects. It grows into the bark and outer sapwood forming a spreading, oozing sore that eventually encircles or 'girdles' the truck or branch. Once girdled, the tree dies because nutrient supplies to and from the roots are stopped. The time from initial attack to death is two to ten years.

The sac fungus is native to Asia. In 1913, it was found on its natural host in Asia – the Chinese chestnut (*Castanea mollissima*) – to which it does no serious harm. It was introduced into New York City by accident, probably in 1904, on imported Asian nursery plants. The first infected trees were found in the Bronx Zoo. The American chestnut is so susceptible to *Cryphonectria* that it was removed from almost its entire range within forty years. All that remained were a few individuals lucky enough to have escaped

the fungal spores, and adventitious shoots spouting from surviving root systems.

In the wake of the chestnut blight, several oak species, beech, hickories, and red maple have become co-dominants: the oak–chestnut forest has turned into oak or oak–hickory forests. This is evident in Watershed 41, western North Carolina (Figure 3.27). Here, the disease first attacked in the late 1920s. The survey carried out in 1934 shows the original forest composition, as the disease had not killed many trees by that time. The American chestnut is plainly the dominant species, occupying over twice the basal area of various hickory species. By 1954, after the forest had borne the brunt of the fungal attack, the tree species composition had altered dramatically. The chestnut had all but disappeared, and hickories (*Carya* spp.), the chestnut oak (*Quercus prinus*), the black oak (*Quercus velutina*), and the yellow poplar (*Liriodendron tulipifera*) had become co-dominants.

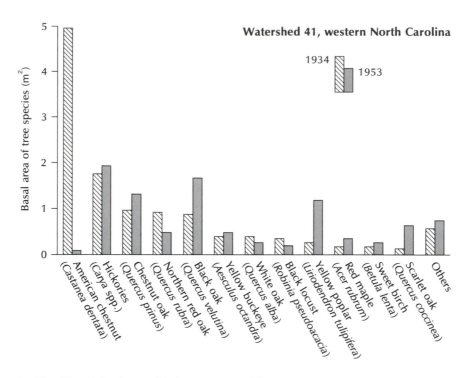

Figure 3.27 The effect of the chestnut blight in Watershed 41, western North Carolina, United States.
Source: From data in Nelson (1955)

Despite the ravages of the chestnut blight, replanting with back-crossed trees derived from blight resistant hybrid chestnuts has led to hopes for a successful replanting programme. Mesic sites (not too dry, not too wet), especially shallow coves, with surviving chestnuts may have the greatest potential for chestnut growth and biocontrol of blight, if effective hypovirulence, blight resistance, and forest management measures are developed. In addition, surviving root systems, which occur throughout the chestnut's natural range, give hope that a genetically engineered, less virulent strain of sac fungus might be released and help to restore a valuable forest species (Choi and Nuss 1992).

SUMMARY

Species, genera, and families display three basic distributional patterns – large or small, widespread or restricted, and continuous or broken. Relict groups are remnants of erstwhile widespread groups that have suffered extinction over much of their former range, owing to climatic or evolutionary changes. Range size and shape display relatively consistent relationships with latitude and altitude. Individuals have home ranges and, sometimes, territories; many select the habitat that they live in, choosing not to occupy all suitable habitat. Distributional limits to species are determined partly by ecological factors, partly by history, and partly by behaviour. Organisms disperse. They may do so actively (by walking, swimming, flying, or whatever) and passively (carried by wind, water, or other organisms). Dispersal ability varies enormously throughout the living world. Ease of passage along dispersal routes varies from almost effortless to nigh on impossible. Dispersal occurs at present, much of it resulting from human introductions. Dispersal in the past forms the subject matter of traditional historical biogeography, where centres-of-origin and dispersal to other parts of the globe are the prime concern. Vicariance is a rival idea. It emphasizes the splitting of ancestral and widespread species distributions by geological or ecological factors. A good example is the breakup of Pangaea, which appears to have dictated the current distributions of several animal and plant groups. The joining of two or more continents leads to faunal mixing. A classic example is the Great American Interchange.

ESSAY QUESTIONS

1 Why has the centre-of-origin and dispersal theory proved so popular in explaining the distribution of organisms?

2 To what extent is the present distribution of animal and plant groups shaped by the breakup of Pangaea?

3 To what extent have humans 'homogenized' the world biota?

FURTHER READING

Briggs, J. C. (1995) *Global Biogeography* (Developments in Palaeontology and Stratigraphy, 14), Amsterdam: Elsevier.

Cox, C. B. and Moore, P. D. (1993) *Biogeography: An Ecological and Evolutionary Approach*, 5th edn, Oxford: Blackwell.

Cronk, Q. C. B. and Fuller, J. L. (1995) *Plant Invaders: The Threat to Natural Ecosystems*, London: Chapman & Hall.

Darlington, P. J., Jr (1957) *Zoogeography: The Geographical Distribution of Animals*, New York: John Wiley & Sons.

Elton, C. S. (1958) *The Ecology of Invasions by Animals and Plants*, London: Chapman & Hall.

Hallam, A. (1995) *An Outline of Phanerozoic Biogeography*, Oxford: Oxford University Press.

Nelson, G. and Rosen, D. E. (eds) *Vicariance Biogeography: A Critique* (Symposium of the Systematics Discussion Group of the American Museum of Natural History, 2–4 May 1979), New York: Columbia University Press.

Udvardy, M. D. F. (1969) *Dynamic Zoogeography, with Special Reference to Land Animals*, New York: Van Nostrand Reinhold.

Wallace, A. R. (1876) *The Geographical Distribution of Animals; with A Study of the Relations of Living and Extinct Faunas as Elucidating the Past Changes of the Earth's Surface*, 2 vols, London: Macmillan.

Woods, C. A. (ed.) (1989) *Biogeography of the West Indies: Past, Present, and Future*, Gainesville, Florida: Sandhill Crane Press.

4

POPULATIONS

Most species exist as groups of interbreeding individuals – populations. This chapter covers:

- the form and function of single populations
- ways in which populations survive
- human exploitation and control of populations

BIRTH, SEX, AND DEATH: THE DEMOGRAPHY OF SINGLE POPULATIONS

Populations

Organisms are all to some extent sociable beings, interacting with one another to survive. Their interactions create populations and communities. **Populations** are loose collections of individuals belonging the same species. Red deer (*Cervus elaphas*) in Britain constitute a population. All of them could interbreed, should the opportunity arise. In practice, most populations, including the red deer population in Britain, exist as sets of **local populations** or **demes**. A local red deer population lives in the grounds of Lyme Park, Cheshire. Its members form a tightly linked, interbreeding group and display features typical of many populations (Box 4.1).

Population growth

A population changes due to natality and immigration, mortality and emigration. **Natality** covers a variety of ways of 'coming into the world' – live birth, hatching, germinating, and fission. In mammals, natality is the same thing as the birth rate. **Mortality** covers the only way of 'leaving the world' – dying. Death may occur through old age (senescence). It is more likely to result from disease, starvation, or predation. Two aspects of reproduction are distinguished. **Fertility** measures the actual number of new arrivals in a population. In humans, the fertility rate averages about one birth every eight years per female of childbearing age. **Fecundity** measures the potential number of new arrivals. In humans, the fecundity rate is about one birth per nine to eleven months per female of childbearing age. Not many women fulfil that particular potential!

Most classic population studies focused on natality and mortality, and assumed that immigration and emigration balance another or are too small to contribute greatly to population change. Obviously, **emigration** and **immigration** are important components of long-term biogeographical change. They are also important in more recent population studies that emphasize small-scale comings and goings within a species range. This idea, in the guise of metapopulation theory, will be discussed later.

Exponential growth

Populations grow **exponentially** when the population increase per unit time is proportional to population

Box 4.1

THE RED DEER POPULATION IN LYME PARK, CHESHIRE, ENGLAND

Lyme Park lies on the western flanks of the Pennines. Its 535 ha consist of park grassland, moorland, and woodland. Red deer (*Cervus elaphas*) have lived in Lyme Park for over 400 years. They are a remnant of a larger population once present in Macclesfield Forest, which used to extend many kilometres along the west side of the Pennines. Up to 1946, while the park was privately owned, the population numbered 170 to 242. Some deer were shot for sport and some for the dinner table. After ownership was passed to the National Trust and Stockport Metropolitan Borough Council, the population rose to over 500. Culling was introduced in 1975 and the population is now maintained at 250 to 300.

A study of the red deer population was carried out from 1975 to 1983 (Goldspink 1987). The population was dominated by hinds. There were over three times more hinds than stags. Hinds were recorded in three main areas – the park, the moorland, and Cluse Hay (which is a subdivision of the moorland) (Figure 4.1). All tagged hinds remained close to their areas of capture. Hinds tended to form large groups of 30 or more throughout the study area. Most calves were born in the first two weeks of June. Maximum densities of hinds and calves per km² varied from 30 in the park to 77 on the moorland. Three bachelor herds of stags were present. Moor stags spent much of the summer on a small area (30 ha) of ground above Cluse Hay. They dispersed more during winter, probably to avoid exposure and windchill. The Knott group, which lives in the park, consisted of prime stags, more than 6 years old, during the winter. Younger stags were recorded on Cage Hill. Stags in the park area were more tolerant of disturbance than those living on the moorland. The rut started in mid-September and went on well into November. The median date of observed copulations was 16 October. The apparent gestation period (to the median date of calving) was about 237 days. Stags from the Knott rutted on Cluse Hay, while those from Cage Hill preferred the moor or the park. Moor stags appeared to play but a small part in the rut, despite their numerical abundance. Dominant stags, which commonly had 40-hind harems for two to three weeks during the rut, were usually replaced within two years.

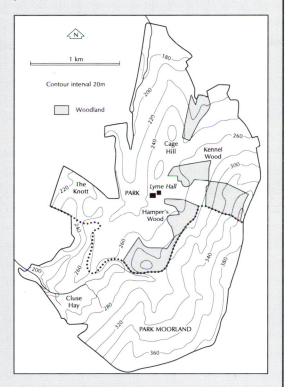

Figure 4.1 The main areas occupied by the red deer (*Cervus elaphas*) population, Lyme Park, Cheshire, England.
Source: After Goldspink (1987)

Growth rates of stags and hinds were low on the moorland. Calf-to-hind ratios were generally low and within the range 0.20–0.34; maxima of 0.55–0.60 were recorded in woodland sites. Conditions on the moorland were severe during the winter due to the poor quality of herbage, mainly purple moor-grass (*Molinia caerulea*), and a lack of shelter. Annual calf mortality varied from 10 to 30 per cent. It was particularly high during the cold winters of 1978–9 and 1981–2. Population density and winter food were probably principal factors limiting the perfor- mance of deer on the moor. Growth rates of stags and hinds in the park were high, but calf-to-hind ratios were low at 0.15–0.26. Culling has reduced levels of natural mortality but further improvements in performance are unlikely to be achieved without a reduction in sheep stocks, some improvement in habitat, and the provision of shelter.

size and is unconstrained. This is readily appreciated by a simple example. Take two individuals. After a year they have produced four offspring. Reproducing at the same rate, those four offspring will themselves have produced eight offspring after a further year. The growth of the population follows a geometric progression – 2, 4, 8, 16, 32, 64, and so on. This exponential growth rapidly produces huge numbers. A female common housefly (*Musca domestica*) lays about 120 eggs at a time (Leland Ossian Howard, cited in Kormondy 1996: 194). Around half of these develop into females, each of which potentially lays 120 eggs. The number of offspring in the second generation would therefore be 7,200. There are seven generations per year. If all individuals of the reproducing generation should die off after producing, the first fertile female will be a great, great, great, great, great grandmother to over 5.5 trillion flies.

Growth in populations with overlapping generations and prolonged or continuous breeding seasons may be described by the equation:

$$dN/dt = rN$$

Notice that the growth rate, dN/dt, is directly proportional to population size, N, and to the intrinsic rate of increase (per capita rate of population growth), r. The intrinsic rate of increase, sometimes called the **Malthusian parameter**, is defined as the specific birth rate, b, minus the specific death rate, d. In symbols, $r = b - d$. (Specific rates are rates per individual per unit time.) Exponential growth, then, simply depends upon how many more births there are than deaths, and on how large the population is. No other factors restrict population growth. Unhindered growth of this kind is the biotic potential of a popu-lation (R. N. Chapman 1928). It is described by the solution to the population growth equation:

$$N = N_0 e^{rt}$$

This equation shows that the starting population, N_0, increases exponentially (e is the base of the natural logarithms) at rate determined by the intrinsic rate of natural increase, r. It describes a J-shaped curve (Figure 4.2). The **maximum intrinsic rate of natural increase** varies enormously between species. In units of per capita per day, it is about 60 for the bacterium *Escherichia coli*; 1.24 for the protozoan *Paramecium aurelia*; 0.015 for the Norway rat (*Rattus norvegicus*); 0.009 for the domestic dog (*Canis domesticus*); and 0.0003 for humans. For the red deer population in Lyme Park, from 1969 to 1975, it was 0.07696.

The **doubling time** is the time taken for a population to double its size. It may be calculated from values of r. The doubling times for the species mentioned above, assuming geometric increase prevails, are: *Escherichia coli* – 0.01 days; *Paramecium aurelia* – 0.56 days; *Rattus norvegicus* – 46 days; *Canis domesticus* – 77 days; and humans – 6.3 years. For Lyme Park red deer, it is 9.0 years.

Logistic growth

The English house sparrow (*Passer domesticus*) was introduced to the United States around 1899. Concern was expressed that in a decade, a single pair could lead to 275,716,983,698 descendants, and by 1916 to 1920 there would be about 575 birds per 40 ha (Kormondy 1996: 194). In the event, there were only 18 to 26 birds per 40 ha, less than 5 per cent of the

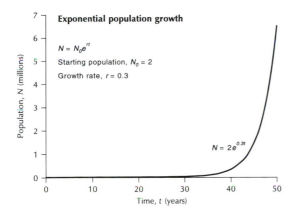

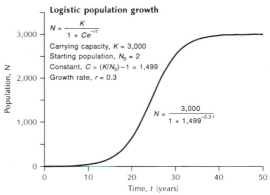

Figure 4.2 Exponential growth of a hypothetical population. Notice that, with a bit of artistic licence, the population growth describes a J-shaped curve

Figure 4.3 Logistic growth of a hypothetical population. Notice that the population growth curve is a stretched S-shape

expected density. Patently, something had prevented the unfettered geometric increase in the immigrant house sparrow population.

A growing population cannot increase indefinitely at a geometric rate. There is a ceiling or **carrying capacity**, a limit to the number of individuals that a given habitat can support. The **biotic potential** of a population is curtailed by an environmental resistance (R. N. Chapman 1928). After a period of rapid growth, the house sparrow population reached its population ceiling, partly because native hawks and owls took to the new English dish on the menu.

The population ceiling or carrying capacity is usually denoted by the letter K. It alters the population growth equation in this way:

$$dN/dt = rN(1 - N/K)$$

This equation for growth rate is called the **logistic equation**. It describes a sigmoid or S-shaped curve (with the top of the 'S' being pushed to the right) where population grows fast initially but tapers off towards the carrying capacity (Figure 4.3). When the population size is small, compared with the carrying capacity, the environmental resistance to growth is minimal. The population responds by virtually unbounded exponential growth. This is the bottom part of the 'S'. When the population size is large and

approaching the carrying capacity, environmental resistance is great. The population barely grows. This is the distorted top part of the 'S'. If the population should exceed the carrying capacity, negative growth occurs; in other words, the population falls until it drops below the carrying capacity.

Population growth (or decline) rates are affected by **density-dependent factors** and by **density-independent factors**. The effects of density-independent factors do not vary with population density and they will affect the same proportion of organisms at any density. Weather, pests, pathogens, and humans commonly affect populations in that way. The effects of density-independent factors do vary with population density, so that the portion of organisms affected increases or decreases with population density. Birth rates and death rates both depend on population density: birth rates tend to decrease with increasing density, while death rates tend to increase with increasing density.

Population irruptions

Irruptions are common features of many populations. They involve a population surging to a high density and then sharply declining. This boom-and-bust pattern occurs in some introduced ungulates. It happened to the Himalayan tahr (*Hemitragus jemlahicus*) in

New Zealand (Caughley 1970). The tahr is a goat-like ungulate from Asia. Introduced into New Zealand in 1904, it spread through a large part of the Southern Alps. After its introduction, the population steadily rose. The birth rate fell slightly but the death rate increased, largely due to greater juvenile mortality. After a period when the population was large, a decline set in. This decline was caused by reduced adult fecundity and further juvenile losses. Changes in the tahr population may have resulted from changes in food supply. As the tahr grazed, they altered the character of the vegetation that they fed on. In particular, snow tussocks (*Chionochloa* spp.), which are evergreen perennial grasses, were an important food source late in winter. They cannot abide even moderate grazing pressure and disappeared from land occupied by the tahr. With no snow tussocks to eat, the tahr browsed on shrubs in winter and managed to kill some of these.

Another irruption occurred on Southampton Island, Northwest Territories, Canada (D. C. Heard and Ouellet 1994). By 1953, Caribou (*Rangifer tarandus groenlandicus*) had been hunted to extinction on Southampton Island. In 1967, 48 caribou were captured on neighbouring Coat's Island and released on Southampton Island. The population of caribou older than 1 year grew from 38 in 1967 to 13,700 in 1991. The annual growth rate during this period was 27.6 per cent and did not decrease with increasing population density. On Southampton Island, caribou did not suffer high winter mortality in some years, but they did on Coat's Island. Caribou density was higher on Coat's, which suggests that adverse weather has a minimal effect when animal density is low.

Irruptions also occur because of unusually high immigration. In the winter of 1986–7, many more rough-legged buzzards (*Buteo lagopus*) moved into Baden-Württemberg, south-west Germany, than in any other winter during at least the preceding century (Dobler *et al.* 1991). From 14 January to 7 April, rough-legged buzzards were observed daily. The highest numbers observed were 109 individuals on 1 February and 110 on 8 February. The buzzard influx was caused by high snow cover and cold spells in the eastern parts of Central Europe. Depressions over Southern Europe may have blocked the route south with cloud and snowfall.

Population crashes

Some populations often **crash**. In early 1989, two-thirds of the Soay sheep (*Ovis aries*) population on St Kilda, in the Outer Hebrides, Scotland, died within a twelve-week period. The cause of the crash was investigated by post-mortem examination and laboratory experiments (Gulland 1992). Post-mortem examination showed emaciated carcasses with a large number of nematode (*Ostertagia circumcincta*) parasites, and that the probable cause of death was protein-energy malnutrition. However, well-nourished Soay sheep artificially infected with the parasite in the laboratory showed no clinical signs or mortality, even when their parasite burdens were the same as those recorded in the dead St Kilda sheep. Thus, parasites probably contributed to mortality only in malnourished hosts, exacerbating the effects of food shortage.

Marine iguanas (*Amblyrhynchus cristatus*) living on Sante Fe, in the Galápagos Islands, suffered a 60 to 70 per cent mortality due to starvation during the 1982–3 El Niño–Southern Oscillation event (Laurie and Brown 1990a, b). Adult males suffered a higher mortality than adult females while food was short, but size explained most of the mortality differences between the sexes. Almost no females bred after the event. In the next year, the frequency of reproduction doubled, the age of first breeding decreased, and mean clutch size increased from two to three. These demographic changes are examples of density-dependent adjustments of population parameters that help to compensate for the crash.

Population crashes may have repercussions within communities. In Africa, savannahs are maintained as grassland partly because of browsing pressures by large and medium herbivores. In Lake Manyara National Park, northern Tanzania, bush encroached on the grassland between 1985 and 1991, shrub cover increasing by around 20 per cent (Prins and Van der Jeugd 1993). Since 1987, poaching has caused a steep decline in the African elephant (*Loxodonta africana*) population in the Park. However, shrub establishment preceded the elephant population decline, and is not attributable to a reduction in browsing. In two areas of the Park, shrub establishment coincided with anthrax epidemics that decimated the impala (*Aepyceros melampus*) population. In the northern section of the Park the epidemic was in 1984; in the southern section it was in 1977. An even-aged stand of umbrella thorn (*Acacia tortilis*) was established within the grassland in 1961, which date coincided with another anthrax outbreak among impala. Similarly, another even-aged stand of umbrella thorn was established at the end of the 1880s, probably following a rinderpest pandemic. The evidence suggests that umbrella-thorn seedling establishment is a rare event, largely owing to high browsing pressures by such ungulates as impala. Punctuated disturbances by epidemics, which cause population crashes in ungulate populations, create narrow windows for seedling establishment. This process may explain the occurrence of even-aged umbrella thorn stands.

Chaotic growth

Irregular fluctuations observed in natural populations, including explosions and crashes, were traditionally attributed to external random influences such as climate and disease. In the 1970s, Robert M. May recognized that these wild fluctuations could result from intrinsic non-linearities in population dynamics (May 1976). For instance, a relatively simple population growth equation of the form

$$N_{t+1} = N_t e[r(1 - N_t/K)]$$

which is applicable to population that suffer epidemics at high densities, possesses an amazing range of dynamic behaviour – stable points, stable cycles, and chaos – depending on the growth rate, *r* (Figure 4.4).

Much of the work on **chaotic population dynamics** is theoretical. Some studies have shown that chaos does occur in natural populations. The ticklegrass or hairgrass (*Agrostis scabra*) displays oscillations and chaos in experimental plots (Tilman

and Wedin 1991). Monocultures of ticklegrass were planted at two different densities on ten different soil mixtures. Populations growing in unproductive soils, low in nitrogen, maintained fairly stable aboveground biomass. Populations growing on richer soil showed biomass oscillations. Populations growing on the richest soils displayed as 6,000-fold crash (Figure 4.5). No other species growing in the garden where the experiment was conducted crashed in 1988, which suggests that environmental agencies were not responsible. The dynamics for the crashing populations were shown to be chaotic. The chaotic regime resulted from the growth inhibition by litter accumulation. The litter causes a one-year delay between growth and the inhibition of future growth. Litter production is greater in more productive plots. The magnitude of the time-delayed inhibitory effect therefore increases with productivity. This may lead to an oscillation between a year of low litter and high living biomass, and a year of high litter and low living biomass. The implications of this study are profound – litter feedback may cause population oscillations and possibly chaos in productive habitats. However, litter accumulates where it falls, and litter-driven chaotic dynamics occurs at small spatial scales – it may avoid detection if looked for at medium and large scales.

Since the late 1980s, part of theoretical ecology has focused on the spatio-temporal dynamics generated by simple ecological models. To a large extent, the results obtained have changed views of complexity (Bascompte and Sole 1995). Specifically, simple rules are able to produce complex spatio-temporal patterns. Consequently, some of the complexity underlying Nature does not necessarily have complex causes. The emerging framework has far-reaching implications in ecology and evolution. It is improving the comprehension of such topics as the scale problem, the response to habitat fragmentation, the relationships between chaos and extinction, and how higher diversity levels are supported in Nature.

Age and sex structure

The aggregate population size, *N*, obscures the fact that individuals differ in age and in ability to

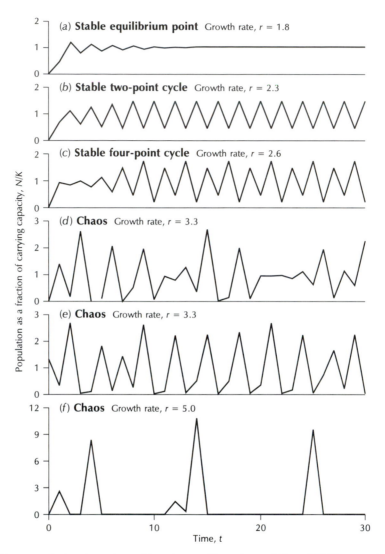

Figure 4.4 Internally driven population dynamics: stable points, stable cycles, and chaos. NB The growth rates, *r*, in (d) and (e) are the same, but the ratios of the initial population to the carrying capacity, N_0/k, are different: in (d), N_0/k = 0.075; in (e), N_0/k = 1.5.
Source: After May (1981)

reproduce. The population age-structure is the proportion of individuals in each age class. A stable age distribution with a zero growth rate has about the same number of individuals in each age class, except for fewer in post-reproductive classes. In contrast, a growing population has a pyramid-shaped structure, with large numbers of pre-reproductive individuals.

Life tables

A **life table**, like an actuarial table, shows the number of individuals expected in each age class based upon fecundity and survival rates. Survival rates, *s*, are related to death rates, *d*, in the following way: $s = 1 - d$. Life-table statistics are normally computed

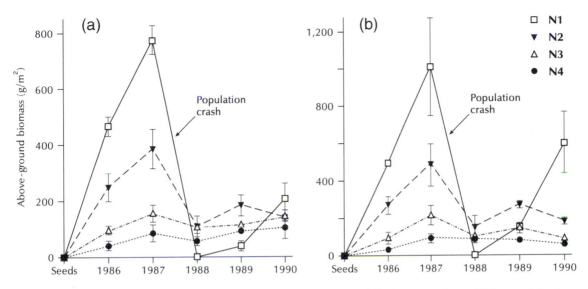

Figure 4.5 A population crash in the ticklegrass (*Agrostis scabra*). (a) High seed density monocultures. (b) Low seed density monocultures. The soil mixtures are divided into four groups, N1–N4. Group N1 is poorest in nitrogen, and Group N4 is the richest in nitrogen.
Source: After Tilman and Wedin (1991)

for females. This is because fecundity for males is very difficult to determine where mating occurs promiscuously. A life table for a hypothetical population of females is shown in Table 4.1. Column 1 lists the age classes. Column 2 gives the survivorship expressed as the fraction surviving at age x; this is the probability that an average new-born will survive to that age. Column 3 gives the fecundity; this is the number of offspring produced by an average individual of age *x* during that period. The sum of fecundity terms over all ages, or the total number of offspring that would be produced by an average organism with no mortality, is called the gross reproductive rate. It is 9.8 in the example. Column 4 shows the number of expected offspring by age class. The figures are simply the product of survivorship and fecundity. The sum of expected offspring gives the net reproductive rate, *R*. This is the expected number of female offspring to which a new-born female will give birth in her lifetime. It is 7.322 in the example. Column 5 shows the product of age and expected offspring. The total of these products divided by the net reproductive rate defines the mean generation time,

T. This is the average age at which females produce offspring. It is 5.26714 years in the example. The earlier young are born, the earlier they, in turn, will have offspring and the more rapid population growth will be. The calculations show that the population will increase by a factor of 7.322 times every 5.26714 years. Plainly, growth rates expressed on a generation basis would be difficult to compare. To avoid this problem, growth rates are normally converted to an annual figure and designated by the Greek letter lambda (λ). The annual growth rate of the hypothetical population is 1.459.

Survivorship curves

A **survivorship curve** shows the fraction of a cohort of new-born (or newly hatched) alive individuals in subsequent years. Natural populations have a great range of survivorship curves. Three basic types are recognized (Pearl 1928) (Figure 4.6):

1 **Type I survivorship curves** are 'rectangular' or convex on semi-logarithmic plots. They show low

mortality initially that lasts for more than half the life-span, after which time mortality increases steeply. This survivorship pattern is common amongst mammals, including the Dall mountain sheep (*Ovis dalli dalli*) and humans. It is also displayed by such reptiles as the desert night lizard (*Xantusia vigilis*).

2 **Type II survivorship curves** are 'diagonal' or straight on semi-logarithmic plots. They show a reasonably constant mortality with age. This survivorship pattern is common in most birds, including the American robin (*Turdus migratorius migratorius*), and reptiles.

3 **Type III survivorship curves** are concave on semi-logarithmic plots. They show extremely high juvenile morality and relatively low mortality thereafter. This survivorship pattern is common in many fish, marine invertebrates, most insects, and plants. It is also characteristic of the British robin (*Erithacus rubecula melophilus*).

These three basic survivorship curves do not exhaust all the possibilities – many intermediate types occur.

Cohort-survival models

Models of population disaggregated by age and sex are called **cohort-survival models** (Leslie 1945, 1948). A cohort-survival model starts with the age structure of a female population at a given time, and with the birth rates and survival rates per individual in each age group. From this information it predicts the age structure of the female population in future years.

Cohort-survival models are used to study population change. One application considered the effects of

Table 4.1 Life table of a hypothetical population

Column 1 Age, x (years)	Column 2 Survivorship, l_x	Column 3 Fecundity, b_x	Column 4 Expected offspring, $l_x b_x$	Column 5 Product of age and expected offspring, $x l_x b_x$
0	1	0	0	0
1	0.99	0.2	0.2	0.2
2	0.98	0.4	0.392	0.784
3	0.96	0.8	0.768	2.304
4	0.92	1.2	1.104	4.416
5	0.86	1.6	1.376	6.880
6	0.78	2.2	1.716	10.296
7	0.68	1.4	0.952	6.664
8	0.56	0.8	0.448	3.584
9	0.42	0.6	0.252	2.268
10	0.26	0.4	0.104	1.040
11	0.06	0.2	0.012	0.132
12	0	0	0	0
		9.8	**7.322**	**38.566**
		This is the gross reproductive rate, GRR	This is the net reproductive rate, R	This is the total weighted age

Mean generation time, $T = 38.566/7.322 = 5.26714$ years
Annual growth rate, λ, is given by $\lambda = R^{1/T}$. It is $7.322^{1/5.26714} = 1.459$

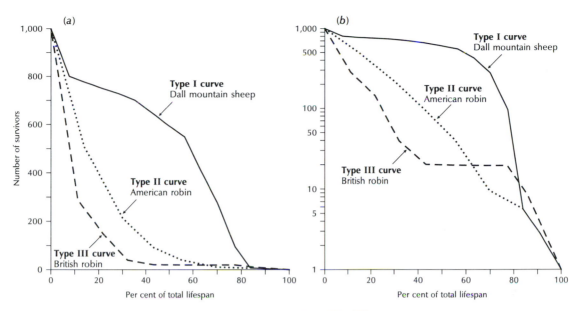

Figure 4.6 Surivorship curves of the Dall mountain sheep (*Ovis dalli dalli*), the American robin (*Turdus migratorius migratorius*), and the British robin (*Erithacus rubecula melophilus*). (a) Arithmetic plot. (b) Semi-logarithmic plot.

over-exploitation on the blue whale (*Balaenoptera musculus*) population (Usher 1972). It used data on the blue whale population collected in the 1930s, before the near extinction of the species in the early 1940s. Age-specific birth rates and survival rates for each of seven age groups were calculated (Table 4.2). The birth terms (fecundity terms in this study) show that female blue whales do not breed in the first four years of life. At full breeding, each cow produces one calf every two years (since the sex ratio is 1:1, this calf has a 50 per cent chance of being female). The survival terms, all but one of which is 0.87, is based on the best available estimate of natural mortality. The survival rate of the 12-plus age class is 0.8. This is because not all members of the oldest age-group die within a two-year period; indeed, some live to be 40. The survival rate of 0.8 per two years gives a life expectancy of 7.9 years for old whales.

The parameters in the table may be arranged as a growth matrix, from which the growth characteristics of the whale population may be extracted. The dominant latent root, λ_1, of the growth matrix is 1.0986. This shows that the population is capable

of growing. The intrinsic rate of natural increase of the blue whale population is given by the natural logarithm of the dominant latent root; it is ln 1.0986 = 0.094. The dominant latent root may be used to estimate the harvest of whales that can be taken without causing a decline in the population. The population size increases from N to $\lambda_1 N$ over a two-year period. The harvest that may be taken, H, expressed as percentage of the total population, is

$$H = 100 \{(\lambda_1 - 1)/\lambda_1\}$$

This is about 4.5 per cent of the total population per year. If the harvest rate were exceeded, the population would decrease, unless homeostatic mechanisms should come into play and alter fecundity and survival rates. Evidence suggests that, when under pressure, the blue whale population shows a slight increase in the pregnancy rate and that individuals reach maturity earlier in life, and that they may generally grow more rapidly. The big question is: can these responses offset the effects of exploitation? For a trustworthy answer to this question, long-term

Table 4.2 Parameters for a cohort-survival model of the blue whale (*Balaenoptera musculus*) population

Parameters (per cow per 2 years)	Age groups						
	0–1	*2–3*	*4–5*	*6–7*	*8–9*	*10–11*	*12+*
Birth rate	0	0	0.19	0.44	0.50	0.50	0.45
Survival rate	0.87	0.87	0.87	0.87	0.87	0.87	0.80

Source: After Usher (1972)

studies involving data collection and modelling are needed.

The wandering albatross (*Diomedia exulans*) is a splendid bird, with a wing span of over three metres. It breeds on southern cool-temperate and sub-Antarctic islands, and forages exclusively in the Southern Hemisphere. Wandering albatross populations around the world have declined since the mid-1960s. The major factor contributing to this decline is accidental deaths associated with longline fishing, which has increased rapidly since the 1960s. Entanglement and collisions with netsonde monitor cables on trawlers may also contribute to albatross mortality. Longline operations involve baited hooks attached to weighted lines being thrown overboard. Seabirds lured to the bait swallow the hooks and die by drowning when they are dragged beneath the surface as the weighted line sinks, or die when the lines are hauled. Some 44,000 albatrosses, including about 10,000 wandering albatrosses, are killed in this manner each year (Brothers 1991). An age-structured model of a wandering albatross population, based on demographic data from the population at Possession Island, Crozets, simulated population trends over time (Moloney *et al.* 1994). The aim was to investigate the potential impacts of new longline fisheries, such as that for the Patagonian toothfish (*Dissostichus eleginoides*) in Antarctica. The model consisted of twelve year age-classes, comprising four age categories: chicks (prior to fledging, 0–1 years), juveniles (before the first return to the breeding colony, 1–4 years), immatures (after returning to the colony but before the first breeding, 5–10 years), and breeding adults (more than 10 years). The model kept track of the

population in each age class, N_i. The populations change owing to eggs being laid (fecundity) and survival from one age class to the next. Fecundity and survival terms are summarized in Table 4.3, together with starting numbers for a stable population. The simulations predicted a decreasing population. The rate of decrease is constant at 2.29 per cent per year and is associated with a stable age structure of 14 per cent chicks, 24 per cent juveniles, 15 per cent immatures, and 47 per cent adults. Further simulations were carried out to test the sensitivity of the wandering albatross population to altered survival rates. On the assumption that longline fishing operations affect juveniles more than adults, there is a time lag of five to ten years before further decreases in population numbers are affected in the breeding populations. In addition, because wandering albatrosses are long-lived, population growth rates take about thirty to fifty years to stabilize after a perturbation, such as the introduction of a new longline fishery.

Metapopulations

There are four chief types of **metapopulations**, normally styled broadly defined, narrowly defined, extinction and colonization, and mainland and island. The broad and narrow categories are more concisely named loose and tight metapopulations, respectively.

Loose metapopulations

A **loose metapopulation** is simply a set of sub-populations of the same species. Rates of mating, competition, and other interactions are much higher

Table 4.3 Demographic parameters for a wandering albatross (*Diomedia exulans*) population

Age class (years)	Age category	Survival (per year)	Fecundity (per year)	Starting population (for a stable age structure)
0	Eggs and chicks	0.640	0	313
1	Juveniles	0.715	0	201
2	Juveniles	0.715	0	147
3	Juveniles	0.715	0	107
4	Juveniles	0.715	0	78
5	Immatures	0.980	0	57
6	Immatures	0.980	0	58
7	Immatures	0.980	0	58
8	Immatures	0.980	0	59
9	Immatures	0.980	0	59
10	Immatures	0.980	0	60
11	Adults	0.922	0.294	1,066

Source: After Moloney *et al.* (1994)

within the subpopulations than they are between the subpopulations. The subpopulations may arise through an accident of geography – the habitat patches in which the subpopulations live lie farther apart than the normal dispersal distance of the species. The conspecific subpopulations that comprise such a metapopulation may or may not be interconnected. By this definition, most species with large and discontinuous distributions form metapopulations.

Where habitat patches are so close together that most individuals visit many patches in their lifetime, the subpopulations behave as a single population – all individuals effectively live together and interact. Where habitat patches are so scattered that dispersal between them almost never happens, the subpopulations behave effectively as separate populations. This situation applies to mammal populations living on desert mountain-tops in the American Southwest (Brown 1971).

Tight metapopulations

A **tight metapopulation** is a set of conspecific subpopulations living in a mosaic of habitat patches, with a significant exchange of individuals between the patches. Migration or dispersal among the sub-

populations stabilizes local population fluctuations. This 'rescue effect' helps to prevent local extinctions, through the recolonization of new habitats or habitats left vacant by local extinctions (or both). In theory, a tight metapopulation structure occurs where the distance between habitat patches is shorter than the species is physically capable of travelling, but longer than the distance most individuals move within their lifetimes. An example is the European nuthatch (*Sitta europaea*) population. The nuthatch inhabits mature deciduous and mixed forest and has a fragmented distribution in the agricultural landscapes of western Europe (Verboom *et al.* 1991). It shows all the characteristics of a tight metapopulation: the distribution is dynamic in space and time; the extinction rate depends on patch size and habitat quality; and the patch colonization rate depends on the density of surrounding patches occupied by nuthatches.

Extinction-and-colonization metapopulations

In the narrowest possible sense, and as originally defined (Levins 1970), a metapopulation consists of subpopulations characterized by frequent turnover. In these **'extinction-and-colonization'** metapopulations

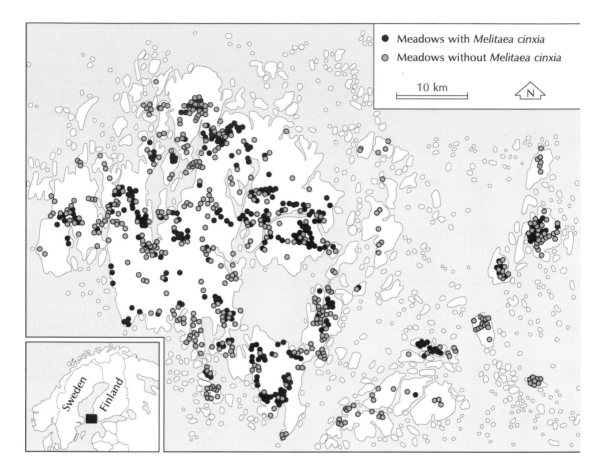

Figure 4.7 The Granville fritillary (*Melitaea cinxia*) population in Åland, Finland, in 1993. Shaded circles are empty meadows; solid circles are occupied meadows.
Source: After Hanski *et al*. (1996b)

(Gutiérrez and Harrison 1996), all subpopulations are equally susceptible to local extinction. Species persistence therefore depends on there being enough subpopulations, habitat patches, and dispersal to guarantee an adequate rate of recolonization. This metapopulation structure seems to apply to the pool frog (*Rana lessonae*) on the Baltic coast of east-central Sweden (Sjögren 1991). Frog immigration mitigates against large population fluctuations and inbreeding, and 'rescues' local populations from extinction. Extinction-and-colonization metapopulation structure also applies to an endangered butterfly, the Glanville

fritillary (*Melitaea cinxia*) in a fragmented landscape (Hanski *et al*. 1995; 1996b). This butterfly became extinct on the Finnish mainland in the late 1970s. It is now restricted to the Åland islands. It has two larval host plants on the islands – ribwort plantain (*Plantago lanceolata*) and spiked speedwell (*Veronica spicata*) – both of which grow on dry meadows. The metapopulation was studied on a network of 1,502 discrete habitat patches (dry meadows), comprising the entire distribution of this butterfly species in Finland (Figure 4.7). Survey of the easily detected larval groups revealed a local population in 536 patches. The butterfly metapopulation satisfied the necessary conditions for a species to persist in a balance between stochastic local extinctions and re-colonizations.

Mainland–island metapopulations

In some cases, there may be a mixture of small sub-populations prone to extinction and a large persistent population. The viability of the 'mainland–island' metapopulations (Gutiérrez and Harrison 1996) is less sensitive to landscape structure than in other types of metapopulation. An example is the natter-jack toad (*Bufo calamita*) metapopulation at four neighbouring breeding sites in the North Rhineland, Germany (Sinsch 1992). Over 90 per cent of all reproductive males showed a lifelong fidelity to the site of first breeding; females did not prefer certain breeding sites. Owing to the female-biased exchange of individuals among neighbouring sites, the genetic distance between local populations was generally low but increased with geographical distance. This spatial pattern is consistent with the structure of a mainland–island metapopulation. Up to three mass immigrations of males per breeding period, replacing previously reproductive individuals, suggested the existence of temporal populations successively repro-ducing at the same locality. Genetic distances were considerably greater between temporal populations than between local ones, indicating partial repro-ductive isolation. An exchange of reproductive individuals between the temporal populations at each site was not detected, but gene flow due to the recruitment of first-breeders originating from offspring other than their own seemed probable. Thus, natterjack metapopulations consist of inter-acting local and temporal populations. Three out of four local populations had low reproductive success, as well as the latest temporal population. The persis-tence of these populations depended entirely on the recruitment of juveniles from the only self-sustaining local population. This 'rescue-effect' impeded local extinction.

Metapopulations and conservation

Metapopulation theory is now widely applied in conservation biology. It is especially important where **habitat fragmentation** is the central concern. The 'classic' example is the northern spotted owl in the United States, for which bird the metapopulation approach to management has been thoroughly devel-oped. Other examples abound, including the euro and Leadbeater's possum, both from Australia.

The spotted owl in north-western North America

The northern spotted owl (*Strix occidentalis caurina*) is closely associated with mature and old-growth con-iferous forests in the Pacific Northwest. These forests once formed a continuous habitat, but they have shrunk over the past half century. They now survive as small remnants occupying just 20 per cent of their former extent. This fragmentation of the spotted owl habitat reduces dispersal success, which may jeopar-dize the long-term survival of the species. Northern spotted owls require territories of around 1,000 ha. They avoid open areas, where they are vulnerable to predation by the larger great-horned owl (*Bubo virginianus*), and will cross only small cleared strips. Their survival therefore depends on the connectivity of the forest habitat (Ripple *et al*. 1991). The preser-vation of a viable owl population is required by the Endangered Species Act and by laws governing the United States Forest Service, which owns 70 per cent of the owl's remaining habitat. The United States government commissioned scientists to discover how much forest, and in what form, was required to ensure the owl's survival. The result was a spatial model of the owl population (Lamberson *et al*. 1992; see also S. Harrison 1994; Gutiérrez and Harrison 1996).

The model let a single spatial element represent the territory of a pair of owls. The dynamics of each pair were modelled using rates of pair-formation, fecundity, and survival measured in the field. Dispersal was estimated from juvenile owls tagged with radio transmitters. Owl demography was linked to forest fragmentation by juvenile dispersal success. To join the breeding population, a newly fledged owl needed to find a vacant and suitable territory. Its chances of doing so depended upon the proportion of forested landscape. The results predicted a sharp threshold, below which the owl population could not persist. With less than 20 per cent forest, the juvenile recruitment to the breeding population was too low to balance mortality of adult territory holders, and the population collapsed. In addition, when juveniles were required to find mates as well as empty territories, a second survival threshold emerged, below which extinction soon ensued. The recommendation was that 7.7 million acres (about 3 million ha) of old-growth forest should be preserved in a system of patches, each patch large enough for more than twenty owl territories within 12 miles (about 20 km) of the nearest patch.

The euro

The common wallaroo or euro (*Macropus robustus*), called the biggada by aborigines, is a large, rather shaggy-haired kangaroo (Colour plate 5). Its range includes most of Australia, except the extreme south and the western side of the Cape York Peninsula, Queensland. Its habitat usually features steep escarpments, rocky hills, or stony rises; and, in areas where extreme heat is experienced for long periods, caves, overhanging rocks, and ledges for shelter.

Members of a euro population living in a fragmented landscape occupied a very large (1,196 ha) patch of remnant vegetation (G. W. Arnold *et al*. 1993). They were sedentary and did not use any other vegetation remnants. However, some young males from this remnant dispersed up to 18 km to other remnants. In areas that had a large remnant (more than 100 ha), individuals were also sedentary. Some individuals included in their home ranges smaller remnants within 700 m. To access these, they usually moved along connecting corridors between remnants. Occasionally, a few individuals made excursions of several kilometres outside their normal home ranges, either overnight or over several months. In areas where all remnants were small (less than 30 ha),

individuals lived alone or in small groups, moving frequently between several remnants. Overall, the euros in the 1,680-km^2 study area appeared to be separated into a number of metapopulations, some of which had very small numbers of animals. Within the metapopulations, movements between populations appear to be dependent on the availability of 'stepping stones' and corridors. In two populations that had low densities of euros, the numbers of juveniles per adult female were significantly lower than in systems with higher densities. Long-term survival of these two populations is questionable.

Leadbeater's possum

Leadbeater's possum (*Gymnobelideus leadbeateri*) lives in the cool, misty mountain forests of the Central Highlands, Victoria, Australia (Colour plate 6). These forests are timber-producing areas. Forest fragmentation is threatening the survival of the possum, which is an endangered species.

The viability of possum metapopulations in the fragmented old-growth forests was assessed using a population model (Lindenmayer and Lacy 1995). Computer simulations with subpopulations of twenty or fewer possums were characterized by very rapid rates of extinction, most metapopulations typically failing to persist for longer than fifty years. Increases in either the migration rate or the number of small subpopulations worsened the metapopulations' demographic instability, at least when subpopulations contained fewer than twenty individuals and when migration rates were kept within realistic values for Leadbeater's possums dispersing between disjunct habitat patches. The worsening demographic situation was reflected in lower rates of population growth and depressed probabilities of metapopulation persistence. These effects appeared to be associated with substantial impacts of chance demographic events on very small subpopulations, together with possum dispersal into either empty patches or functionally extinct (single-sex) subpopulations. Increased migration rates and the addition of subpopulations containing forty possums produced higher rates of population growth, lower probabilities of extinction, and longer persistence times. Extinctions in these scenarios were also more likely to be reversed through recolonization by dispersing individuals.

A common problem with fragmented populations is that genetic information within the gene pool becomes less well mixed. At the highest rates of migration, subpopulations of forty possums were thoroughly mixed and behaved genetically as a single larger population. However, over a century, the gene-pool mixing would decline and lead to a significant loss in genetic variability. This loss would occur even with highest migration rates among five, forty-animal subpopulations. While demographic stability might occur in metapopulations of two hundred individuals, considerably more individuals than this might be required to avoid a significant decline in genetic variability over a century.

A QUESTION OF SURVIVAL: POPULATION STRATEGIES

Animals and plants possess primary ecological strategies. These strategies are recurrent patterns of specialization for life in particular habitats or niches. They involve the fundamental activities of an organism, including resource capture, growth, and reproduction. There are two main schemes for population strategies. The first involves two basic strategies – opportunism and competitiveness; the second involves three basic strategies – competitiveness, stress toleration, and ruderalism.

Opportunists and competitors

Two basic population strategies depend on how long a habitat is favourable. A crucial factor is a population's generation time compared with the 'life-span' of a stable habitat. One extreme case is where the generation time is about the same as a stable habitat's 'life-span'. This would apply, for example, to fruit flies (*Drosophila*) living in a tropical forest. A fruit fly lives on ripe fruits, which plainly are temporary habitats. If a fruit fly should attain adulthood, it simply migrates to another ripe fruit. Under these circum-

stances, there is nothing to be lost by exceeding the habitat's carrying capacity because the resources left for future generations will not be jeopardized. For this reason, short-lived habitats tend to be occupied by exploiters or opportunists with boom-and-bust population dynamics. Such opportunistic species are called *r*-strategists, after the intrinsic growth rate, *r*, in population growth equations.

Another extreme case occurs where the generation time is tiny compared with a stable habitat's 'life-span'. This would apply, for instance, to an orang-utan (*Pongo pygmaeus*), to whom an entire forest is a stable and permanent habitat. Under these circumstances, it pays species to look to the future. Long-lived habitats with a fairly stable carrying capacity would be degraded if population density should overshoot the carrying capacity. This would reduce the habitat's carrying capacity for future generations. Species in these habitats are adapted to harvesting food in a crowded environment. They are highly competitive and they tend to have low population growth rates. They are called *K*-strategists, after the carrying capacity parameter, *K*, in population growth equations.

The r-strategists

The *r*-strategists continually colonize temporary habitats. Their strategy is essentially opportunistic. They have evolved high population growth rates, produced by a high fecundity and a short generation time. Migration is a major component of their population dynamics, and may occur as often as once per generation. The habitats they colonize are commonly free from rivals. This favours a lack of competitive abilities and small size. They defend themselves against predators by synchronizing generations (hence satiating predators) and by being mobile, which enables them to play a game of hide and seek. The greatest *r*-strategists are insects and bacteria. The African armyworm (*Spodoptera exempta*) is a good example. This migrant moth is a crop pest in East Africa. It lays up to 600 eggs 'at a sitting', has a generation time of a little over three weeks, and large population outbreaks occur on the young growth of grasses, including maize (*Zea mays*). Some birds are *r*-strategists. The quail (*Coturnix coturnix coturnix*) is a case in point (Puigcerver *et al.* 1992).

The *K*-strategists

K-strategists occupy stable habitats. They maintain populations at stable levels, and are intensely competitive. They are often selected for large size. Many vertebrates are *K*-strategists. They are characteristically large, long-lived, very competitive, have low birth and death rates, and invest much time and effort in raising offspring. Their low growth rate helps to avoid the habitat degradation that would follow an overshooting of the carrying capacity. Animal and plant *K*-strategists alike tend to invest much energy in defence mechanisms. Animal *K*-strategists often have small litters or clutches and parents are very careful with their young. An extreme example is the wandering albatross (*Diomedia exulans*). This bird breeds in alternate years (when successful), lays just one egg, and is immature for nine to eleven years, the longest period of immaturity for any bird (see p. 94).

Not all vertebrates are *K*-strategists. The nomadic Australian zebra finch (*Taeniopygia castanotis*) is *r*-selected, and the blue tit (*Parus caerulus*), with the largest recorded clutch size for a passerine bird, is an opportunist. If their populations should fall dramatically, then the birth rate will often increase to bring the population bouncing back to its stable level. This happens by increasing the litter or clutch size. Closely related species may display different population strategies. Two trefoil species (*Trifolium*) in Europe – the pale trefoil (*Trifolium pallescens*) and Thal's trefoil (*Trifolium thalii*) – prefer distinct habitats and reproduce only by seeds (Hilligardt 1993). *Trifolium pallescens* grows on alpine screes, often in morainic areas; *Trifolium thalii* is predominantly found in subalpine and alpine pastures. *Trifolium thalii*, a plant of more competitor-influenced habitats, is a *K*-strategist; *Trifolium pallescens* has characteristics of an invasive, pioneer *r*-strategist. It has less developed vegetative structures, higher seed production, higher seed-bank reserves, smaller seeds, and higher rates of seedling mortality.

Competitors, stress-tolerators, and ruderals

Some populations are finely adjusted to habitats that vary considerably in time. Where temporal habitat variations are pronounced, populations may experience adverse conditions or stress. A three-strategy model, which adds a stress-toleration strategy to competitive and opportunistic strategies, has been applied to plants (Grime 1977; Grime *et al.* 1988).

External environmental factors that affect plants fall into two chief categories – stress and disturbance. Stress involves all factors that restrict photosynthesis (mainly shortages of light, water, or nutrients; and suboptimal temperatures). Disturbance involves partial or total destruction of plant biomass arising from herbivores, pathogens, and humans (trampling, mowing, harvesting, and ploughing), and from wind-damage, frosting, droughting, soil erosion, and fire. Combining high and low stress with high and low disturbance yields four possible contingencies to which vegetation adapts (Table 4.4). Notice that there is no practicable strategy in plants for living in a stressful and highly disturbed environment. This is

Table 4.4 Environmental contingencies and ecological strategies in plants

Disturbance intensity	Stress intensity	
	Low	High
Low	Competitors	Stress-tolerators
High	Ruderals	(No practicable strategy)

Source: After Grime (1977)

Table 4.5 A key for identifying ecological strategies of herbaceous plants

	Levels of dichotomous key				Strategy
1	2	3	4	5	
Annual	⇒ Potentially fast-growing	⇒ Flowering precocious			Ruderal
		⇒ Flowering delayed			Competitive–ruderal
	⇒ Small and slow-growing				Stress-tolerant–ruderal
Biennial	⇒ Potentially large and fast-growing				Competitive–ruderal
	⇒ Small and slow-growing				Stress-tolerant–ruderal
Perennial	⇒ Vernal geophyte				Stress-tolerant–ruderal
	⇒ Not vernal geophyte	⇒ Rapid leaf turnover	⇒ Rapid proliferation and fragmentation of shoots		Competitive–ruderal
			⇒ Shoots not fragmenting rapidly	⇒ Shoots tall and laterally extensive	Competitor
				⇒ Shoots short or creeping	Competitive–stress-tolerant–ruderal (C–S–R)
		⇒ Slow leaf turnover	⇒ Shoots tall, or laterally extensive, or both		Stress-tolerant competitor
			⇒ Shoots short and not laterally extensive		Stress-tolerator

Source: After a diagram in Grime et al. (1988)

because continuous and severe stress in highly disturbed habitats prevents a sufficiently swift recovery or re-establishment of the vegetation. (Some humans – teachers come to mind – manage to survive under those conditions.) The remaining three environmental contingencies appear to have produced three basic ecological strategies in plants – **competitors, stress-tolerators**, and **ruderals**. Competitors correspond to K-strategists, and ruderals to r-strategists. Stress-tolerators are an additional group whose strategy is a response to the temporal variability and adversity of the physical environment. In addition, there are four intermediate ecological strategies – competitive ruderal, stress-tolerant competitor, stress-tolerant ruderal, and competitor–stress-tolerator–ruderal strategy (or C–S–R). Strategies of different herbaceous plants may be found using a dichotomous key (Table 4.5).

The primary and intermediate ecological strategies may be displayed on a triangular diagram (Figure 4.8). The plants associated with each strategy share the same adaptive characteristics. The following examples are all British plants:

1 **Competitors** exploit low stress and low disturbance environments. The rosebay willow-herb (*Chamerion angustifolium*) occurs in a wide variety of undisturbed habitats. It is notably common in fertile, derelict environments (urban clearance sites, cinders, building rubble, mining and quarry waste, and other spoil heaps). It is abundant in woodland glades, wood margins, scrub, and young plantations; it is widespread in open habitats such as walls, cliffs and rock outcrops, waysides and waste ground, and river banks; and seedlings are sometimes found in wetland (but not submerged areas), arable fields, and pastures.

2 **Stress-tolerators** exploit high stress, low disturbance, low competition environments. The heath grass (*Danthonia decumbens*) favours mountain grassland. It is also found in pastures and heathland, and occasionally in road verges, grassy paths, scree slopes, rock outcrops, and wetland.

3 **Ruderals** exploit high disturbance, low stress, and low competition environments. The shepherd's purse (*Capsella bursa-pastoris*) lives on disturbed, fertile ground. It is especially abundant as a weed on arable land and in gardens, and occurs in a range of disturbed artificial habitats such as demolition sites, paths, spoil heaps, and manure heaps.

4 **Competitive ruderals** exploit low stress, medium competition, and medium disturbance environments. The policeman's helmet (*Impatiens glandulifera*) is abundant on river banks, shaded marshland, and disturbed, lightly shaded areas in woodland and scrub.

5 **Stress-tolerant competitors** exploit medium competition, medium stress, and low disturbance environments. The male fern (*Dryopteris filix-mas*) is restricted to two very different habitats – woodland habitats and skeletal habitats (mainly cliffs and walls, but also quarry spoil, rock outcrops and screes, river banks, and some cinder tips).

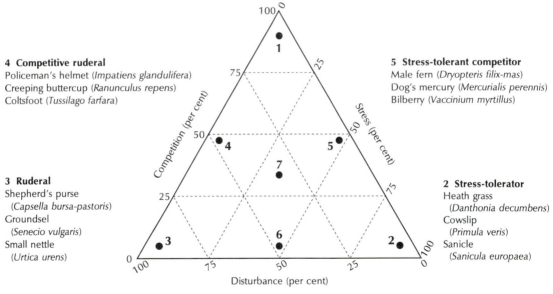

1 Competitor
Rosebay willow-herb (*Chamerion angustifolium*)
Stinging nettle (*Urtica dioica*)
Reed grass (*Phalaris arundinacea*)

4 Competitive ruderal
Policeman's helmet (*Impatiens glandulifera*)
Creeping buttercup (*Ranunculus repens*)
Coltsfoot (*Tussilago farfara*)

5 Stress-tolerant competitor
Male fern (*Dryopteris filix-mas*)
Dog's mercury (*Mercurialis perennis*)
Bilberry (*Vaccinium myrtillus*)

3 Ruderal
Shepherd's purse
  (*Capsella bursa-pastoris*)
Groundsel
  (*Senecio vulgaris*)
Small nettle
  (*Urtica urens*)

2 Stress-tolerator
Heath grass
  (*Danthonia decumbens*)
Cowslip
  (*Primula veris*)
Sanicle
  (*Sanicula europaea*)

6 Stress-tolerant ruderal
Carline thistle (*Carlina vulgaris*)
Hard poa (*Desmazeria rigida*)
Purging flax (*Linum catharticum*)

7 'C–S–R' strategist
Yorkshire fog (*Holcus lanatus*)
Cat's ear (*Hypochaeris radicata*)
Common sorrel (*Rumex acetosa*)

Figure 4.8 Ecological strategies in plants. The triangular diagram is used to define various mixes of three basic strategies – competitor, stress tolerator, and ruderal. Pure competitors lie at the top corner of the triangle, pure stress-tolerators at the bottom right corner, and pure ruderals at the bottom left corner. Intermediate strategies are shown. Example species are all British.
Source: After Grime *et al.* (1988)

6 **Stress-tolerant ruderals** exploit medium disturbance, medium stress, and low competition environments. The carline thistle (*Carlina vulgaris*) occurs in areas of discontinuous turf associated with calcareous pastures, rock outcrops, lead-mine heaps, and derelict wasteland. It is also frequent on sand dunes.

7 **Competitor–stress-tolerator–ruderal (C–S–R) strategists** occupy a middle-of-the-road position (medium competition, medium stress, medium disturbance) that enables them to live in a wide range of conditions. The Yorkshire fog (*Holcus lanatus*) is recorded as seedlings in every type of habitat. It is most abundant in meadows and pastures, and is prominent in spoil heaps, waste ground, grass verges, paths, and in hedgerows. It is less common on stream banks, arable land, marsh ground, and in skeletal habitats such as walls and rock outcrops. It occurs in low frequency in scrub and woodland clearings, and is found also in grassy habitats near the sea and on mountains.

Different life-forms tend to have a narrow range of ecological strategies. Trees and shrubs centre on a point between the competitor–stress-tolerator–ruderal and competitive–ruderal positions. Herbs have a wide range of strategies, though competitor–stress-tolerator–ruderal is common. Bryophytes (mosses and liverworts) are mainly ruderals and stress-tolerator–ruderals. Lichens are mainly stress-tolerator–ruderals and stress-tolerators.

Migration strategies

Migration rates vary considerably between species. In plants, three basic migration strategies seem to exist and are analogous to ruderal, opportunist, and competitive population strategies.

Fugitive species

These are the 'weeds' of the plant kingdom. They colonize temporary, disturbed habitats, reproduce rapidly, and soon depart before the habitat disappears or before competition with other organisms overwhelms them. The common dandelion (*Taraxacum officinale*) is a fine example. A **fugitive-migration strategist** occupies a patchwork of habitat islands. The tamarack (*Larix laricina*) is a fugitive-strategist (Delcourt and Delcourt 1987: 319–22). Its changing distribution mirrors the availability of wetland habitats, and any snapshot of its distribution shows low dominance in most areas, with dominance hot-spots in local wetland sites (e.g. Payette 1993).

Opportunist species

Fast migrators spread rapidly, pushing forwards along a steep migration front, but failing to maintain large populations in its wake. Spruce (*Picea*) is an **r-migration strategist** and displays these characteristic (Delcourt and Delcourt 1987: 306–12). With the retreat of the Laurentide Ice Sheet, spruce spread rapidly onto the newly exposed ground. The rate of migration reached 165 m/yr around 12,000–10,000 years ago. The spruce's northward advance was halted only by the ice. Behind its migration front, populations diminished as climates warmed. Black spruce (*Picea mariana*) was the first conifer species to invade northern Quebec immediately after the ice had melted 6,000 years ago (Desponts and Payette 1993).

Equilibrium species

Slow migrators rise more slowly to dominance after an initial invasion, the highest values of dominance lying well behind the 'front lines' and reflecting an unhurried build-up of populations. Oak (*Quercus*) is a **K-migration** strategist. It was a minor constituent of late glacial forests in eastern North America. It did reach the Great Lakes in late glacial times, but its rise to dominance was a slow process that reached a ceiling in the Holocene epoch in the region now occupied by the eastern deciduous forest (Delcourt and Delcourt 1987: 313–16).

PLAYING WITH NUMBERS: POPULATION EXPLOITATION AND CONTROL

Exploiting populations

The **commercial exploitation** of marine mammals, fish, big game, and birds has caused the **extinction** of several species and the **near-extinction** of several others. The cases of the passenger pigeon, great auk, northern fur seal, and the American bison illustrate this point.

Passenger pigeon

These graceful birds once flourished in North America. On 1 September 1914, the last passenger pigeon (*Ectopistes migratorius*), called 'Martha', died in the Cincinnati Zoological Garden. The rapid extinction of such a flourishing population resulted from three misfortunate circumstances. First, the pigeons were very tasty when roasted or stewed. Second, they migrated in huge and dense flocks, perhaps containing 2 billion birds, that darkened the sky as they passed overhead. Third, they gathered in vast numbers, perhaps several hundred million birds, to roost and to nest. A colony in Michigan was estimated to be 45 km long and an average of around 5.5 km wide. Professional hunters – pigeoners – formed large groups to trap and kill the pigeons. Nesting trees were felled to collect the squabs (young, unfledged pigeons). The clearing of forest for farmland and the expansion of the railroad made huge tracts of the pigeon's range accessible and the persecution intensified. By about 1850, several thousand people were employed in catching and marketing passenger pigeons. Every means imaginable was used to obtain the tasty bird. Special firearms, cannons, and forerunners to the machine gun were built. In 1855 alone, one New York handler had a daily turnover of 18,000 pigeons. In 1869, 7.5 million birds were captured at one spot. In 1879, 1 billion birds were captured in the state of Michigan. The harvesting rate disrupted breeding and the population size began to plummet. Even by 1870, large breeding assemblies were confined to the Great Lake States, and the northern end of the pigeon's former range. The last nest was found in 1894. The last bird was seen in the wild in 1899.

Auks

The 'penguin of the Arctic', the great auk (*Pinguinus impennis*), was an early casualty of uncontrolled harvesting. This flightless bird was the largest of the Alcidae, which family includes guillemots, murres, and puffins. It used to nest around the North Atlantic Ocean, in north-eastern North America, Greenland, Iceland, and the north-western British Isles (Figure 4.9). The last pair died in Iceland in 1844. A few museum specimens survive.

The razorbill (*Alca torda*) is the largest living auk. For many years, this species was killed for its feathers, and by fishermen for use as bait. This exploitation caused the species to decline. The razorbill is now protected and is staging a comeback, so it may avoid the fate of its close relative, the great auk.

Fur seals

Millions of northern fur seals (*Callorhinus ursinus*) used to breed on the Pribilof Islands, in the Bering Sea, Alaska. Between 1908 and 1910, Japanese seal hunters slaughtered 4 million of them. The Fur Seal Treaty of 1911 was signed by Japan, Russia, Canada, and the United States. This was the first international agreement to protect marine resources. The signatories agreed rates of harvesting for pelts. The happy result was that the northern fur seal population bounced back. New colonies occasionally appear. Northern fur seal pups were first observed on Bogoslof Island, in the south-east Bering Sea, in 1980 (Loughlin and Miller 1989). By 1988 the population had grown to more than 400 individuals, including 80 pups, 159 adult females, 22 territorial males, and 188 juvenile males. Some animals originated from rookeries of the Commander Island; others were probably from the Pribilof Islands.

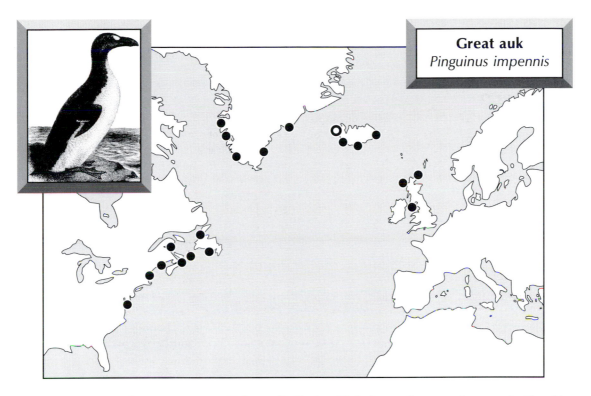

Figure 4.9 The great auk's (*Pinguinus impennis*) former distribution. Black dots are former nesting grounds. The white circle marks the site where the last two birds died in 1844.
Source: Map after Ziswiler (1967); picture from Saunders (1889)

American bison

Humans have hunted terrestrial, as well as marine, mammal species to extinction or near extinction. A spectacular, if shameful, example concerns the American bison or 'buffalo' (*Bison bison*) (Figure 4.10). In 1700, the buffalo population on the Great Plains was around 60 million. Native Indians had hunted this population sustainably for thousands of years. A new era of buffalo hunting started in the 1860s, when the Union Pacific Railway was opened. At first, special hunting parties were hired to supply the railway workers with meat. William 'Buffalo Bill' Cody rose to fame at this time for having killed 250 buffalo in a single day! The tracks of the Union Pacific cut right across the heart of the buffalo country. Once the railway was operating, the railway company

encouraged buffalo shoot-outs from the trains. Thus was perpetrated a senseless and bloodthirsty slaughter. Tens of thousands of animals were shot and left to rot alongside the tracks. By the 1890s, only a few hundred buffaloes were still alive in the United States, and a few hundred more of the wood buffalo variety in Canada. In 1902, the population was less than a hundred. Happily, buffalo breed well under animal husbandry. Thanks to William T. Hornaday of the New York Zoological Society, the American Bison Society (founded in 1905), and the United States and Canadian governments, the species was saved. There was a vigorous programme of captive breeding and releasing of animals onto protected ranges, such as that in Yellowstone Park. The bison has now recovered to a population size of tens of thousands and has been introduced into Alaska.

Controlling populations

Much conservation effort is currently expended on **reintroducing species** to at least some of their former domain, and keeping successfully introduced species under control. These two aspects of population management are exemplified by beaver reintroductions in Sweden and by managing feral goats in New Zealand and feral cats on Marion Island.

Swedish beavers

The European beaver (*Castor fiber*) is Europe's largest rodent. It lives beside shores of watercourses and lakes, in excavated burrows or constructed lodges. The beaver was heavily hunted in Sweden during the nineteenth century (Bjärvall and Ullström 1986: 77). By the time it received protection in Sweden, it was too late – the last animals had been shot two years earlier. In 1922, reintroductions were begun using animals captured in southern Norway. The population grew to about 400 animals in 1939 and 2,000 in 1961. During the 1960s the population tripled and

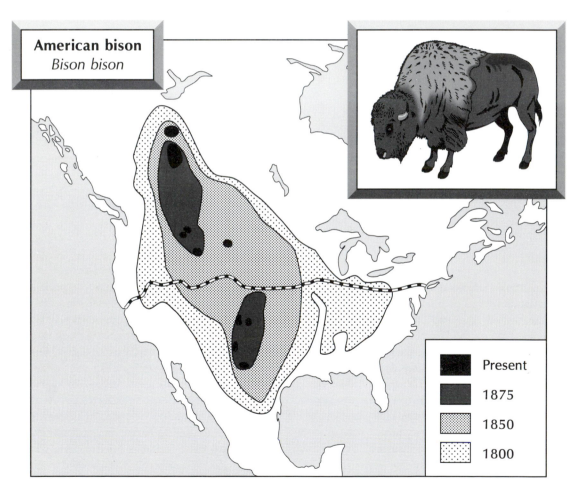

Figure 4.10 The shrinking range of the American bison (*Bison bison*).
Source: After Ziswiler (1967)

increased to 40,000 animals by 1980. Given such a large increase, some hunting was permitted from 1973. The hunting appears to have been justified. The reintroduced beaver population in Sweden has behaved as introduced ungulate populations – it 'exploded' and then started to decline (Hartman 1994). In two study areas, the population growth-rate turned negative after 34 and 25 years and at densities of 0.25 and 0.20 colonies per km^2, respectively. Management policy for an irruptive species should probably allow hunting during the irruptive phase. Hunting keeps the irruptive population in check and so maintains food resources and avoids uncontrolled population decline.

New Zealand goats

Feral goats (*Capra hircus*) live in about 11 per cent of New Zealand, mostly on land reserved for conservation of the indigenous biota. They were introduced from Europe before 1850. They were recognized as a threat to native vegetation from the 1890s onward and soon attained the rank of a pest. The effect of their browsing is most noticeable on islands, where subtropical forest has in some cases been converted to rank grassland (Rudge 1990). Uncontrolled goat densities are usually less than 1 per ha, but have reached 10 per ha in one area (Parkes 1993). The total population is at least 300,000. Goats are highly selective in what they eat, but switch their diets as the more palatable food species are eliminated from their habitat.

Unharvested feral populations in New Zealand form stably dynamic *K*-strategist relationships with their resources. Their impacts are thus chronic, but the relationship can be manipulated to favour the resource if management can be applied 'chronically', that is, as sustained control. The Department of Conservation has attempted to organize its control actions in the highest priority areas either to eradicate goats where possible or to sustain control where eradication is not presently possible. In 1992–3, 6,160 hunter-days of ground-control effort, and 230 hours of helicopter hunting effort, were expended against feral goats, and 10.2 km of goat fencing was constructed (though goats are notoriously hard to fence in). If the present effort is maintained through to the first decade of the new millennium, about half the feral goat populations now on the conservation estate are expected to be eradicated or controlled. However, goats are a valuable farming resource, and this may assure the survival of feral populations in hobby farms and hill country farms bordering forests (Rudge 1990).

Marion Island cats

Marion Island is the largest of the Prince Edward Islands and lies in the southern Indian Ocean, about midway between Antarctica and Madagascar. It is a volcanic island with an area is 29,000 ha that supports a tundra biota. Domestic cats (*Felis catus*) were introduced to Marion Island as pets in 1949. By 1975, the population was about 2,139 and had risen to 3,405 in 1977. As on other sub-Antarctic islands where cats were introduced, the feral cats played havoc with the local bird populations. They probably caused the local extinction by 1965 of the common diving petrel (*Pelecanoides urinatrix*), and reduced the populations and breeding successes of winter-breeding great-winged petrels (*Pterodroma macroptera*) and grey petrels (*Procellaria cinerea*). Control of the cat population was begun in 1977. The viral disease feline panleucopaenia was chosen as the most efficient and cost-effective method. The population fell from 3,405 cats in 1977 to 615 cats in 1982. However, the decline slowed down and evidence suggested that **biological control** was becoming ineffective. A full-scale hunting effort was launched in the southern spring of 1986 (Bloomer and Bester 1992). The aim was to eradicate the cat population from the island. In 14,725 hours of hunting, 872 cats were shot dead and 80 were trapped. The number of cats sighted per hour of night hunting decreased, which was strong indication that the population had declined. But, by the end of the third season, it became evident that hunting alone was no longer removing enough animals to sustain the population decline, and **trapping** was incorporated into the eradication programme. Trapping with baited walk-in cage traps has a drawback: other species, such as sub-Antarctic skuas (*Catharacta antarctica*), also become ensnared. However, a small loss of non-target species has to be offset against the long-term benefits of the cat trapping programme to the bird population.

SUMMARY

Populations are interbreeding groups of individuals (demes). Unconstrained population growth describes a rising exponential curve; when constrained, it describes a logistic curve, or may behave chaotically. Population irruptions and crashes are common. Age and sex are important components of population growth. They are summarized in life tables and survivorship curves, and they are studied using cohort-survival models. Metapopulations are defined according to the geographical location of individuals within a population – whether they form one large interacting group, or whether they live in several relatively isolated subpopulations, for example. Recent work on the conservation of threatened populations draws heavily on metapopulation theory. Populations adopt various strategies to maximize their chances of survival. Opportunists (*r*-strategists) and competitors (*K*-strategists) are two basic groups. Stress-tolerators form another strategic option. The three basic migration strategies are fugitives, opportunists, and equilibrists. Wild populations suffer heavily at the hands of humans. Some species were hunted to extinction. Others were hunted to near extinction but managed to recover. Population control is often necessary to prevent damage to agricultural resources and environmental degradation. Control involves culling or eradicating introductions, such as goats and cats. It may also may involve re-establishing a species over vacated parts of its former range.

ESSAY QUESTIONS

1 Why do some populations irrupt and crash?

2 What are the advantages and disadvantages of the different population survival strategies?

3 Why do some populations need controlling?

FURTHER READING

Hanski, I. and Gilpin, M. E. (1997) *Metapopulation Biology: Ecology, Genetics, and Evolution*, New York: Academic Press.

Kormondy, E. J. (1996) *Concepts of Ecology*, 4th edn, Englewood Cliffs, NJ: Prentice Hall.

Kruuk, H. (1995) *Wild Otters: Predation and Populations*, Oxford: Oxford University Press.

Perry, J. H., Woiwod, I. P., Smith, R. H., and Morse, D. (1997) *Chaos in Real Data: Analysis on Non-linear Dynamics from Short Ecological Time Series*, London: Chapman & Hall.

Taylor, V. J. and Dunstone, N. (eds) (1996) *The Exploitation of Mammal Populations*, London: Chapman & Hall.

Tuljapurkar, S. and Caswell, H. (1997) *Structured-Population Models*, New York: Chapman & Hall.

5

INTERACTING POPULATIONS

Populations interact with one another. This chapter covers:

- co-operating populations
- competing populations
- plant-eating populations
- flesh-eating populations
- biological population control

Two populations of the same species, living in the same area, may or may not affect each other. Interactions between pairs of populations take several forms, depending on whether the influences are beneficial or detrimental to the parties concerned (Table 5.1). In mutually beneficial relationships, both populations gain; in mutually detrimental relationships, both lose. Other permutations are possible. One population may gain and the other lose, one may gain and the other be unaffected; one may lose and the other be unaffected.

LIVING TOGETHER: CO-OPERATION

Four types of interactions between populations or species favour peaceful coexistence – neutralism, protocooperation, mutualism, and commensalism. Neutralism involves no interaction between two populations. It is probably very rare, because there are likely to be indirect interactions between all populations in a given ecosystem. The other forms of interaction are common.

Protocooperation

Protocooperation involves both populations benefiting one another in some way from their interaction. However, neither population's survival depends on the interaction (the interaction is not obligatory). Some small birds ride on the backs of water buffalo – the birds obtain food and in doing so rid the buffalo of insect pests. Some birds pick between the teeth of crocodiles – the birds get a meal and the crocodile gets a free dental hygiene session.

Animal–animal protocooperation

A remarkable example of protocooperation is the partnership between an African bird, the greater honey-guide (*Indicator indicator*), and a mammal, the honey badger or ratel (*Mellivora capensis*) (Plate 5.1). The honey-guide seeks out a beehive, then flitters around making a chattering cry to attract the attention of a honey badger, or any other interested mammal, including humans. The honey badger rips open the hive and feeds upon honey and bee larvae. After the honey badger has had its fill, the bird feeds. Its digestive system can even cope with the nest-chamber wax, which is broken down by symbiotic wax-digesting bacteria living in its intestines. In captivity, individuals have been kept alive for up to 32 days feeding exclusively on wax! The honey-guide is adept at finding beehives but is incapable of opening them; the honey badger is good at opening beehives but is not good at finding them. The protocooperation clearly profits both species.

Table 5.1 Interactions between population pairs

Kind of interaction	Species		Comments
	A	B	
Neutralism	No effect	No effect	The populations do not affect one another
Protocooperation	Gains	Gains	Both populations gain from their interaction, but the interaction is not obligatory
Mutualism	Gains	Gains	Both populations gain from their interaction, and the interaction is obligatory (they cannot survive without it)
Commensalism	Gains	No effect	Population A (the commensal) gains; population B (the host) is unaffected
Competition	Loses	Loses	The populations inhibit one another
Amensalism	Loses	No effect	Population A is inhibited; population B is unaffected
Predation	Gains	Loses	Population A (the predator) kills and eats members of population B (the prey)

Plant–plant protocooperation

Protocooperation seems to occur among plants. Two species of bog moss, *Sphagnum magellanicum* and *Sphagnum papillosum*, were grown under laboratory conditions (Li *et al.* 1992). Under dry conditions, *Sphagnum magellanicum* outcompeted *Sphagnum papillosum* for water, largely because it is better designed for transporting and storing water. *Sphagnum magellanicum* is a drought-avoider and prefers peat hummocks, while *Sphagnum papillosum* is a drought-tolerator and prefers lower sites on a hummock. However, *Sphagnum magellanicum* benefits *Sphagnum papillosum* growing at higher hummock positions by encouraging lateral flow of water. Both species grow better when mixed, suggesting a degree of protocooperation.

Animal–plant protocooperation

Plants often form co-operative partnerships with animals. The most common examples are associations between plants and pollinators, and between plants and seed dispersers. Plants are rooted to the place that they grow so they must employ a go-between to carry their pollen to other plants and their seeds to other sites. Colourful flowers with nectar and brightly coloured fruit have presumably evolved to attract animals as potential delivery agents. Some herbivores eat fruits, the seeds in which may pass through the herbivore intestines unharmed and grow into new plants from the droppings. Mammals and birds carry seeds in their fur and feathers. The sheep carries a garden-centre-full of plant seeds around in its fleece (p. 60). Birds, bees, and butterflies are the commonest pollinators. Some mammals are pollinators, including humans, nectar-feeding bats, and the slender-nosed honey possum or noolbender (*Tarsipes rostratus*). The noolbender is a diminutive Australian marsupial that feeds on nectar, pollen and, to some extent, small insects that live in flowers (Colour plate 7).

An unusual example of plant–animal co-operation is the alliance between some species of *Acacia* and epiphytic ants. Species of acacia that normally support ant colonies are highly palatable to herbivorous insects; those that do not normally house an ant fauna to ward off herbivores are less palatable. Acacias guarded by ant gangs produce nectaries and swollen

Plate 5.1 Ratel or honey badger (*Mellivora capensis*)
Photograph by Pat Morris

thorns that attract and benefit the ants. The plants put matter and energy into attracting ants that will protect their leaves, rather than into chemical warfare. This antiherbivore ploy is broad-based, since the ants fiercely attack a wide range of herbivores.

Mimics

Müllerian mimicry (named after Fritz Müller, a nineteenth-century German zoologist) occurs when two or more populations of distasteful or dangerous species come to look similar. An example is bees and wasps, which are normally banded with yellow and black stripes. In Trinidad, the unpalatable butterflies *Hirsutis megara* (from the family Ithomiidae) and *Lycorea ceres* (from the family Danaidae) resemble one another (Figure 5.1). The poisonous passion-flower

butterflies (*Heliconius*) are confined to the Americas. There are some forty species, most of which are restricted to rain forest. In different parts of South America, the various species tend to look much alike and form Müllerian mimicry rings. It is not too difficult to see the advantage of such biological photocopying. Tenderfoot predators presumably learn to avoid unsavoury prey by trial and error, killing and eating at random. If distasteful species varied widely in appearance, then the predators would be forced to kill many of each before learning those to avoid. However, when the unsavoury prey are coloured in the same way, then the predator quickly learns to avoid one basic pattern. Its sample snacks are therefore spread out over many prey species. By resembling one another, unsavoury prey populations keep their losses by predation to a minimum.

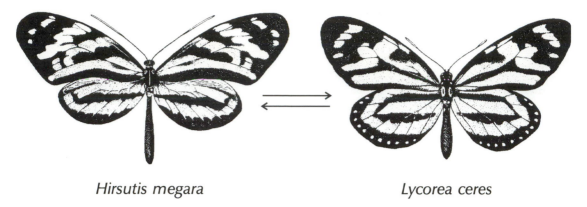

Hirsutis megara   ### *Lycorea ceres*

Figure 5.1 The unpalatable Trinidadian butterflies *Hirsutis megara* (from the family Ithomiidae) and *Lycorea ceres* (from the family Danaidae) resemble one another – they are Müllerian mimics.
Source: From Brower (1969)

Batesian mimicry, named after the nineteenth-century English naturalist Henry Walter Bates, occurs when a palatable animal comes to look like an unpalatable one. The harmless dronefly (*Eristalis tenax*) has warning coloration much like a honey bee's (*Apis mellifera*). The advantage in such mimicry is avoidance by would-be predators mistaking it for a bee with a sting in its tail. The dronefly, and other mimics, is thus a 'sheep in wolf's clothing', bluffing its way through life on the strength of masquerade. Batesian mimicry is quite common in the animal kingdom. Many species of harmless snakes mimic poisonous snakes – harmless and poisonous coral snakes in Central America are so similar that only an expert can tell them apart. Batesian mimicry is disadvantageous to the poisonous 'model' because predators will eat harmless mimics and so take longer to learn to avoid the model's warning colours. So, technically speaking, it is a brand of predation.

Mutualism

In a mutualistic relationship, both populations benefit from interacting. However, they are dependent on their interaction to survive. **Mutualism** is far less common than protocooperation, of which it is a more extreme form. It is the evolutionary equivalent of 'putting all your eggs in one basket' – the reliance of one species on another is complete. Certain Australian termites, for example, cannot produce enzymes to digest the cellulose in wood. They exploit wood as a food source by harbouring a population of protozoans (*Myxotricha paradoxa*) capable of making cellulose-digesting enzymes in their guts. The protozoans are intestinal endosymbionts. Neither the termites nor the protozoans would survive without the other. One generation of termites passes on the protozoans to the next by exchanging intestinal contents.

Lichens are thought to depend on a mutualistic alliance between a fungus and an alga – the fungus provides the shell, the alga the photosynthetic power house. However, the algae in some lichens can be grown without their fungal host, so the relationship may in some cases be protocooperation.

Mutualism is also known in marine environments. Two encrusting sponges (*Suberites rubrus* and *Suberites luridus*) protect the queen scallop (*Chlamys opercularis*) from predation by a starfish, the common crossfish (*Asterias rubens*) (Pond 1992). They probably do so by making it difficult for the common crossfish to grip the scallop with their tube-like feet, and by excluding other organisms settling on the scallop, so hindering its mobility. The sponge benefits from the association by being afforded protection from the shell-less nudibranch-gastropod predator *Archidoris pseudoargus*, and more generally by being carried to favourable locations.

Commensalism

Commensalism occurs when one population (the **commensal**) benefits from an interaction and the other population (the **host**) is unharmed. Bromeliads and orchids grow as epiphytes on trees without doing any harm. The house mouse (*Mus musculus domesticus*) has lived commensally with humans since the first permanent settlements were established in the Fertile Crescent (Boursot *et al.* 1993). The house mouse benefits from the food and shelter inadvertently provided by humans, while the humans suffer no adverse effects. Four other examples are as follows.

Cattle egrets and cattle

Cattle egrets (*Bubulcus ibis*), which are native to Africa and Asia, follow cattle grazing in the Sun, pouncing on crickets, grasshoppers, flies, beetles, lizards, and frogs flushed from the grass as the cattle approach (Colour plate 8). The cattle are unaffected by the birds, while the birds appear to benefit – they feed faster and more efficiently when associated with the cattle. Interestingly, the cattle egret has expanded its range, successfully colonizing Guyana, South America, in the 1930s. It filled a niche created by extensive forest clearance. Without large herbivores to stir up its meal, it has taken to feeding in the company of domestic livestock, which also disturbs insects and other prey living in grass.

Midge and mosquito larvae in pitcher-plant pools

In Newfoundland, Canada, the carnivorous pitcher plant, *Sarracenia purpurea*, has water-filled leaves holding decaying invertebrate carcasses (S. B. Heard 1994). Larvae of a midge (*Metriocnemus knabi*) and a mosquito (*Wyeomyia smithii*) feed on the carcasses. The two insect populations are limited by the carcass supply, but the use the carcasses at different stages of decay and live commensally. The midges feed by chewing upon solid materials, while the mosquitoes filter-feed on tiny bits that break off the decaying matter and on bacteria. Mosquitoes eating particles has no effect on the midges, but midges feeding on solid material creates a supply of organic fragments and, by increasing the surface area of decaying matter, bacteria.

Mynas and king-crows

In India, two bird species, the common myna (*Acridotherus tristis*) and the jungle myna (*Acridotherus fuscus*), forage in pure and mixed flocks on fallow land (Veena and Lokesha 1993). Another bird species, the king-crow (*Dicrurus macrocercus*), a drongo, feeds on insects disturbed by the foraging mynas. King-crows tend to consort with larger myna flocks, comprising twenty or more individuals. The king-crows and mynas eat different foods, so they do not compete. The king-crows benefit from the association, whereas the mynas are unaffected – a case of commensalism.

River otters and beavers

In the boreal forest of north-east Alberta, Canada, a partial commensalism has evolved between river otters (*Lutra canadensis*) and beavers (*Castor canadensis*) (Reid *et al.* 1988; Colour plate 9). In winter, otters dig passages through the beaver dams to guarantee underwater access to adjacent water bodies, which are frozen over. The passages lead to a reduction in pond water-level, which may increase the otters' access to air under the ice, and to concentrate the fishes upon which the otters feed. Beavers repair the passage during times of open water, so otters benefit from the creation and maintenance of fish habitats by the beavers, who themselves are unaffected by the otters' activities.

STAYING APART: COMPETITION

While some species co-operate with one another, others compete for food or territory.

Competing for resources

Competition is a form of population interaction in which both populations suffer. It occurs when two or more populations compete for a common resource, chiefly food or territory, that is limited. It may engage members of the same species (**intraspecific competition**) or members of two or more different species (**interspecific competition**). The **limiting resource** prevents one or both the species (or population) from growing.

Competitive exclusion

The outcome of competing for a limiting resource is not obvious and depends upon a variety of circumstances. It would be tempting to think either that both species may coexist, though at reduced density, or else that one species may displace the other. However, coexistence is exceedingly unlikely. As a very strict rule, no two species can coexist on the same limiting resource. If they should try to do so, one species will outcompete the other and cause its extinction. This finding is encapsulated in **Gause's principle**, named after Georgii Frantsevich Gause. It is also called the **competitive exclusion principle** (Hardin 1960). It was first established in mathematical models by Alfred James Lotka (1925) and Vito Volterra (1928), and later in laboratory experiments by Gause (1934).

Gause used a culture medium to grow two *Paramecium* species – *Paramecium aurelia* and *Paramecium caudatum*. Each population showed a logistic growth curve when cultured separately. When grown together, they displayed a logistic growth curve for the first six to eight days, but the *Paramecium caudatum* population slowly fell and the *Paramecium aurelia* population slowly rose, though it never attained quite so high a level as when it was cultured separately. Many other laboratory experiments using various competing populations give comparable results. Competitors used include protozoa, yeasts, hydras, *Daphnia*, grain beetles, fruit flies, and duckweed (*Lemna*).

Scramble competition

Competition between populations occurs in two ways – **scramble competition** and **contest competition**. First, scramble, exploitation, or resource competition occurs when two or more populations use the same resources (food or territory), and when those resources are in short supply. This competition for resources is indirect. One population affects the other by reducing the amount of food or territory available to the others.

A classic example of scramble competition concerns the native red squirrel (*Sciurus vulgaris*) and the introduced grey squirrel (*Sciurus carolinensis*) in England, Wales, and Scotland (Figure 5.2; Colour plate 10). The red squirrel favours coniferous forest, but also lives in deciduous woodland, particularly beech woods. The grey squirrel, a native of eastern North America, is adapted to life in broad-leaved forests, but is adaptable and will colonize coniferous forests. In Britain, the red squirrel's range has progressively shrunk during the twentieth century as the grey squirrel, introduced from America in the late nineteenth and early twentieth centuries, has extended its range. A dramatic decline in red squirrel range occurred from 1920 to 1925, largely due to disease following great abundance (Shorten 1954). The grey squirrel rapidly advanced into the areas relinquished by the red squirrel, and into areas around their introduction sites. Up to around 1960, the red squirrel still had a few strongholds in the southern parts of England, especially East Anglia and the New Forest. The grey squirrel has advanced into these areas and replaced the red squirrel in much of East Anglia (H. R. Arnold 1993). Red squirrels can still be found on the Isle of Wight and Brownsea Island in Poole Harbour, Dorset.

The success of the grey squirrel may lie primarily in its superior adaptability to the herbivore niche at canopy level in deciduous woodland (C. B. Cox and Moore 1993: 61). However, one study highlights the complexity of squirrel competition (Kenward and Holm 1993). Conservative demographic traits combined with feeding competition could explain the red squirrel's replacement by grey squirrels. In

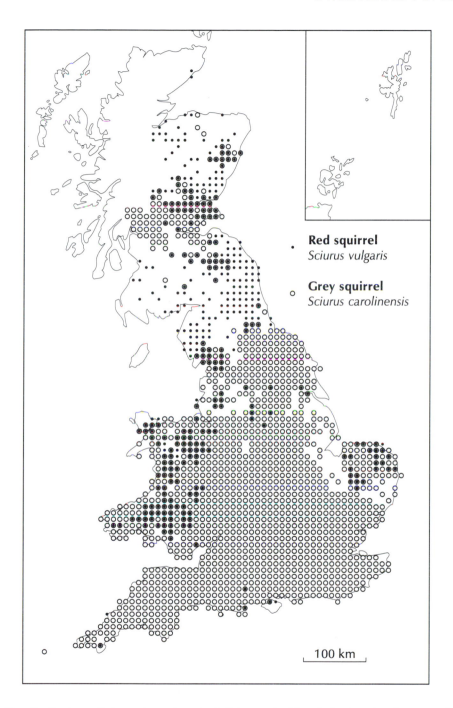

Red squirrel
Sciurus vulgaris

Grey squirrel
Sciurus carolinensis

100 km

Figure 5.2 The distributions of the native red squirrel (*Sciurus vulgaris*) and the introduced grey squirrel (*Sciurus carolinensis*) in England, Wales, and Scotland.
Source: After H. R. Arnold (1993)

oak–hazel woods, grey squirrel foraging, density, and productivity are related to oak (*Quercus* spp.) and acorn abundance. In contrast, red squirrels forage where hazels (*Corylus avellana*) are abundant; their relatively low density and breeding success are related to hazelnut abundance. Red squirrels fail to exploit good acorn crops, although acorns are more abundant than hazels. However, in Scots pine (*Pinus sylvestris*) forests, their density and breeding success is as high as grey squirrels' in deciduous woods. Captive grey squirrels thrive on a diet of acorns, but red squirrels have a comparative digestive efficiency of only 59 per cent, seemingly because they are much less able than grey squirrels to neutralize acorn polyphenols. A model with simple competition for the autumn hazel crop, which is eaten by grey squirrels before the acorn crop, shows that red squirrels are unlikely to persist with grey squirrels in woods with more than 14 per cent oak canopy. Oak trees in most British deciduous woods give grey squirrels a food refuge that red squirrels fail to exploit. Thus, red squirrel replacement by grey squirrels may arise solely from feeding competition, which is exacerbated by the post-war decline in coppiced hazel.

Contest competition

Contest or **interference competition** occurs through direct interaction between individuals. The direct interaction may be subtle, as when plants produce toxins or reduce light available to other plants, or overtly aggressive, as when animal competitors 'lock horns'.

Direct competition occurs in four duckweed species (*Lemna minor*, *Lemna natans*, *Lemna gibba*, and *Lemna polyrhiza*) (Harper 1961). *Lemna minor* grows fastest in uncrowded conditions. Under crowded conditions, when all species are grown together, *Lemna minor* grows the slowest. Nutrient levels have no effects on the growth rates, so the changes are caused by competition for light – the *Lemna* species interfere with one another.

Several animal and plant species interfere with competitors by chemical means. The chemicals released have an inhibiting, or allelopathic, impact on rivals. The nodding thistle (*Carduus nutans*) is allelopathic, releasing soluble inhibitors that discourage the growth of pasture grasses and legumes, at two phases of its development – at the early bolting phase when larger rosette leaves are decomposing and releasing soluble inhibitors, and at the phase when bolting plants are dying (Wardle *et al.* 1993). Moreover, nodding thistle seedlings seem to be stimulated by thistle-tissue additions to the soil. The thistle plants may weaken pasture and, at the same time, encourage recruitment of their own kind. Crowberry (*Empetrum hermaphroditum*) inhibits the growth of Scots pine (*Pinus sylvestris*) and aspen (*Populus tremula*) (Zackrisson and Nilsson 1992). It releases water-soluble phytotoxic substances from secretory glands on the leaf surface. These toxins interfere with Scots pine and aspen seed germination on the forest floor.

Aggressive competition is not so common as is popularly believed. As a rule, organisms avoid potentially dangerous encounters. But aggression does occur. A classic example is two acorn barnacles – *Chthamalus stellatus* and *Balanus balanoides* – that grow intertidal-zone rocks around Scotland. A study on the shore at Millport (Figure 5.3) showed that, when free-swimming larvae of *Chthamalus* decide to settle, they could attach themselves to rocks down to mean tide level (Connell 1961). Where *Balanus* is present, they attach only down to mean high neap tide. During neap tides, the range between low water mark and high water mark is at its narrowest. Young *Balanus* grow much faster than *Chthamalus* larvae – they simply smother them or even prise them off the rocks. This behaviour is direct, aggressive action to secure the available space. When *Balanus* is removed, *Chthamalus* colonizes the intertidal zone down to mean tide level. However, without *Chthamalus*, *Balanus* is unable to establish population above mean high neap tide, seemingly because of adverse weather (warm and calm conditions) promoting desiccation.

Mechanisms of coexistence

Species coexistence is the rule in Nature. It depends upon avoiding competition and is achieved

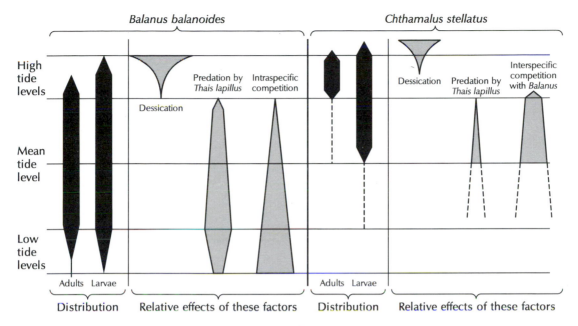

Figure 5.3 The vertical distribution of larvae and adult acorn barnacles on intertidal-zone rocks, Millport, Scotland. Two competing species are shown – *Balanus balanoides* and *Chthamalus stellatus*. *Thais lapillus* is a predatory whelk.
Source: After Connell (1961)

through several mechanisms: the environment is normally diverse and contains hiding places and food patches of various sizes; many species use a range of resources rather than one; animals have varying food preferences and commonly switch diets. The result is that competition is often diffuse: a species competes with many other species for a variety of resources, and each resource represents a small portion of the total resource requirements.

Resource partitioning

Species avoid niche overlap by partitioning their available resources according to size and form, chemical composition, and seasonal availability. Five warbler species (genus *Dendroica*) live in spruce forests in Maine, United States. They each feed in a different part of the trees, use somewhat different foraging techniques to find insects among the branches and leaves, and have slightly different nesting dates (Figure 5.4) (MacArthur 1958). The differences in feeding zones are large enough to account for

coexistence of the blackburnian warbler (*Dendroica fusca*), black-throated green warbler (*Dendroica virens*), and bay-breasted warbler (*Dendroica castanea*). The Cape May warbler (*Dendroica tigrina*) is different. Its survival depends upon outbreaks of forest insects to provide a glut of food. The myrtle warbler (*Dendroica coronata coronata*) is not so common as the other species and is less specialized.

Some species partition resources solely according to prey size. Among such predatory species as hawks, larger species normally eat larger prey than smaller species. In North America, the goshawk (*Accipiter gentilis*) eats prey weighing up to 1,500 g (Storer 1966). The smaller Cooper's hawk (*Accipiter cooperi*) takes prey almost as large, but only very occasionally. It mainly eats prey in the 10–200 g range. The smaller-still sharp-shinned hawk (*Accipiter striatus*) will eat prey up to 100 g, and sometimes a little more, but most of its prey are in the 10–50 g range. In these species, the goshawk is 1.3 times longer than Cooper's hawk, which is itself 1.3 times longer than the sharp-shinned hawk.

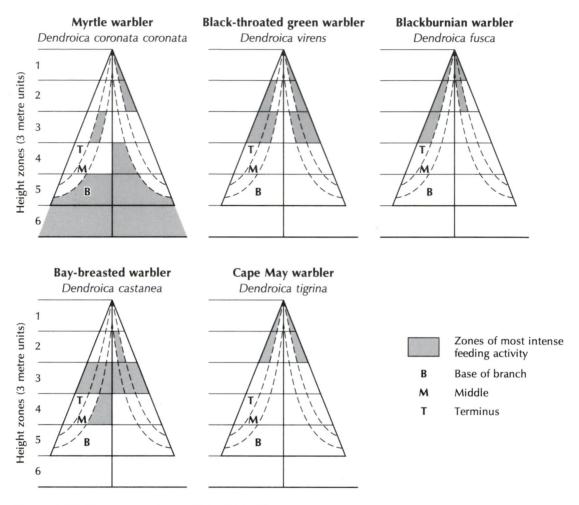

Figure 5.4 Warblers in spruce forests, Maine, United States.
Source: After MacArthur (1958)

Character displacement

Competition between two species may be avoided by evolutionary changes in both. Two predators may eat the same size prey and therefore, all other factors being constant, they are competitors. Under these circumstances, it would not be uncommon for one species gradually to tackle slightly large prey and the other species to tackle slightly smaller prey. The two species would thus diverge by evolutionary changes.

Evolutionary divergence can be inferred only from the phenomenon of **character displacement**. This phenomenon occurs when a species' appearance or behaviour differs when a competitor is present from when it is absent. An example is afforded by the beak size of ground finches (Geospizinae) living on the Galápagos Islands (Figure 5.5) (Lack 1947). Abingdon and Bindloe Islands have three species of *Geospiza* that partition seed resources according to size – the large ground finch (*Geospiza magnirostris*), the medium ground finch (*Geospiza fortis*), and the

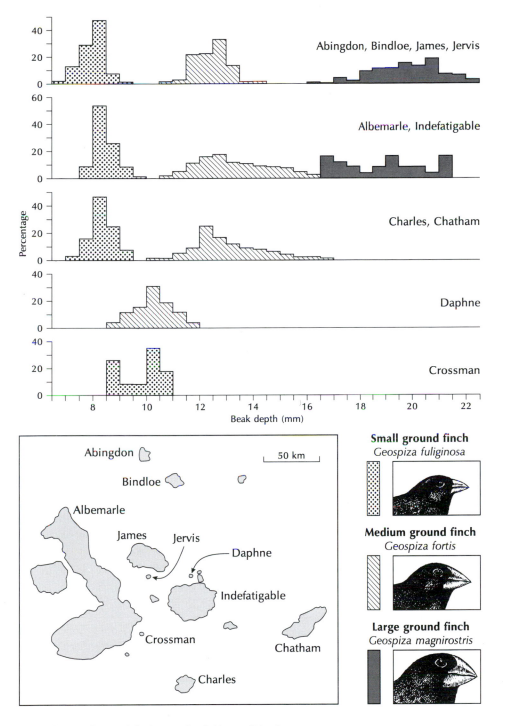

Figure 5.5 Beak size of ground finches on the Galápagos Islands.
Source: After Lack (1947)

small ground finch (*Geospiza fuliginosa*). *Geospiza magnirostris* has a large beak adapted to husk large seeds that smaller finches would fail to break open. *Geospiza fuliginosa* has a small beak that can husk smaller seeds more efficiently than the larger species, *Geospiza fortis*. *Geospiza fortis* feeds on intermediate-sized seeds. Now, *Geospiza magnirostris* does not reside on Charles or Chatham Islands. On these islands, *Geospiza fortis* tends to have a heavier beak, on average, than on Abingdon or Bindloe Islands. On Daphne Island, where *Geospiza fortis* lives without *Geospiza fuliginosa*, its beak is intermediate in size between the two species on Charles and Chatham Islands. On Crossman Islands, *Geospiza fuliginosa* lives without the other two finches, and its beak is intermediate in size there. The habitats on all the islands are very similar, and the explanation for this divergence in beak size is that competition has caused character displacement.

Spatial competition

All organisms, and particularly terrestrial plants and other sessile species, interact mainly with their neighbours. Neighbourhoods may differ considerably in composition, largely because of colonization limitation (propagules have not yet arrived at some sites). In areas of apparent uniformity, there is a great number of 'botanical ghettos'. This spatial variety enables more species to coexist than in a uniform environment. Traditional competition theory holds that there can be no more consumer species than there are limiting resources. The **spatial competition hypothesis** suggests that neighbourhood competition, coupled with random dispersal among sites, allows an unlimited number of species to survive on a single resource (Tilman 1994). Such rich coexistence occurs because species with sufficiently high dispersal rates persist in sites that are unoccupied by superior competitors.

The spatial competition hypothesis appears to explain the coexistence of numerous grassland plant species in the Cedar Creek Natural History Area, Minnesota, United States. The grassland plants compete strongly below ground for nitrogen, which is the only limiting resource. The best competitor is the little bluestem (*Schizachyrium scoparium*), a grass. According to classical competition theory, as there is just one limiting resource (soil nitrogen), this species should outcompete all others and take over. In garden plots it does just that. But in the natural grassland it shares the environment with over a hundred other species. The answer to this enigma may lie in the relative energy allocation for root growing and for reproduction. Species investing in root growth are good competitors where nitrogen levels are limiting, but they are also poor dispersers. Species investing in reproduction, such as the tickle-grass (*Agrostis scabra*) and the quackgrass (*Agropyron repens*), though not the best nitrogen competitors, are good dispersers. So, the 'good-dispersers-but-poor-competitors' invade abandoned fields immediately, and are not dominated by the 'slow-dispersers-but-good-competitors' until thirty or forty years later. It is possible, therefore, that superior competitors are prevented, by their poor colonizing abilities, from occupying the entire landscape, and that this provides sites in which numerous species of inferior competitors can persist.

VEGETARIAN WILDLIFE: HERBIVORY

Three forms of population interaction benefit one species but are detrimental to the other. **Predation** occurs when a predatory population (the winner) exploits a prey population (the loser). Normally, a predator has a seriously harmful effect on its prey – it kills it and eats it, but not necessarily in that order! This is classic carnivory. Plants lose leaves to herbivores. Herbivory is thus a form of predation. Parasitism is a relatively mild form of predation, though it may still lead to death. **Parasitism** is like predation but the host (a member of the population adversely affected), rather than being consumed immediately, is exploited over a period. **Amensalism** occurs when one population is adversely affected by an interaction but the other species is unaffected.

Plant eaters

Types of herbivore

Herbivores are animals that eat live plants. **Herbivory** is a special kind of predation where plants are the prey. All parts of a plant – flowers, fruits, seeds, sap, leaves, buds, galls, herbaceous stems, bark, wood, and roots – are eaten by something, but no one herbivore eats them all. There are several specialized herbivore niches – grazers, browsers, grass, grain, or seed feeders (graminivores), grain and seed eaters (granivores), leaf eaters (folivores), fruit and berry eaters (frugivores), nut eaters (nucivores), nectar eaters (nectarivores), pollen eaters (pollenivores), and root eaters. Detritivores eat dead plant material and will be discussed in Chapter 6.

A vast army of invertebrate herbivores munches, rasps, sucks, filters, chews, and mines its way through the phytosphere. A smaller but potent force of vertebrate herbivores grazes and browses its way through the phytosphere (Table 5.2). Some mammalian herbivore niches are unusual. The giant panda (*Ailuropoda melanoleuca*) is a purely herbivorous member of the Carnivora. The palm-nut vulture (*Gypohierax angolensis*) is the only herbivorous member of the Falconiformes. It eats nuts of the oil palm (*Elaeis guinensis*) and raphia palm (*Raphia ruffia*), as well as fish, molluscs, and crabs. Even invertebrate niches can be surprising. An adult tiger beetle (*Cicindela repanda*), a common and widespread water-edge species, was recently seen feeding on fallen sassafras fruits lying on a Maryland beach, United States (Hill and Knisley 1992). This was the first report of frugivory by a tiger beetle, which may be an opportunistic frugivore, especially in the autumn just after emergence, when fruits would provide a valuable energy resource before overwintering.

Browsers and **grazers** are chiefly mammals, but a few large reptiles and several bird species occupy this niche, too. Some of the geese and goose-like waterfowl crop grass. The South American hoatzin (*Opisthocomus hoazin*), a highly aberrant cuckoo, mainly eats the tough rubbery leaves, flowers, and fruits of the tall, cane-like water-arum (*Montri-*

chardia) and the white mangrove (*Avicenna*). The curious New Zealand kakapo (*Strigops habroptilus*) is a ground-dwelling parrot. It extracts juices from leaves, twigs, and young shoots by chewing on them and detaching them from the plant, or it fills its large crop with browse and retires to a roost where it chews up the plant material, swallows the juices, and 'spits out' the fibre as dry balls. Members of the grouse family (Tetraonidae) browse on buds, leaves, and twigs of willows and other plants of low nutritive value. Their digestive systems house symbiotic bacteria that digest the vegetable materials and are comparable to the digestive system of ruminants. A few birds eat roots and tubers. Pheasants dig into the soil with their beaks, while others scratch with their feet. Some cranes, geese, and ducks are adapted for obtaining the roots, rhizomes, and bulbous parts of aquatic plants.

Grain, seed, nectar, fruit, and nut eating are specialities of many birds. Sometimes, the relationship between a fruit-producing plant and its 'predator' is close (see Box 5.1). Many mammals eat fruit, but very few are specialized fruit eaters. An exception is the fruit-eating bats, which include nearly all members of the Pteropodidae. These animals have reduced cheek teeth as an adaptation for their fluid diet.

Defence mechanisms in plants

As prey species, plants have a distinct disadvantage over the animal herbivore counterparts – they cannot move and escape from predators; they have to stand firm and cope with herbivore attack as best they may. The world is green, so obviously plants can survive the ravages of plant-eating animals. They do so in two main ways. First, some herbivore populations evolve self-regulating mechanisms that prevent their destroying their food supply; or other mechanisms, particularly predation, may hold herbivore numbers in check. Second, all that is green may not be edible – plants have evolved a battery of defences against herbivores.

Plant defences are formidable. Some discourage herbivores by structural adaptations such as thorns and spikes. Some engage in chemical warfare,

Table 5.2 Herbivorous tetrapods

Order	Name	Herbivores within Order	Example	Example of plant tissue eaten
Amphibians Anura	Frogs, toads	Few	Common frog (*Rana temporaria*) tadpoles	Water weed
Reptiles Chelonia	Tortoises and turtles	Some	Giant tortoise (*Testudo gigantea*)	Grasses, sedges
Squamata	Snakes and lizards	Few	Marine iguana (*Amblyrhynchus cristatus*)	Seaweeds
Birds Anseriformes	Ducks and geese	Many	Southern screamer (*Chauna torquata*)	Aquatic plants, grasses, and seeds
Falconiformes	Eagles and hawks	Few	Palm-nut vulture (*Gypohierax angolensis*)	Palm nuts, fish, molluscs, crabs
Galliformes	Game birds	Most	Red grouse (*Lagopus lagopus*)	Heather shoots, insects
Columbiformes	Pigeons	All	Wood pigeon (*Columba palumbus*)	Legume leaves, seeds
Psittaciformes	Parrots	Most	Blue-and-yellow macaw (*Ara ararauna*)	Fruits, kernels
Apodiformes	Hummingbirds and swifts	Some	Sword-billed hummingbird (*Ensifera ensifera*)	Nectar, insects
Passeriformes	Perching birds	Many	Plantcutter (*Phytotoma rutila*)	Fruit, leaves, shoots, buds, seeds
Mammals Marsupialia	Kangaroos, etc.	Many	Honey possum or noolbender (*Tarsipes spencerae*)	Nectar, pollen, some insects
Chiroptera	Bats	Few	Long-tongued bat (*Glossophaga soricina*)	Fruit, nectar, some insects
Dermoptera	Flying lemurs	All	Flying lemurs (*Cynocephalus* spp.)	Leaves, buds, flowers, fruit
Edentata (Xenarthra)	Sloths and armadillos	Some	Three-toed sloths (*Bradypus* spp.)	Leaves, fruit
Primates	Apes, monkeys, and lemurs	Most	Human (*Homo sapiens*)	Everything except wood
Rodentia	Voles, mice, squirrels, etc.	Most	Alpine marmot (*Marmota marmota*)	Grass, plants, roots

Table 5.2 (continued)

Order	Name	Herbivores within Order	Example	Example of plant tissue eaten
Lagomorpha	Rabbits, hares, and pikas	All	Mountain hare (*Lepus timidus*)	Fine twigs, sprigs of bilberry and heather
Carnivora	Dogs, cats, bears, etc.	Few	Giant panda (*Ailuropoda melanoleuca*)	Bamboo
Hyracoidea	Hyraxes (conies)	All	Rock hyraxes (*Heterohyrax* spp.)	Grass
Proboscidea	Elephants	All	African elephant (*Loxodonta africana*)	Grass, leaves, twigs
Sirenia	Sea cows	All	Manatee (*Trichechus manatus*)	Aquatic plants
Perissodactyla	Odd-toed ungulates[a]	All	Brazilian tapir (*Tapirus terrestris*)	Succulent vegetation, fruit
Artiodactyla	Even-toed ungulates[b]	All	Goats (*Capra* spp.)	Everything

Source: Partly after Crawley (1983)
Notes: [a] Horses and zebras, tapirs, rhinoceroses
   [b] Pigs, peccaries, hippopotamuses, camels, deer, cows, sheep, goats, antelope

employing by-products of primary metabolic pathways. Many of these chemical by-products, including terpenoids, steroids, acetogenins (juglone in walnut trees), phenylpropanes (in cinnamon and cloves), and alkaloids (nicotine, morphine, caffeine) are distasteful or poisonous and are very effective deterrents (see Larcher 1995: 21).

Herbivores have developed ways of side-stepping plant defence systems. Some have evolved enzymes to detoxify plant chemicals. Others time their life cycles to avoid the noxious chemicals in the plants. Two examples will illustrate these points – cardiac glycosides in milkweed and tannins in oaks.

Poisonous milkweed

The milkweed *Asclepias curassavica* is abundant in Costa Rica and other parts of Central America. It contains secondary plant substances called cardiac glycosides (or cardenolides) that affect the vertebrate heartbeat and are poisonous to mammals and birds (Brower 1969). Cattle will not eat the milkweed, even though it grows abundantly in grass. They are wise not to do so, for it causes sickness and occasionally death. However, certain insects, including the larvae of danaid butterflies (Danainae), eat it without any deleterious effects. The danaid butterflies are distasteful to insect-eating birds and serve as models in several mimicry complexes (p. 111). The danaids have evolved biochemical mechanisms for eating the milkweed and storing the poison in their tissues. Thus, they acquire protection from predators from the plants that they eat.

Unpalatable oak-leaves

The common oak (*Quercus robur*) in western Europe is attacked by the larvae of over 200 species of butterflies and moths (Feeny 1970). It protects itself against this massive assault by using tannins for chemical and structural defences. The tannins lock proteins in complexes that insects cannot digest and utilize, and they also make leaves tough and unpalatable. The herbivorous attackers partly get round these defences by concentrating feeding in early spring, when the leaves are young and, because they contain less tannin, less tough. They also alter their life cycles in summer and autumn – many late-feeding insects overwinter as larvae and complete their development on the soft and tasty spring leaves. Some insect species do feed on oak leaves in summer and autumn. These tend to grow very slowly, which may be an adaptation to a low-nitrogen diet when proteins are locked away in older leaves.

Herbivore and plant interactions

Herbivores interact with plants in two chief ways. First, in **non-interactive herbivore–plant systems**, herbivores do not affect vegetation growth, even when there are very large numbers of them. Second, in **interactive herbivore–plant systems**, herbivores affect vegetation growth. Two examples will illustrate these different systems – seed predation and grazing.

Seed predation

Herbivores that feed upon plant seeds (seed predators) do not normally influence plant growth. Birds eating plant seeds have no effect on plant growth, though they could influence the long-term survival of a plant population by eating a large portion of the annual seed production. On the other hand, plants do influence herbivore populations – seed predator populations will be low in times of seed shortages. British finches feed either on herb seeds or on tree seeds. The herb-seed feeders have stable populations, whereas the tree-seed feeders have fluctuating populations (Newton 1972). Herbs are usually consistent in the annual number of seeds produced, but tree-seed production is variable and, in some cases, irregular. Variable tree-seed production is illustrated by the North American piñon pine (*Pinus edulis*). The piñon pine yields a heavy cone crop at irregular intervals. The crop satiates its invertebrate seed and cone predators. It also allows efficient foraging by piñon jays (*Gymnorhinus cyanocephalus*), which disperse the seeds and store them in sites suitable for germination and seedling growth (Ligon 1978). Many other trees are mast fruiters. In mast fruiting, all trees of the same species within a large area produce a big seed crop in one year. They then wait a long time before producing anything other than a few seeds. The advantage of such synchronous seeding is that seed predators are satiated and are unlikely to eat the entire crop before germination has begun.

Grazing

In interactive herbivore–plant systems, the herbivores affect the plants, as well as the plants affecting the herbivores. As a rule, plants have more impact on herbivore populations than herbivores have on plant populations – plant abundance, quality, and distribution greatly affect herbivore numbers. Nonetheless, herbivory can affect plant populations. This is the case where vegetation is grazed. Sheep, for example, are 'woolly lawnmowers' that maintain pasture. The effects of grazing on vegetation are plainest to see where a fence separates grazed and ungrazed areas. The grazed area consists mainly of grasses, while the ungrazed area is dominated by shrubs and tall herbs. Given enough time, the ungrazed area would revert to woodland. Domestic livestock sometimes has such a large influence on plant communities that the herbivore may be identified from the vegetation it produces (Crawley 1983: 295). In the United Kingdom, dock (*Rumex obtusifolius*) is abundant in horse pastures; silverweed (*Potentilla anserina*) is common where geese graze. Spiny shrubs and aromatic, sticky herbs flourish where goats graze in Mediterranean maquis.

Patterns of herbivore–plant interaction are complex and generalization is difficult. Population models, similar to those used to study predator–prey relationships, suggest some basic possibilities (Box 5.2).

HUNTER AND HUNTED: CARNIVORY

Flesh eaters

Carnivores are animals that eat animals. **Carnivory** is predation where animals are the prey. Carnivory includes predators eating herbivores as well as top predators eating predators. A predator benefits from interaction with a prey – it gets a meal; the prey does not benefit from interaction with a predator – it normally loses its life. Rare exceptions are lizards that lose their tails to predators but otherwise survive attacks. Predation is plain to see and easy to study. It is evident when one animal eats another, often in a gory spectacle.

Predators come in a variety of shapes and sizes. Highlights from a predator catalogue might be the 'small but deadly' – the least weasel (*Mustela nivalis*)

is the smallest carnivore ever – and the 'large and equally deadly' – the African lion (*Panthera leo*). Some fossil carnivores were awesome in size and possibly in speed. The most famous examples, even before their appearance in Jurassic Park, are the dinosaurs *Tyrannosaurus rex* and *Velociraptor*. But more recent avian predators were just as terrifying. The phorusrhacoids were a group of large, flightless, flesh-eating birds (L. G. Marshall 1994). They lived from 62 million to about 2.5 million years ago in South America and became the dominant carnivores on that continent. They ranged in height from 1 to 3 m, and were able to kill and eat animals the size of small horses.

Carnivorous specialization

All animals, except those at the very top of food chains, are eaten by a predator. The zoological menu is vast, but predators tend to specialize. Carnivore niches include general flesh-eaters (faunavores), fish eaters (piscivores), blood feeders (haematophages or sanguivores), egg eaters (ovavores), insect eaters (insectivores), ant eaters (myremovores), and coral eaters (corallivores). Omnivores are unspecialized carnivores-cum-herbivores. Scavengers feed on dead animal remains. Coprophages feed on dung or faeces.

A few carnivorous niches are highly specialized and rather unusual. There are few sanguivores, the most notorious (apart from Count Dracula) are the New World vampire bats. The sharp-beaked ground finch (*Geospiza difficilis*), one of the Galápagos finches, obtains some its nourishment by biting the bases of growing feathers on boobies (*Sula* sp.) and eating the blood that oozes from the wound. Coprophagy is more widespread than might be supposed. Several birds eat animal dung. The ivory gull (*Pagophila eburnea*) feeds on the dung dumped on ice by polar bears, walruses, and seals. Puffins and petrels eat whale dung floating on the water. Certain vultures and kites feed solely on human and canine faeces around native villages. Myrmecophagy (ant eating) is found in several species in Australia, South America, Africa, and Eurasia that are adapted to feed upon ants and termites (Figure 5.7). Extreme adaptations for

Box 5.2

POPULATION MODELS OF HERBIVORE–PLANT SYSTEMS

Plant–herbivore systems are special cases of predator–prey systems. The plant is the prey and the herbivore is the predator. The equations devised to study predators and their prey (p. 130) may be adapted to study herbivores preying on plants. The equations, one for the plant population and one for the herbivore population, are (Crawley 1983: 249):

$$dP/dt = [aP(K - P)]/K - bHP$$

$$dH/dt = cHP - dH.$$

In these equations, P is the plant population, H is the herbivore population, K is the carrying capacity of plants, and a, b, c, and d are parameters: a is the intrinsic growth rate of plants in the absence of herbivores; b is the depression of the plant population per encounter with a herbivore, a measure of herbivore searching efficiency; c is the increase in the herbivore population per encounter with plants, a measure of herbivore growth efficiency; and d is the decline in herbivores (death rate) in the absence of plants.

The parameters and the carrying capacity affect system stability and steady-state population numbers. In brief, increasing the intrinsic growth rate of plants, a, increases system stability and

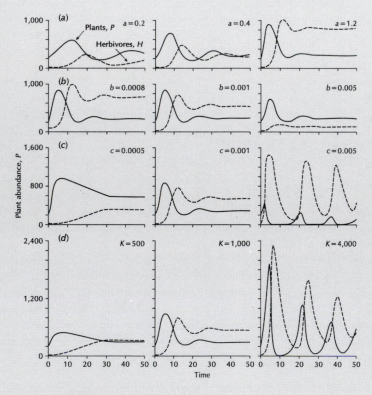

Figure 5.6 Herbivore–plant interactions.
Source: After Crawley (1983)

herbivore steady-state density, but has no effect on plant abundance (Figure 5.6a). Increasing herbivore searching efficiency, *b*, reduces, somewhat surprisingly, herbivore steady-state density, but has no effect on system stability or on steady-state plant abundance (Figure 5.6b). Increasing herbivore growth efficiency, *c*, reduces steady-state plant abundance and so reduces system stability, but increases steady-state herbivore numbers (Figure 5.6c). Increasing the carrying capacity of the environment for plants, *K*, increases herbivore steady-state density and reduces system stability, but, paradoxically, has no effect on steady-state plant abundance (Figure 5.6d).

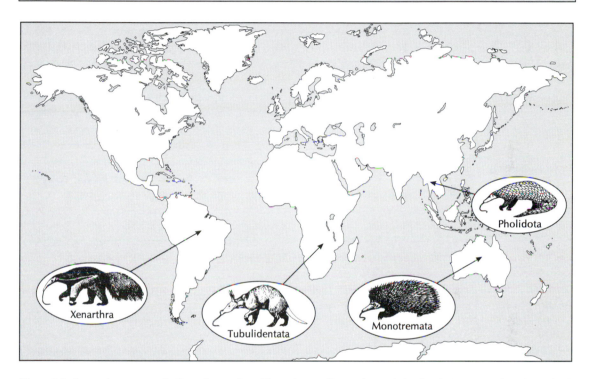

Figure 5.7 Ant-eating mammals from four orders: Xenarthra – all members of the family Myrmecophagidae, which includes the giant or great anteater (*Myrmecophaga tridactyla*); Tubulidentata – the aardvark (*Orycteropus afer*); Pholidota – eight species of pangolin (*Manis*); and Monotremata – the Australian and New Guinean spiny anteaters or echidnas, *Tachyglossus aculeatus* and *Zaglossus bruijni*. The numbat (*Myrmecobius fasciatus*), which is not illustrated, is an Australian marsupial anteater.
Source: Animal drawings from Rodríguez de la Fuente (1975)

this diet are seen in teeth, jaws, and skulls. The teeth are reduced to flat, crushing surfaces, as in the aardvark (*Orycteropus afer*), or they are missing, as in the giant or great anteater (*Myrmecophaga tridactyla*). Tongues are long, mobile, and worm-like, with uncommon protrusibility ('stick-out-ability'). Enlarged salivary glands produce a thick, sticky secretion that coats the tongue. Ant-eating mammals also tend to have elongated snouts and powerful forelimbs for digging and ripping open nests.

Some plants are carnivorous, including the great sundew (*Drosera anglica*) and the greater bladderwort (*Utricularia vulgaris*). Pitcher plants of the genus *Heliamphora* (Sarraceniaceae) in the Guayana Highlands of Venezuela have carnivorous traits (Jaffe *et al.* 1992). *Heliamphora tatei* is a true carnivore. It attracts prey through special visual and chemical signals, traps and kills prey, digests prey through secreted enzymes, contains commensal organisms, and has wax scales that aid prey capture and nutrient absorption. Other species of *Hemiamphora* possess all these traits except they seem neither to secrete proteolytic enzymes nor to produce wax scales. The pattern of the carnivorous syndrome among *Heliamphora* species suggests that the carnivorous habit has evolved in nutrient-poor habitats to improve the absorption of nutrients by capturing mainly ants. Only *Heliamphora tatei* further evolved mechanisms for rapid assimilation of organic nutrients and for capturing a variety of flying insects. The carnivorous traits are lost in low light conditions, indicating that nutrient supply is limiting only under fast growth.

Prey switching

The switching of prey follows the principle of optimal foraging. Predators select the prey that provides the largest net energy gain, considering the energy expended in capturing and consuming prey. This usually means the prey that happened to be most abundant at the time. It also means that predators normally take very young, very old, or physically weakened prey. Some carnivores show local feeding specializations within their geographical range. In Mediterranean environments, the European badger (*Meles meles*), a carnivore species with morphological, physiological, and behavioural traits proper to a feeding generalist, prefers to eat the European rabbit (*Oryctolagus cuniculus*) (Martin *et al.* 1995). It will eat other prey according to their availability when the rabbit kittens are not abundant. The badger is a poor hunter, and its ability to catch relatively immobile baby rabbits may explain the curious specialization.

Prey selection

Where several carnivores compete for the same range of prey species, coexistence is possible if each carnivore selects a different prey. In the tropical forests of Nagarahole, southern India, there is a wide range of large mammalian prey species, mainly ungulates and primates, that are eaten by the tiger (*Panthera tigris*), the leopard (*Panthera pardus*), and the dhole (*Cuon alpinus*) (Karanth and Sunquist 1995). The three predators show significant selectivity among prey species. Gaur (*Bos gaurus*) is preferred by tigers, whereas wild pig (*Sus scrofa*) is underrepresented in leopard diet, and the langur or leaf monkey (*Presbytis entellus*) is underrepresented in dhole diet. Tigers selected prey weighing more than 176 kg, whereas leopard and dhole focused on prey in the 30–175 kg size class. Average weights of principal prey killed by tiger, leopard, and dhole were, respectively, 91.5 kg, 37.6 kg and 43.4 kg. Tiger predation was biased towards adult males in chital or axis deer (*Axis axis*), sambar (*Cervus unicolor*), and wild pig, and towards young gaur. Dholes selectively preyed on adult male chital, whereas leopards did not. If there is choice, large carnivores selectively kill large prey, and non-selective predation patterns reported from other tropical forests may be the result of scarcity of large prey. Because availability of prey in the appropriate size classes is not a limiting resource, selective predation may facilitate large carnivore coexistence in Nagarahole.

Carnivore communities

Carnivore communities normally consist of predators with overlapping tastes. For this reason, they posses complex feeding relationships. A dry tropical forest in the 2,575-km² Huai Kha Khaeng Wildlife Sanctuary, Thailand, contained twenty-one carnivore species from five families (Rabinowitz and Walker 1991). The carnivores fed on at least thirty-four mammal species, as well as birds, lizards, snakes, crabs, fish, insects, and fruits. Nearly half the prey identified in large carnivore faeces, which was mainly deposited by Asiatic leopard (*Panthera pardus*) and clouded leopard (*Neofelis nebulosa*), was barking deer (*Muntiacus muntjak*), with sambar deer (*Cervus unicolor*), macaques (*Macaca* spp.), wild boar (*Sus scrofa*), crestless Himalayan porcupine (*Hystrix hodgsoni*), and hog badger (*Arctonyx collaris*) being important secondary prey items. Murid rodents, especially the yellow rajah-rat (*Maxomys surifer*), and the bay bamboo-rat (*Cannomys badius*), accounted for one-third of identified food items in small carnivore faeces. Non-mammal prey accounted for just over one-fifth, and fruit seeds for one-eighth, of all food items found in small carnivore faeces.

Predators and prey

Commonly, predator and prey populations evolve together. Predators which are better able to find, capture, and eat prey are more likely to survive. Natural selection tends to favour those hunters' traits within prey populations. In turn, predation pushes the prey population to evolve in a direction that favours individuals good at hiding and escaping. Over thousands of generations, evolutionary forces acting upon predator and prey populations lead to elaborate and sophisticated adaptations. These features are seen in the social hunting behaviour of lions and wolves, the folding fangs and venom-injecting apparatus of viperine snakes, and spiders and their webs.

Predator–prey cycles

There are several pairs of predators and prey that appear to display cyclical variations in population density. The pairs include: sparrow–hawk (Europe), muskrat–mink (central North America), hare–lynx (boreal North America), mule deer–mountain lion (Rocky Mountains), white-tailed deer–wolf (Ontario, Canada), moose–wolf (Isle Royale), caribou–wolf (Alaska), and white sheep–wolf (Alaska).

The rationale behind **predator–prey cycles** is deceptively straightforward. Consider a population of weasels, the predator, and a population of voles, the prey. If the vole population should be large, then, with a glut of food scampering around, the weasel population will flourish and increase. As the weasel population grows, so the vole population will shrink. Once vole numbers have fallen low enough, the prey shortage will cause a reduction in weasel numbers. The vole population, enjoying the dearth of predators, will then rise again. And so the cycle continues.

Although the rationale behind predator–prey cycles is plausible, sustained oscillations in predator and prey systems are not always the result of population interactions. Canadian lynx (*Lynx canadensis*) populations show a nine to ten-year peak in density (Figure 5.8). This cycle is revealed by lynx pelt records kept by Hudson's Bay Company in Canada. The Canadian lynx feeds mainly upon the snowshoe hare (*Lepus canadensis*), which also follows a ten-year

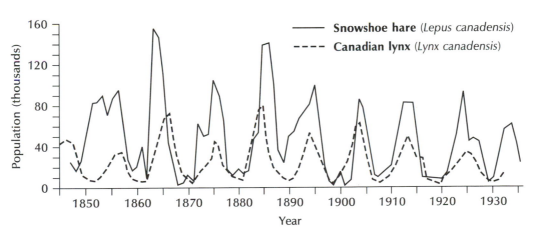

Figure 5.8 Cycles in the population density of the Canadian lynx (*Lynx canadensis*) and the snowshoe hare (*Lepus canadensis*), as seen in the number of pelts received by the Hudson's Bay Company.
Source: After MacLulich (1937)

cycle. Field investigations suggested that the cycle in snowshoe hare density is correlated with the availability of food (the terminal twigs of shrubs and trees). The lynx population, therefore, follows swings in snowshoe hare numbers, and does not drive them.

Oscillations in predator–prey systems were modelled in the 1920s (Lotka 1925; Volterra 1926, 1931). Refined versions of these models have produced three important findings. First, stable population cycles occur only when the prey's carrying capacity (in the presence of predators) is large. This finding goes against intuition. It is called the 'paradox of enrichment' and states that an increase in carrying capacity decreases stability. Second, the population cycles are stable only when the prey population grows faster than the predator population, or when the predators are relatively inefficient at catching their prey. Third, some predator–prey 'cycles' display the features of chaotic dynamics.

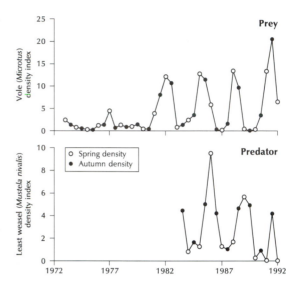

Figure 5.9 Observed cycles in predators and prey, Alajoki, western Finland. The prey is voles (*Microtus* spp.) and the predator is the least weasel (*Mustela nivalis*).
Source: After Hanski *et al*. (1993)

Chaos in Finnish weasel and vole populations

Long-term studies of predator (weasel) and prey (vole) dynamics in Finland have revealed population cycles of three to five years (Figure 5.9). A predator–prey model with seasonal effects included (by allowing weasels to breed only when the vole density exceeds a threshold) predicted population changes that closely resembled the observed changes in weasel and vole populations (Hanski *et al*. 1993). The study suggested that the cycles in vole populations are driven by weasel predation, and are chaotic. These findings are supported by further work on boreal voles and weasels that shows the importance of multispecies predator–prey assemblages with field-vole-type rodents as the keystone species (Hanski and Korpimäki 1995; Hanski and Henttonen 1996).

Geographical effects

Laboratory experiments

Gause (1934) studied a simple predator–prey system in the laboratory (Figure 5.10). He used two micro-

scopic protozoans, *Paramecium caudatum* and *Didinium nasutum*. *Didinium* prey voraciously on *Paramecium*. In one experiment, *Paramecium* and *Didinium* were placed in a test-tube containing a culture of bacteria supported in a clear **homogeneous** oat medium. The bacteria are food for the *Paramecium*. The outcome was that *Didinium* ate all the *Paramecium* and then died of starvation. This result occurred no matter how large the culture vessel was, and no matter how few *Didinium* were introduced. In a second experiment, a little sediment was added to the test-tube. This made the medium **heterogeneous** and afforded some refuges for the prey. The outcome of this experiment was the extinction of the *Didinium* and the growth of the prey population to its carrying capacity. In a third experiment, the prey and predator populations were 'topped up' every three days by adding one *Paramecium* and one *Didinium* to the test-tube. In other words, immigration was included in the experiment. The outcome was two complete cycles of predator and prey populations. Gause concluded that, without some form of external interference such as immigration, predator–prey systems are self-annihilating.

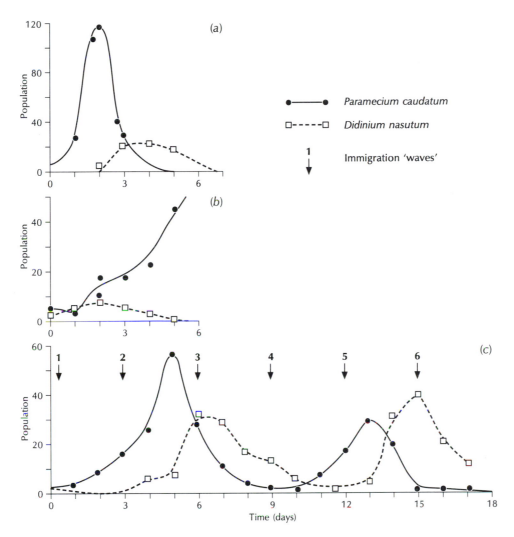

Figure 5.10 Outcomes of Gause's laboratory experiments with the protozoans *Paramecium caudatum* (prey) and *Didinium nasutum* (predator). (a) Oat medium without sediment. (b) Oat medium with sediment. (c) Oat medium without sediment but with immigrations.
Source: After Gause (1934)

Later, laboratory experiments using an orange → herbivorous mite → carnivorous mite food chain showed that coexistence was possible (Huffaker 1958). When the environment was homogeneous (oranges placed close together and spaced evenly), the predatory mite (*Typhlodromus occidentalis*) overate its phytophagous prey (*Eotetranychus sexmaculatus*) and died out, as Gause's *Didinium* had done. Increasing the distance between oranges simply lengthened the time to extinction. But when the environment was made heterogeneous, the system stabilized. Several different 'environments' were constructed. The basic 'landscape' was forty oranges placed on rectangular trays like egg cartons, with some of the oranges partly covered with paraffin or paper to limit the available feeding area. Complications were

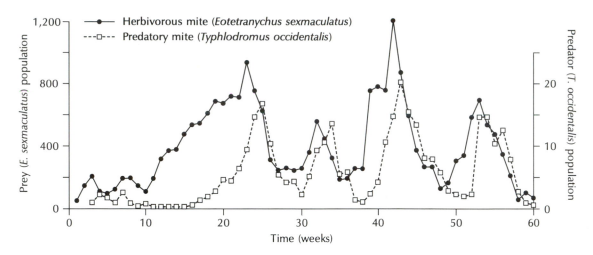

Figure 5.11 Population cycles in a laboratory experiment where a predatory mite, *Typhlodromus occidentalis*, feeds upon another mite, *Eotetranychus sexmaculatus*. The environment is heterogeneous, consisting of 252 oranges with one-twentieth of each orange available to the prey for feeding.
Source: After Huffaker *et al.* (1963)

introduced by using varying numbers of rubber balls as 'substitute' oranges. In some cases, entire new trays of oranges were added and artificial barriers of Vaseline constructed that the mites could not cross. Stability occurred in a 252-orange landscape with complex Vaseline barriers. The prey were able to colonize oranges in a hop, skip, and jump fashion, and to keep one step ahead of the predator, which exterminated each little colony of prey it found. During the seventy weeks before the predators died out and the experiment stopped, three complete predator–prey population cycles occurred (Figure 5.11).

Nearly forty years after Gause's original work, new experiments using *Paramecium aurelia* as prey and *Didinium* as predator managed to stabilize the system (Luckinbill 1973, 1974). When *Paramecium* was grown with *Didinium* in a 6 ml of standard cerophyl medium, *Didinium* ate all the prey in a few hours. When the medium was thickened with methyl cellulose to slow down the movements of predator and prey alike, the populations went through two or three diverging oscillations lasting several days before becoming extinct. When a half-strength cerophyl medium was thickened with methyl cellulose,

the populations maintained sustained oscillations for thirty-three days before the experiment was concluded. A mathematical model of the system suggests that the stability arises because the increase in the cost–benefit ratio of energy spent searching to energy gained capturing prey apparently inhibits the predator searching at low prey densities (G. W. Harrison 1995).

An ambitious set of laboratory experiments studied competition and predation at the same time (Utida 1957). The system contained a beetle, the azuki bean weevil (*Callosobruchus chinensis*), as prey that was provided with an unlimited food supply. But this coleopteran paradise had a 'sting in the tail' in the form of two competing species of predatory wasp – *Neocatolaccus mamezophagus* and *Heterospilus prosopidis*. The wasp species had similar life histories and depended upon the beetle as a food source. During the four years of the experiment, which represented seventy generations, all three populations fluctuated wildly but managed to coexist (Figure 5.12). The wasp population fluctuations were out of phase with one another. This was because *Heterospilus* was more efficient at finding and exploiting the beetle when it was at low densities, while *Neocatolaccus* was more

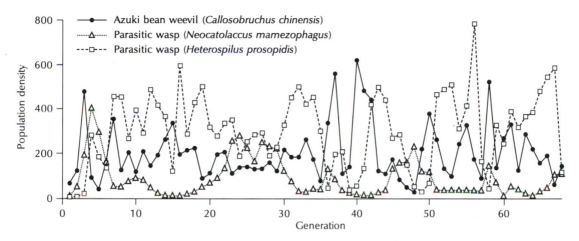

Figure 5.12 Population changes in a laboratory experiment with a beetle host, the azuki bean weevil (*Callosobruchus chinensis*), and two parasitic wasps, *Neocatolaccus mamezophagus* and *Heterospilus prosopidis*. The experiment ran for four years. All three populations fluctuated considerably, but they did survive.
Source: After Utida (1957)

efficient at higher prey densities. The competitive edge thus shifted between the two wasp species as the beetle density changed with time (owing to density-dependent changes in reproduction rate and the effects of the two wasp populations). The stability of the system was thus the purely result of predatory and competitive biotic interactions.

Mathematical 'experiments'

In theory, geography is a crucial factor in the co-existence of predator and prey populations. A simple mathematical model showed that when predators and prey interact within a landscape, coexistence is rather easily attained (J. M. Smith 1974: 72–83). It is favoured by prey with a high migration capacity, by cover or refuge for the prey, by predator migration during a restricted period, and by a large number of landscape 'cells'. Furthermore, if the predator's migration ability is too low, it will become extinct. If the predator's migration ability is high, co-existence is possible if the prey is equally mobile.

Simple mathematical models of predator–prey interactions in a landscape have evolved into sophisticated metapopulation models. These, too, stress the vital role of refuges for prey species and migration

rates between landscape patches. They also reveal the curious fact that some metapopulations may persist with only 'sink' populations, in which population growth is negative in the absence of migration. However, long-term persistence requires some local populations becoming large occasionally (Hanski *et al.* 1996a).

LIFE AGAINST LIFE: ALTERNATIVES TO CHEMICAL CONTROL

Pests are organisms that interfere with human activities, and especially with agriculture. They are unwelcome competitors, parasites, or predators. The chief agricultural pests are insects (that feed mainly on the leaves and stems of plants), nematodes (small worms that live mainly in the soil, feeding on roots and other plant tissues), bacterial and viral diseases, and vertebrates (mainly rodents and birds feeding on grain and fruit). Weeds are a major problem for potential crop loss. A typical field is infested by ten to fifty weed species that complete with the crop for light, water, and nutrients.

Pest control is used to reduce pest damage, but even with the weight of modern technology behind it, pest control is not enormously successful. In the United States, one-third of the potential harvest and one-tenth of the harvested crop is lost to pests. A control operation is successful if the pest does not cause excessive damage. It is a failure if excessive damage is caused. Just how much damage is tolerable depends on the enterprise and the pest. An insect that destroys 5 per cent of a pear crop may be insignificant in ecological terms, but it may be disastrous for a farmer's margin of profit. On the other hand, a forest insect may strip vast areas of trees of their leaves, but the lumber industry will not go bankrupt.

Pests are controlled in several different ways. The blanket application of toxic chemicals called pesticides has inimical side-effects on the environment. Several other options for pest control are available (Table 5.3).

Biological control

Biological pest control pits predator species against prey species – parasites, predators, and pathogens are used to regulate pest populations. Two approaches exist – inundative biocontrol and classical biocontrol (Harris 1993). In **inundative control**, an organism is applied in the manner of a herbicide. Like a herbicide, the control agent is usually marketed by industry. In **classical biocontrol**, an organism (or possibly a virus) is established from another region. The pest is kept in check indefinitely, usually by government agencies acting in the public interest. Both biocontrol approaches, in the right circumstances, are effective means of pest control and bring few harmful environmental impacts, as the following examples will show.

Prickly-pear cactus in Australia

The prickly pear (*Opuntia stricta*), a cactus, is native to North and South America. It was brought to eastern Australia in 1839 as a hedge plant. It spread fast, forming dense stands, some 1–2 m high and too thick for anybody to walk through. By 1900, it infested 4 million ha. Control by poisoning the cactus was not

Table 5.3 Main pest control techniques

Control method	r-strategist pests	Intermediate pests	K-strategist pests
Pesticides (use chemical compounds to kill pests directly)	Early widespread application based on forecasting	Selective pesticides	Precisely targeted applications based on monitoring
Biological (use natural enemies, viruses, bacteria, or fungi)		Introduction or enhancement of natural enemies	
Cultural (change agricultural or other practices to alter the pest's habitat)	Timing, cultivation, and rotation	→	← Changes in agronomic practice, destruction of alternative hosts
Resistance (breed animal and crop-plant varieties resistant to pests)	General, polygenic resistance	→	← Specific monogenic resistance
Genetic (sterilize pest population to reduce its growth rate)			Sterile mating techniques

Source: After Conway (1981)

economically feasible – clearing infested land with poison was far more costly than the worth of the land. Searches in native habitats of the prickly pear were begun in 1912 to find a possible biological control agent. Eventually one was found – a moth, *Cactoblastis cactorum*, native to northern Argentina, the larvae of which burrow into the pads of the cactus, causing physical damage and introducing bacterial and fungal infections. Between 1930 and 1931, when the moth population had become enormous, the prickly pear stands were ravaged. By 1940, the prickly pear was still found here and there, but in very few places was it a pest.

A different control agent, *Dactylopius opuntiae*, was used in an area of New South Wales, Australia, where *Cactoblastis cactorum* had not been not a successful biological control agent (Hosking *et al.* 1994). *Dactylopius opuntiae* reduced a prickly pear (*Opuntia stricta* var. *stricta*) population in the central tablelands of New South Wales.

False ragweed in Australia and India

The false ragweed (*Parthenium hysterophorus*), a native of Central America, is a problem weed of Australian rangeland, particularly in Queensland. Following field surveys in Mexico, the rust fungus *Puccinia abrupta* var. *partheniicola* was selected as a potential biological control agent (Parker *et al.* 1994). One isolate was chosen for further investigation. Infection with the rust hastened leaf senescence, significantly decreased the life-span and dry weight of false ragweed plants, and led to a tenfold reduction in flower production. Subsequent studies showed that the rust was sufficiently host-specific to be considered for introduction. In Bangalore, India, *Cassia uniflora*, a leguminous undershrub of some economic value, has replaced, over a five-year period, more than 90 per cent of the false ragweed on a 4,800 m^2 site (Joshi 1991). Leachates from *Cassia uniflora* are allelopathic, inhibiting false-ragweed seed germination and hampering the establishment of a summer false-ragweed generation. The colonies of *Cassia uniflora* are robust enough to prevent a false-ragweed generation from forming below them in winter.

Pests in the Mediterranean

The whitefly, *Parabemisia myricae*, is one of the most serious citrus pests in the eastern Mediterranean region of Turkey. In 1986, a host-specific parasitoid of *Parabemisia myricae*, the aphelinid *Eretmocerus debachi*, was imported from California (Sengonca *et al.* 1993). In the following years *Parabemisia myricae* populations were rapidly reduced in all citrus orchards where the parasitoids were released. *Eretmocerus debachi* was a good disperser, well adapted to the climatic conditions. Since its successful colonization, the whitefly is no longer a serious pest.

In Cyprus, the black scale (*Saissetia oleae*) is a pest primarily of olive tree (*Olea europaea*). It attacks several other plants, including citrus trees and oleander (*Nerium oleander*). Two parasitoids (*Metaphycus bartletti* and *Metaphycus helvolus*) were imported, mass-reared, and permanently established in the island (Orphanides 1993). Following limited releases of these parasitoids, black scale populations fell from outbreak levels to almost non-existence. Black scale populations have stayed low since parasitoid releases were discontinued.

Genetic control

Another alternative to chemical control of pests is **genetic control**. One method of genetic control is to breed sterile organisms. For example, if gypsy moth (*Lymantria dispar*) pupae are irradiated with a sterilizing dose of gamma radiation and mated with normal females, the first generation male moths are sterile (Schwalbe *et al.* 1991). However, difficulties associated with large-scale deployment of partially sterile males renders genetic control impractical on a large scale.

Genetic control is also achieved by breeding crop plants that are more resistant to pests. It was this control technique that, in 1891, came to the rescue of the French wine industry. *Phylloxera* is an aphid native to America. It lives in galls on leaves and roots of vines where it cannot be reached by sprays. It multiplies prodigiously. Vines infected by *Phylloxera* become stunted and die. In 1861, it was accidentally introduced into Languedoc, France. Two decades later, four-fifths of the Languedoc vineyards had been devastated and every wine-growing area of France was infected; no remedy had been found; and the outlook was bleak for wine drinkers. In 1891, it was discovered that the American vine (*Vitis labrusca*) was almost immune to *Phylloxera*. Scions of the European vine (*Vitis vinifera*) were grafted onto American root stocks to produce a hybrid vine that, if not entirely immune, was affected far less seriously.

Integrated pest management

Modern pest control involves **integrated pest management**. This is an ecological approach to pest management that brings together at least four techniques. First, it uses natural pest enemies, including parasites, diseases, competitors, and predators (biological control). Second, it advocates the planting of a greater diversity of crops to lessen the possibility that a pest will find a host. Third, it advocates no or little ploughing so that natural enemies of some pests have a chance to build up in the soil. Fourth, it allows the application of a set of highly specific chemicals, used sparingly and judiciously (unlike the old application method that tended to be profligate). Integrated pest management involves the use of chemicals, the development of genetically resistant stock, biological control, and land culture. Land culture is the physical management of the land — whether and how it is ploughed, what kind of crop rotation used, the dates of planting, and basic means of handling crop harvests to reduce presence of pest in residues and products sold.

Integrated pest management has been used to tackle the oriental fruit moth (*Grapholitha molesta*), which attacks several fruit crops (Barfield and Stimac 1980). The moth is prey to a species of braconid wasp, *Macrocentrus ancylivrous*. The introduction of the wasp into fields and orchards helped to control the moth population. But an interesting discovery was made: the efficacy of the wasp in peach orchards was increased when strawberry fields lay nearby. The strawberry fields are an alternative habitat for the wasp and help it to overwinter.

SUMMARY

No population can exist isolation — it needs to interact with others. Population interactions take several forms but fall into two categories — co-operation and competition. Co-operation occurs to varying degrees. Weak co-operation (protocooperation) is beneficial to both populations, but it is not obligatory — the interacting species would survive without one another. It includes mimicry. Mutualism is an extreme form of protocooperation in which the interaction is obligatory. Commensalism occurs when one of the co-operating species benefits and the other is unharmed by the association. Competition occurs where species interactions are detrimental to at least one of the interactants. The competitive exclusion principle states that no two species may occupy exactly the same niche. If they should occupy similar niches, competition will occur. Competition takes the form of scramble competition (for resources) or contest competition (sometimes involving aggressive action between individuals). Coexistence of competing species is achieved by resource partitioning, character displacement, and spatial heterogeneity. Herbivory is the predation of plants by plant-eaters. The plant kingdom has evolved defences to foil their herbivorous adversaries. The defences include structural deterrents and chemical arsenals. Herbivores interact with plants in several characteristic ways. This is seen in patterns of seed predation and grazing. Carnivory is the predation of one animal by another. There are several carnivorous specializations, some rather bizarre. Evolution has finely tuned carnivore communities to promote coexistence through prey switching and prey selection. Predators and their prey sometimes display cycles in population numbers. Such cycles are sometimes the result of food availability, though chaotic behaviour in some predator–prey systems does appear to rest within the interacting populations themselves. Stability in predator–prey systems is more readily attained in heterogeneous environments. Life may be pitted against life in an attempt to control agricultural pests. Biological control is a useful alternative to pesticide application. Genetic control is another option. Biological control is usually applied within a broader system of integrated pest management.

ESSAY QUESTIONS

1 What are the results of interspecific competition?

2 What are the advantages and disadvantages of species co-operation?

3 What are the pros and cons of biological control?

FURTHER READING

Crawley, M. J. (1983) *Herbivory: The Dynamics of Animal–Plant Interactions* (Studies in Ecology, Vol. 10), Oxford: Blackwell Scientific Publications.

Grover, J. P. (1997) *Resource Competition*, New York: Chapman & Hall.

Jackson, A. R. W. and Jackson, J. M. (1996) *Environmental Science: The Natural Environment and Human Impact*, Harlow: Longman.

Kingsland, S. E. (1985) *Modeling Nature: Episodes in the History of Population Biology*, Chicago and London: University of Chicago Press.

Kormondy, E. J. (1996) *Concepts of Ecology*, 4th edn, Englewood Cliffs, NJ: Prentice Hall.

MacDonald, D. (1992) *The Velvet Claw: A Natural History of the Carnivores*, London: BBC Books.

6

COMMUNITIES

Communities consist of several interacting populations. Ecosystems are communities together with the physical environment that sustains them. This chapter covers:

- the nature of communities and ecosystems
- ecological roles in communities
- feeding relationships
- biological diversity

SOCIAL AND PHYSICAL CONNECTIONS: COMMUNITIES AND ECOSYSTEMS

An **ecosystem** is a space in which organisms interact with one another and with the physical environment. A **community** is the assemblage of interacting organisms within an ecosystem. Communities are sometimes called **biocoenoses**, with each part given a separate name – phytocoenose (plants), zoocoenose (animals), micro-biocoenose (micro-organisms) (Sukachev and Dylis 1964: 27). Where the physical environment supporting the community is included, the term biogeocoenose is used (and is a vowel-rich and somewhat awkward alternative to the term ecosystem). Ecosystems and communities range in size from a cubic centimetres to the entire world.

A local ecosystem

The Northaw Great Wood, Hertfordshire, England, is a deciduous wood, with a small plantation of conifers. It occupies 217 ha in the headwaters of the Cuffley Brook drainage basin. The eastwards-draining streams have cut into a Pebble Gravel plateau that sits at around 400 m. They have eroded valleys in London Clay, Reading Beds (sands), and, at the lowest points,

chalk (Figure 6.1). A variety of soils have formed upon these rocks. The soils support many trees (Figure 6.2). The main ones are oak (chiefly *Quercus robur*), hornbeam (*Carpinus betulus*), and silver birch (*Betula pendula*). The herb layer is varied, the main communities corresponding to the chief habitats – the forest floor, streams and other damp places, paths, clearings, the woodland edge, a marshy area, and a chalky area. It consists of flowering plants, ferns and horsetails, mosses and liverworts, and fungi. There are 283 plant species, at least 3 liverwort and 12 moss species, and 302 fungi species. A few lichens grow on tree trunks and buildings.

The plants and fungi support a great diversity of insects and woodlice. There are predators (ground beetles, burying beetles, ladybirds, and others), defoliators (leaf beetles and weevils), seed borers (weevils), bark borers (weevils and bark beetles), wood borers ('longhorn' beetles, 'ambrosia' beetles, and click beetles), root feeders (weevils and click beetle larvae – wireworms), and woodlice. There are butterflies and moths, snails and slugs. There are two amphibians – the common toad (*Bufo bufo*) and the common frog (*Rana temporaria*); and three reptiles – the warty newt (*Triturus cristatus*), the slow-worm (*Anguis fragilis*), and the ringed or grass snake (*Natrix natrix*). There are fifty species of breeding birds and some twenty-two

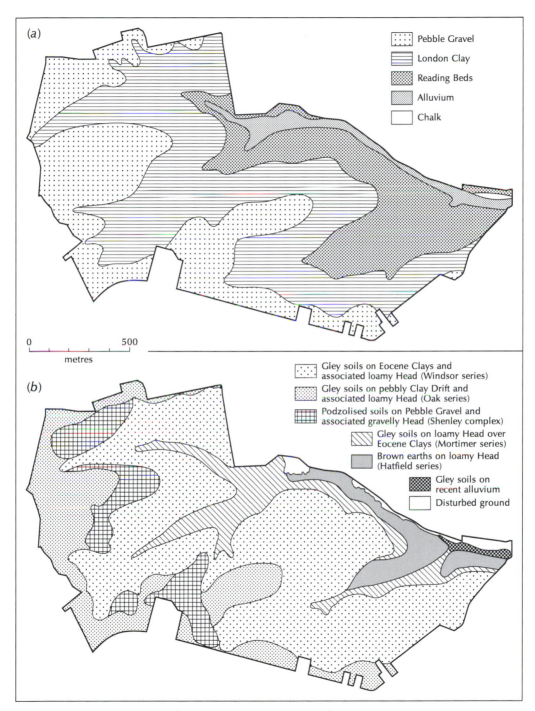

Figure 6.1 Northaw Great Wood, Hertfordshire, England. (a) Geology. (b) Soils.
Source: (a) After Sage (1966b). (b) After D. W. King (1966)

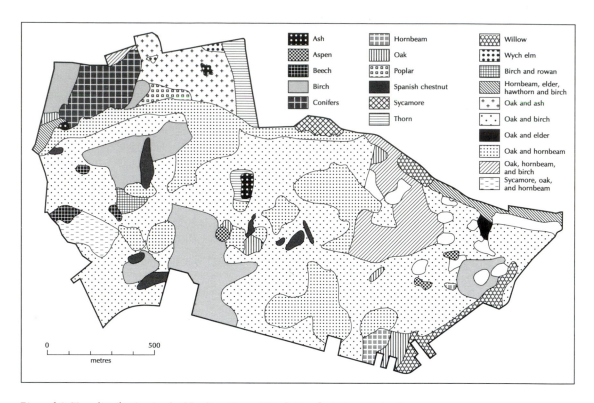

Figure 6.2 Tree distribution in the Northaw Great Wood, Hertfordshire, England.
Source: After Horsley (1966)

visitors. In summer, the dominant species in the breeding community are chaffinch (*Fringilla coelebs*), willow warbler (*Phylloscopus trochilus*), blue tit (*Parus caerulus*), great tit (*Parus major*), robin (*Erithacus rubecula*), blackbird (*Turdus merula*), wren (*Troglodytes troglodytes*), dunnock or hedge sparrow (*Prunella modularis*), blackcap (*Sylvia atricapilla*), garden warbler (*Sylvia bocin*), chiffchaff (*Phylloscopus collybita*), greater spotted woodpecker (*Dendrocopos major*), nuthatch (*Sitta europaea*), redstart (*Phoenicurus phoenicurus*), and tree pipit (*Anthus trivialis*). Twenty-three wild species of mammal are recorded (Table 6.1). Six mammalian orders are represented – Insectivora (shrews, hedgehogs, and voles), Chiroptera (pipistrelle and noctule bats), Lagomorpha (hares and rabbits), Rodentia (squirrels, rats, mice, and voles), Carnivora (foxes, badgers, stoats, and weasels), and Artiodactyla (fallow deer and muntjac deer).

Most ecosystems are heterogeneous and involve a mosaic of individuals, populations, and habitats. The Northaw Great Wood is mainly an oak–hornbeam woodland, with stands of silver birch on more acidic and better drained soils. However, it contains patches and corridors. The corridors are formed by paths and the streams. The patches include cleared areas and areas where other tree species dominate. There are, for example, three stands of ash (*Fraxinus excelsior*), three stands of beech (*Fagus sylvatica*), three stands of aspen (*Populus tremula*), and seven stands of sweet chestnut (*Castanea sativa*).

Within the wood, several communities exist. A good example is the bracken (*Pteridium aquilinium*), wood anemone (*Anemone nemorosa*), and bluebell (*Hyacinthoides non-scripta*) community (Sage 1966a: 14). These species manage to live together as a plant association by dividing the habitat to avoid undue

Table 6.1 Mammals in Northaw Great Wood, Hertfordshire, England (to 1966)

Order	Family	Species
Insectivora	Erinaceidae	Hedgehog (*Erinaceus europaeus*)
	Talpidae	Mole (*Talpa europaea*)
	Soricidae	Common shrew (*Sorex araneus*), pygmy shrew (*Sorex minutus*), water shrew (*Neomys fodiens*)
Chiroptera	Vespertilionidae	Common pipistrelle (*Pipistrellus pipistrellus*), noctule bat (*Nyctalus noctula*)
Lagomorpha	Leporidae	Rabbit (*Oryctolagus cuniculus*), brown hare (*Lepus europaeus*)
Rodentia	Sciuridae	Grey squirrel (*Sciurus carolinensis*)
		Bank vole (*Clethrionomys glareolus*), field vole (*Microtus agrestis*), water vole (*Arvicola terrestris*)
		Wood mouse (*Apodemus agrestis*), yellow-necked mouse (*Apodemus flavicollis*), brown rat (*Rattus norvegicus*)
		Dormouse (*Muscardinus avellanarius*)
Carnivora	Canidae	Red fox (*Vulpes vulpes*)
		Stoat (*Mustela erminea*), weasel (*Mustela nivalis*), badger (*Meles meles*)
Artiodactyla	Cervidae	Fallow deer (*Dama dama*), Reeve's muntjac (*Muntiacus reevesi*)

competition for resources. Bracken rhizomes may penetrate about 60 cm. The bluebell bulbs are not nearly so deep. The wood anemone taps the top 6 cm of soil. The wood anemone is a pre-vernal flowerer (March to May), whereas the bluebell is a vernal flowerer (April to May). Both complete assimilation before the bracken canopy develops in June, so avoiding competition for light. The bluebell is the only species that can withstand bracken, but if bracken growth is especially vigorous, the accumulations of fallen fronds exclude even the bluebell.

Communities possess vertical layers. This **stratification** results primarily from differences in light intensity. In some communities, distinct layers of leaves are evident – canopy, understorey, shrub layer, and ground layer. Animal species are sorted among the layers according to their food and cover requirements. In the Northaw Great Wood, tree, shrub, field, and ground layers are present. Each layer acts as a kind of semi-transparent blanket and modifies the conditions below. Light levels are reduced, as is precipitation intensity, from the canopy to the ground.

The horizontal distribution of species within communities is more complicated. Two extreme situations exist. First, a **gradient** is a gradual change in a species assemblage across an area. In the Northaw Great Wood, the dominant tree species change along a toposequence. Silver birch on summits gives way to oak and hornbeam in midslope positions and along the valley floor. Second, a **patch** is a clustering of species into somewhat distinctive groups. Although tree communities do gradually alter along environmental gradients in the Northaw Great Wood, there are also stands of trees that form distinct patches (Figure 6.2). Gradients and patches are two ends of a continuum and the subject of a lively debate in vegetation science.

The global ecosystem

All living things form the **biosphere**. The biosphere interacts with non-living things in its surroundings (air, water, soils, and sediments) to win materials and energy. The interaction creates the **ecosphere**, which is defined as life plus life-support systems. It consists of ecosystems – individuals, populations, or communities interacting with their physical environment. Indeed, the ecosphere is the global ecosystem (Huggett 1995: 8–11; 1997).

Communities and ecosystems possess properties that emerge from the individual organisms. These **emergent properties** cannot be measured in individuals – they are community properties. Four important community properties are biodiversity, production and consumption, nutrient cycles, and food webs. These community properties enable the biosphere to perform three important tasks (Stoltz *et al.* 1989). First, the biosphere harnesses energy to power itself and build up reserves of organic material. Second, it garners elements essential to life from the atmosphere, hydrosphere, and pedosphere. Third, it is able to respond to cosmic, geological, and biological perturbations by adjusting or reconstructing food webs.

PRODUCING AND CONSUMING: COMMUNITY ROLES

All organisms have a community and an ecological role. Some organisms produce organic compounds, and some break them down. The chief roles are as follows:

1   **Producers (autotrophs)**. These include all photo-autotrophic organisms: green plants, eukaryotic algae, blue-green algae, and purple and green sulphur bacteria. In some ecosystems, chemo-autotrophs produce organic materials by oxidation of inorganic compounds and do not require sunlight.
2   **Consumers (heterotrophs)**. These are organisms that obtain their food and thus energy from the tissues of other organisms, either plants or animals or both – herbivores, carnivores, and top carnivores. Consumers are divided into several trophic levels. Consumers that eat living plants are primary consumers or herbivores. Consumers of herbivores are the flesh-eating secondary consumers or carnivores. In some ecosystems, there are carnivore-eating carnivores; these are tertiary consumers or top carnivores.
3   **Decomposers (saprophages) and detritivores (saprovores)**. Decomposers dissolve organic matter, while detritivores break it into smaller pieces and partly digest it.

Autotrophs produce organic material, the consumers eat it, and the decomposers and detritivores clean up the mess (excrement and organic remains).

Producer society: community production

Primary producers

Green plants use solar energy, in conjunction with carbon dioxide, minerals, and water, to build organic matter. The organic matter so manufactured contains chemical energy. **Photosynthesis** is the process by which radiant energy is converted into chemical energy. **Production** is the total biomass produced by photosynthesis within a community. Part of the photosynthetic process requires light, so it occurs only during daylight hours. At night, the stored energy is consumed by slow oxidation, a process called **respiration** in individuals and **consumption** in a community.

Sunlight comes from above, so ecosystems tend to have a **vertical structure** (Figure 6.3). The upper **production zone** is rich in oxygen. The lower **consumption zone**, especially that part in the soil, is rich in carbon dioxide. Oxygen is deficient in the

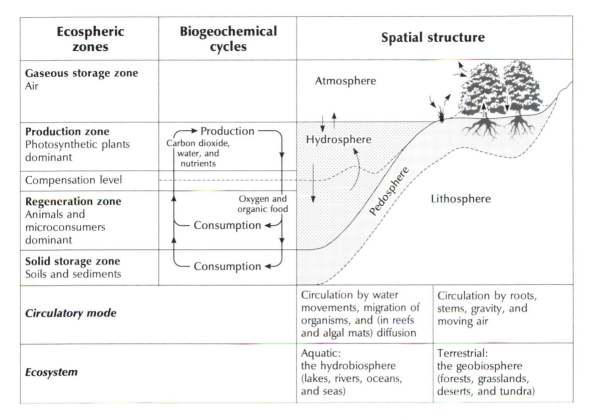

Ecospheric zones	Biogeochemical cycles	Spatial structure	
Gaseous storage zone Air		Atmosphere	
Production zone Photosynthetic plants dominant	Production Carbon dioxide, water, and nutrients	Hydrosphere	
Compensation level			
Regeneration zone Animals and microconsumers dominant	Oxygen and organic food Consumption	Pedosphere Lithosphere	
Solid storage zone Soils and sediments	Consumption		
Circulatory mode		Circulation by water movements, migration of organisms, and (in reefs and algal mats) diffusion	Circulation by roots, stems, gravity, and moving air
Ecosystem		Aquatic: the hydrobiosphere (lakes, rivers, oceans, and seas)	Terrestrial: the geobiosphere (forests, grasslands, deserts, and tundra)

Figure 6.3 Production zones, consumption zones, and biogeochemical cycles in ecosystems.
Source: From Huggett (1997) partly after Odum (1971)

consumption zone, and may be absent. Such gases as hydrogen sulphide, ammonia, methane, and hydrogen are liberated where reduced chemical states prevail. The boundary between the production zone and the consumption zone, which is known as the **compensation level**, lies at the point where there is just enough light for plants to balance organic matter production against organic matter utilization.

Phytomass is the living material in producers. It normally excludes dead plant material, such as tree bark, dead supporting tissue, and dead branches and roots. Where dead bits of plants are included, the term **standing crop** is applied. Evergreen trees in tropical forests are largely made of dead supporting structures (branches, trunks, roots); about 2 per cent is green plant biomass (leaves) (Figure 6.4). The relative proportion of leaves, branches, trunks, and roots

is similar in temperate forests, though the absolute values are lower. Grassland plants are all leaf and root – there is negligible branch and trunk. Tundra and semi-desert plants largely made of green plant biomass and contain little supporting tissue.

Primary production

The green plants that form the production zone, because they produce their own food from solar energy and raw materials, are called **photo-autotrophs**. The amount of organic matter that they synthesize per unit time is **gross primary production**. Most of this matter is created in the plant leaves. Some of it is transported through the phloem to other parts of plants, and especially to the roots, to drive metabolic and growth processes.

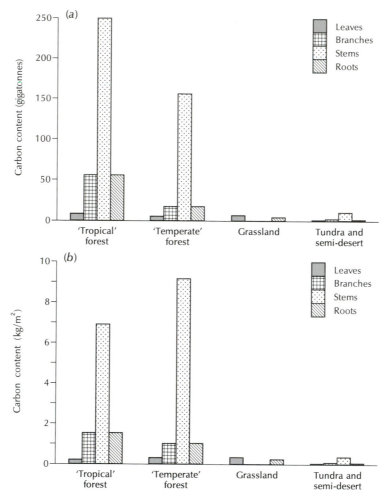

Figure 6.4 The distribution of biomass in different biomes. (a) Total carbon (gigatonnes, Gg) stored in leaves, branches, stems, and roots. (b) Carbon stored per unit area (kg/m²) in leaves, branches, stems, and roots. 'Tropical forest' includes tropical forest, forest plantations, shrub-dominated savannah, and chaparral. 'Temperate forest' includes temperate forest, boreal forest, and woodland.
Source: After data in Goudriaan and Ketner (1984)

Net primary production is the gross primary production less the chemical energy burnt in all the activities that constitute plant respiration. Net primary production is usually about 80 to 90 per cent of gross primary production. The mean net global primary production is 440 g/m²/yr (Vitousek *et al.* 1986), or about 224 petagrammes (1 Pg = 224 × 10¹⁵ grammes = 224 billion tonnes) (Figure 6.5). It is difficult to comprehend such vast figures. A cubic

kilometre of ice (assuming the density is 1 g/cm³) would weigh 1 billion tonnes (1 petagramme). A block of ice 1 km high and resting on an area of 15 km × 15 km would have about the same mass as the global net primary production.

In terrestrial ecosystems, the main producers are green plants (megaphytes), with the lower plants playing a minor role. Land plants produce on average 899 g/m²/yr (Vitousek *et al.* 1986). That is a total of

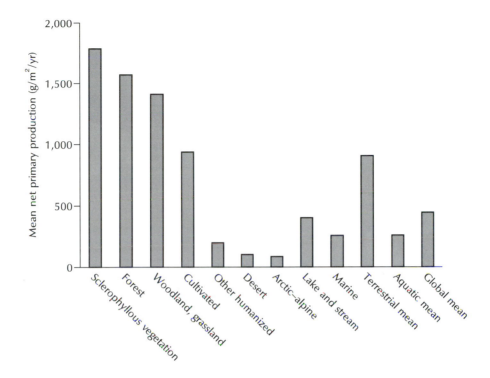

Figure 6.5 Mean net primary production for the biosphere and its component biomes.
Source: Data from Vitousek *et al.* (1986)

about 132 billion tonnes for the entire land area. In aquatic ecosystems, the main producers are autotrophic algae. These unicellular and colonial organisms are suspended in the water and are part of the plankton. Aquatic plants produce on average 225 g/m²/yr (Vitousek *et al.* 1986). That is a total of 92.4 billion tonnes for the entire water area. Of this total, 91.6 billion tonnes are produced by marine plants, and a mere 0.8 billion tonnes by freshwater plants. A very tiny part of global net primary production is contributed by chemolithotrophic bacteria using chemical reactions around vents in the sea floor or in soils.

The world pattern of terrestrial net primary production is shown in Figure 6.6. It mirrors the simultaneous availability of heat and moisture. Production is high in the tropics, warm temperate zone, and typical temperate zone, and low in the arid subtropics, continental temperate regions, and polar zones.

Consumer society: community consumption

Consumers

The material produced by photosynthesis serves as a basic larder for an entire ecosystem – it is used as an energy-rich reserve of organic substances and nutrients that is transferred through the rest of the system. The potential chemical energy of net primary production is available to organisms that eat plants, both living plant tissue and dead plant tissue, and indirectly therefore to animals (and the few plants) that eat other animals. All these organisms depending upon other organisms for their food are **heterotrophs** or **consumers**. They are browsers, grazers, predators, or scavengers. They include microscopic organisms, such as protozoans, and large forms, such as vertebrates. The majority are chemoheterotrophs, but a few specialized photosynthetic bacteria are

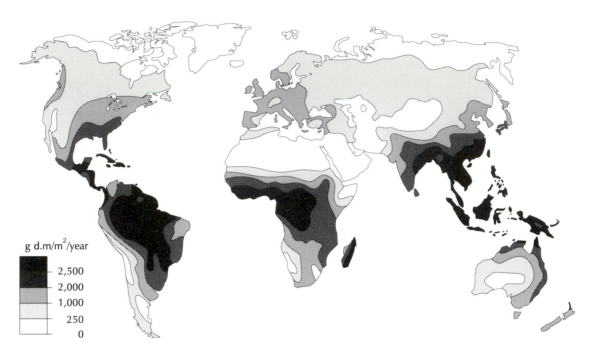

g d.m/m^2/year

- 2,500
- 2,000
- 1,000
- 250
- 0

Figure 6.6 The world pattern of terrestrial net primary production. Units are grammes of dry matter per square metre per year (g d.m/m²/year).
Source: From Huggett (1997) after Box and Meentemeyer (1991)

photoheterotrophs. The stored chemical energy in consumers is called secondary production. There are two broad groups of consumers. **Macroconsumers** or **biophages** eat living plant tissues. **Microconsumers** or **saprophages** slowly decompose and disintegrate the waste products and dead organic matter of the biosphere.

The available phytomass in water is small, but the primary production (the amount of dry organic matter produced in a year) is relatively large because aquatic plants multiply fast. Animals eat the plants and incorporate much of the primary production in their bodies. For this reason, the **secondary production** or **zoomass** (animal biomass) in aquatic ecosystems is commonly more than fifteen times larger than the phytomass. The situation is different in terrestrial ecosystems. Much of the phytomass consists of non-photosynthetic tissue, such as buds, roots, and trunks. For this reason, the standing phytomass is always very large, especially in woodlands. Primary

production, on the other hand, is relatively small – a modest amount of dry organic matter is created each year. A mere 1 per cent, or thereabouts, of the phytomass is eaten by animals living above the ground. The zoomass is therefore small, being about 1 to 0.1 per cent of the phytomass. Biomass is the weight or mass of living tissue in an ecosystem – phytomass plus zoomass. It has an energy content, which may be thought of as bioenergy.

The human harvest of plants for food, fuel, and shelter accounts for about 4.5 per cent of global terrestrial net production (Table 6.2). Land used in agriculture or converted to other land uses accounts for about 32 per cent of terrestrial net primary production. All human activities have reduced terrestrial net primary production by around 45 per cent. This probably amounts to the largest ever diversion of primary production to support a single species (Vitousek *et al*. 1986).

Ecosystems also contain 'geomass' – soils (including

Table 6.2 Human appropriation of net primary production in the 1980s

Manner of consumption	Net primary production	
	Total mass (billion tonnes)	Percentage of terrestrial net primary production
Used by humans		
Plants eaten	0.8	0.48
Plants fed to domestic animals	2.2	1.67
Fish eaten by humans and domestic animals	1.2	–
Wood for paper and timber	1.2	0.91
Fuel wood	1.0	0.76
Total	7.2	4.54[a]
Used or diverted		
Cropland	15.0	11.35
Converted pastures	9.8	7.42
Other (cities, deforested)	17.8	13.48
Total	42.6	32.25
Used, diverted, or reduced		
Used or diverted	42.6	32.25
Reduced by conversion	17.5	13.25
Total	60.1	45.50

Source: After Diamond (1987)
Note: [a] Excluding fish

litter and dead organic matter), sediments, air, and water that harbour a supply of water and nutrients (macronutrients, micronutrients, and other trace elements).

Decomposers and detritivores

Saprophages include decomposers (or saprophytes) and detritivores (or saprovores). **Decomposers** are organisms that feed on dead organic matter and waste products of an ecosystem. They do so by secreting enzymes to digest organic matter in their surroundings, and soaking up the dissolved products. They are mainly aerobic and anaerobic bacteria, protozoa, and fungi. An example is the shelf fungus (*Trametes versicolor*). This grows on rotting trees and is important in decomposing wood. A very few animals, such as tapeworms, and some green plants,

such as the Indian pipe, also obtain their food by diffusion from the outside.

Detritivores are organisms that feed upon detritus. They include beetles, centipedes, earthworms, nematodes, and woodlice. They are all microconsumers. Detritivores assist the breakdown of organic matter. By chewing and grinding dead organic matter before ingestion, they comminute it and render it more digestible. When egested, the carbon–nitrogen (C/N) ratio is a little lower, and the acidity (pH) a little higher, than in the ingested food. These changes mean that the faeces provide a better substrate for renewed decomposer (and notably bacterial) growth. Successively smaller fragments of dead organic matter are passed through successively smaller detritivores, after having been subject to decomposer attack at each stage. The result is a comminution spiral (Figure 6.7).

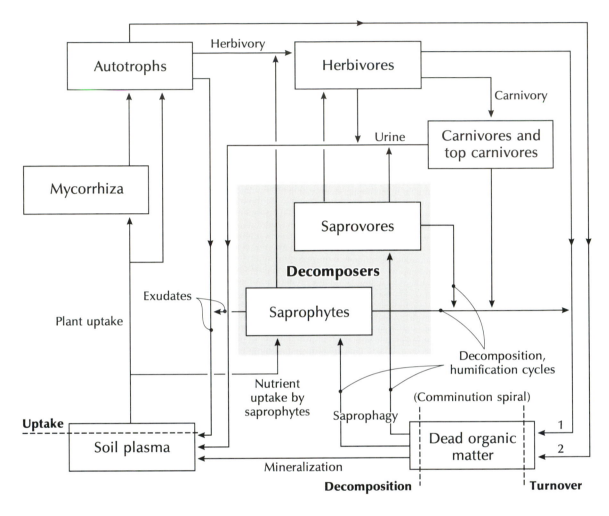

Figure 6.7 Grazing and detrital feeding relations in an ecosystem, showing comminution spirals.
Source: After Huggett (1980)

Humification accompanies comminution and decomposition to produce a group of organic compounds called **humus**. Humus is in a sense the final stage of organic decomposition, but it is also the product of microbial synthesis and is always subject to slow microbial decomposition, the end product of which is stable humus charcoal. Humus decomposition in the temperate zones is incomplete and humus is brown or black. Decomposition is usually more advanced in the tropics and humus may be colourless. Mineral nutrients are released into the soil solution during decomposition, a process styled **mineralization**.

Recycling machines: ecosystem turnover

Biogeochemicals

The biosphere is made of three main elements – hydrogen (49.8 per cent by weight), oxygen (24.9 per cent), and carbon (24.9 per cent). Several other elements are found in the biosphere, and some of them are essential to biological processes – nitrogen (0.27 per cent), calcium (0.073 per cent), potassium (0.046 per cent), silicon (0.033 per cent), magnesium (0.031 per cent), phosphorus (0.03 per cent), sulphur (0.017 per cent), and aluminium (0.016 per cent). These elements, except aluminium, are the basic ingredients for organic compounds, around which biochemistry revolves. Carbon, hydrogen, nitrogen, oxygen, sulphur, and phosphorus are needed to build nucleic acids (RNA and DNA), amino acids (proteins), carbohydrates (sugars, starches, and cellulose), and lipids (fats and fat-like materials). Calcium, magnesium, and potassium are required in moderate amounts. Chemical elements required in moderate and large quantities are **macronutrients**. More than a dozen elements are required in trace amounts, including chlorine, chromium, copper, cobalt, iodine, iron, manganese, molybdenum, nickel, selenium, sodium, vanadium, and zinc. These are **micronutrients**. Functional nutrients play some role in the metabolism of plants but seem not to be indispensable. Other mineral elements cycle through living systems but have no known metabolic role. The biosphere has to obtain all macronutrients and micronutrients from its surroundings.

Biogeochemical cycles

There is in the ecosphere a constant turnover of chemicals. The motive force behind these chemical cycles is life. In addition, on geological time-scales, the cycles are influenced by forces in the geosphere producing and consuming rocks. **Biogeochemical cycles**, as they are called, involve the storage and flux of all terrestrial elements and compounds except the inert ones. Material exchanges between life and life-support systems are a part of biogeochemical cycles.

At their grandest scale, biogeochemical cycles involve the entire Earth. An exogenic cycle, involving the transport and transformation of materials near the Earth's surface, is normally distinguished from a slower and less well understood endogenic cycle involving the lower crust and mantle. Cycles of carbon, hydrogen, oxygen, and nitrogen are gaseous cycles – their component chemical species are gaseous for a leg of the cycle. Other chemical species follow a sedimentary cycle because they do not readily volatilize and are exchanged between the biosphere and its environment in solution.

Minerals cycle through ecosystems, the driving force being the flow of energy. The circulation of mineral elements through ecosystems involves three stages – uptake, turnover, and decomposition. Solutes and gases are taken up by green plants, the rate of uptake broadly matching biomass production rate, and incorporated into phytomass. Oxygen is released in photosynthesis. The remaining minerals are either passed on to consumers or else returned to soil and water bodies when plants die or bits fall off. The minerals in the consumers eventually return to the soil, sea, or atmosphere. Figure 6.8 summarizes some major **mineral cycles**.

The distribution of biogeochemicals within ecosystems varies among biomes. Figure 6.9 shows the amount of carbon stored in biomass, litter, humus, and stable humus charcoal in the major ecozones. This gives a good indication of how organic matter is apportioned in different ecosystems. Biomass carbon is twenty to forty times greater in forests than in other ecosystems. Humus carbon is highest in temperate forests and grasslands. It is lower in tropical forests because it is rapidly decomposed in the year-round, hot and humid conditions. Stable-humus-charcoal carbon, which degrades exceedingly slowly, is highest in tropical forests, and is still significant in temperate forests and grasslands.

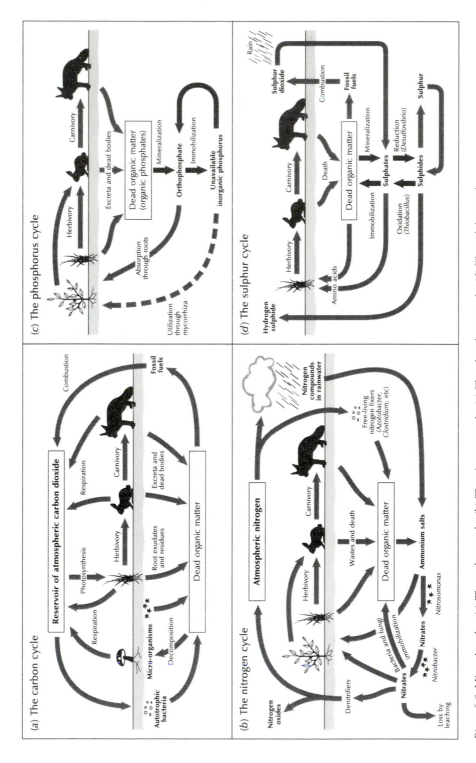

Figure 6.8 Mineral cycles. (a) The carbon cycle. (b) The nitrogen cycle. (c) The phosphorus cycle. (d) The sulphur cycle.
Source: Partly after Jackson and Raw (1966)

EATERS AND THE EATEN: FOOD CHAINS AND FOOD WEBS

Food webs

An ecosystem contains two chief types of food webs: a grazing food web (plants, herbivores, carnivores, top carnivores) and a decomposer or detrital food web (Figure 6.7).

Grazing food chains and webs

This simple feeding sequence

plant → herbivore → carnivore → top carnivore

is a **grazing food chain**. An example is

leaf → caterpillar → bird → weasel.

Usually though, food and feeding relations in an ecosystem are more complex because there is commonly a wide variety of plants available, different herbivores prefer different plant species, and carnivores are likewise selective about which herbivore they will consume. Complications also arise because some animals, the omnivores, eat both plant and animal tissues. For all these reasons, the energy flow through an ecosystem is in most cases better described as a food web.

Figure 6.10 shows the food web for Wytham Wood, Oxfordshire, England. The incoming solar energy supports a variety of trees, shrubs, and herbs. Oak (*Quercus* spp.) is the dominant plant. The plants are eaten by a variety of herbivores. Several invertebrates feed on plant material, and especially leaves. The winter moth (*Operophtera brumata*) and the pea-green oak twist (*Tortrix viridana*) are examples. The herbivores are prey to carnivores, including spiders, parasites, beetles, birds, and small mammals. *Cyzenis albicans* is a tachinid fly and a specific parasite of the winter moth. There are several predatory beetles in the litter layer, the commonest large ones being *Philonthus decorus*, *Feronia madida*, *Felonia melanaria*, and *Abax parallelopipedus*. Titmice feed partly on plants (e.g. beech mast), partly on insects, and partly on spiders. The main species are the great

tit (*Parus major*) and the blue tit (*Parus caeruleus*). Small mammals include the bank vole (*Clethrionomys glareolus*) and the wood mouse (*Apodemus sylvaticus*). Top carnivores include parasites and hyperparasites, shrews, moles, weasels, and owls. The common shrew (*Sorex araneus*), pygmy shrew (*Sorex minutus*), and mole (*Talpa europaea*) dominate the ranks of top carnivore. They are all eaten by the tawny owl (*Strix aluco*).

Decomposer food chains and webs

Dead organic matter and other waste products generally lie upon and within the soil. As they are decomposed, minerals are slowly released that are reused by plants. There are complex food and feeding relations among the decomposers and a decomposer or detrital food chain is recognized, the organisms in which are all microconsumers (decomposers and detritivores).

On land, litter and soil organic matter support a **detrital food web**. In Wytham Wood, decomposers (mainly bacteria and fungi) slowly digest the litter. Litter fragments are eaten by detritivores – earthworms, soil insects, and mites. Predator beetles eat the detritivores, thus linking the grazing and detrital food webs. In aquatic ecosystems, the producers (mainly phytoplankton) live in the upper illuminated areas of water bodies. They are eaten by animal microplankton and macroplankton. In turn, the animal plankton are eaten by fishes, aquatic mammals, and birds that take their prey out of water. All the organisms in the detrital aquatic food chain are eventually decomposed in the water and in sediments on lake, river, or sea bottoms.

Ecological pyramids

Energy and biomass are distributed among producers, herbivores, predators, top predators, and decomposers. Their distributions both resemble a pyramid – energy and biomass become less in moving from producer, to herbivore, to carnivore, to top carnivore. This is because at each higher trophic level, there is successively less energy and biomass available to eat. In

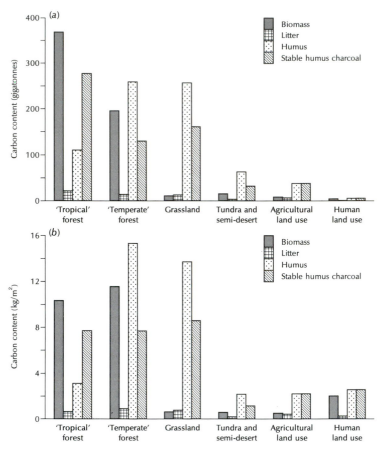

Figure 6.9 Carbon stored in biomass, litter, humus, and charcoal in the major ecozones. (a) Total carbon (gigatonnes, Gg). (b) Carbon per unit area (kg/m^2). 'Tropical forest' includes tropical forest, forest plantations, shrub-dominated savannah, and chaparral. 'Temperate forest' includes temperate forest, boreal forest, and woodland.
Source: After data in Goudriaan and Ketner (1984)

Cedar Bog Lake, Minnesota, the net production figures (in kcal/m^2/yr) are: producers 879, herbivores 104, and carnivores 13 (Lindemann 1942). The biomass of each trophic level could be measured in the field, or could be derived by converting the production figures to biomass. Each gramme of dry organic matter is equivalent to about 4.5 kcal, so each kilocalorie unit is equivalent to about 0.222 kg. The biomasses for Cedar Lake Bog (in kg/m^2) are therefore approximately: producers 195, herbivores 23, and carnivores 3. This conversion to biomass does not change the shape of the pyramid.

Some ecosystems have an inverted pyramidal distribution of biomass – narrow at the bottom and wide at the top. In the English Channel, there are 4 g of phytoplankton per m^2 and 21 g of zooplankton per m^2. Clear-water aquatic ecosystems have a lozenge-shaped biomass distribution – narrow at the top and bottom and wide in the middle.

A **pyramid of numbers** represents the number of organisms at each trophic level – the number of plants, the number of herbivores, and the number of carnivores (Figure 6.11). This information can be troublesome when comparing two different ecosystems – it is not very informative to equate 'a diatom with a tree, or an elephant with a vole' (Phillipson

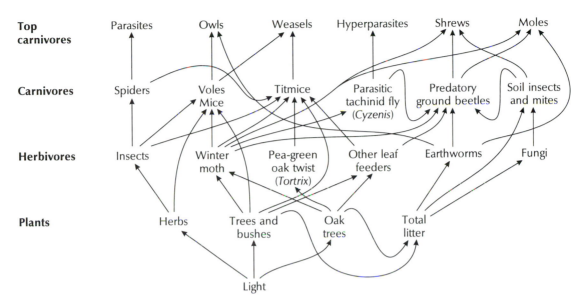

Figure 6.10 A food web, Wytham Wood, Oxfordshire, England.
Source: After Varley (1970)

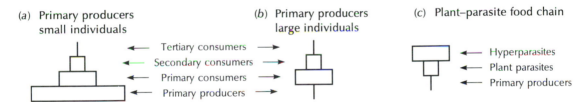

Figure 6.11 Ecological pyramids. (a) Primary producers are small organisms. (b) Primary producers are large organisms. (c) A plant–parasite–hyperparasite food chain.
Source: After Phillipson (1966)

1966: 13). The typical pyramid of numbers applies when producers are small, as they are in aquatic ecosystems. In forests, the producers – mainly trees – are large, and a pyramid of the various consumer levels perches on a thin base. In plant–parasite–hyperparasite food chains, the pyramid of numbers is inverted.

Keystone species

Keystone species are species central to an ecosystem, species upon which nearly all other species depend. Several keystone species have been identified in the wild, but it is not easy to predict which species will be keystone because the connections between species in food webs are often complex and obscure. For instance, large cats act as keystone predators in Neotropical forests. They limit the number of medium-sized terrestrial mammals, which in turn control forest regeneration. On Barro Colorado Island, Panama, jaguars (*Felis onca*), pumas (*Felis concolor*), and ocelots (*Felis pardalis*) have been removed. The populations of big-cat prey – including the red coati (*Nasua nasua*), the agouti (*Dasyprocta variegata*), and the paca (*Agouti paca*) – are about ten times higher than on Cocha Cashu, Peru, where big cats still live (Terborgh 1988). However, this increase may result from natural population variability rather than the lack of jaguars and pumas (Wright *et al.* 1994). The extreme removal of herbivores and frugivorous mammals would drastically affect forest regeneration, altering tree species composition, but the effects of modest changes in densities are less clear.

Several examples of keystone species will be considered, and then some general ideas about keystone-species removal will be examined.

Keystone predators

The sea otter (*Enhydra lutris*) is a keystone predator *par excellence* (Duggins 1980). A population of around 200,000 once thrived on the kelp beds lying close to shore from northern Japan, through Alaska, to southern California and Mexico (Figure 6.12). In 1741, Vitus Bering, the Danish explorer, reported seeing great numbers of sea otters on his voyage among the islands of the Bering Sea and the North Pacific Ocean. Some furs were taken back to Russia and soon this new commodity was highly prized for coats. Hunting began. In 1857, Russia sold Alaska to the United States for $7,200,000. This cost was recouped in forty years by selling sea otter pelts. In 1885 alone, 118,000 sea otter pelts were sold. By 1910, the sea otter was close to extinction, with a world-wide population of fewer than 2,000. It was hardly ever seen along the Californian coast from 1911 until 1938.

The inshore marine ecosystem changed where the sea otter disappeared. Sea urchins, which were eaten by the otters, underwent a population explosion. They consumed large portions of the kelp and other seaweeds. While the otters were present, the kelp formed a luxuriant underwater forest, reaching from the sea bed, where it was anchored, to the sea surface. With no otters to keep sea-urchins in check, the kelp vanished. Stretches of the shallow ocean floor were turned into sea-urchin barrens, which were a sort of submarine desert.

Happily, a few pairs of sea otters had managed to survive in the outer Aleutian Islands and at a few localities along the southern Californian coast (Figure 6.12). Some of these were taken to intermediate sites in the United States and Canada where they were protected by strict measures. With a little help, the sea otters staged a comeback and the sea urchins declined. The lush kelp forest grew back and many lesser algae moved in, along with crustaceans, squids, fishes, and other organisms. Grey whales migrated closer to shore to park their young in breaks along the kelp edge while feeding on the dense concentrations of animal plankton.

Keystone predators sometimes are more effective within certain parts of their range. The sea star (*Pisaster ochraceus*) is a keystone predator of rocky intertidal communities in western North America (Paine 1974). This starfish preys primarily on two mussels – *Mytilus californianus* and *Mytilus trossulus*. A study along the central Oregon coast showed that three distinct predation regimes exist (Menge *et al.* 1994). Strong keystone predation occurs along wave-exposed headlands. Less strong predation by sea stars, whelks, and possibly other predators occurs in a wave-protected cove. Weak predation occurs at a wave-protected site regularly buried by sand.

Keystone herbivores and omnivores

Some keystone herbivores and omnivores are so dominant that they help to structure the ecosystems in which they live. Beavers, elephants, and humans are cases in point. The beaver (*Castor canadensis*) builds dams, so creating ponds and raising water tables (Naiman *et al.* 1988). The wetter conditions produce wet meadows, convert streamside forest into shrubby coppice areas, and open gaps in woods some distance away by toppling trees. The elongated area fashioned by a beaver family may exceed 1 km in length. Beavers also cause changes to the biogeochemical characteristics of boreal forest drainage networks (Naiman *et al.* 1994).

Elephants, rhinoceroses, and other big herbivores play a keystone role in the savannahs and dry woodlands of Africa (Laws 1970; Owen-Smith 1989).

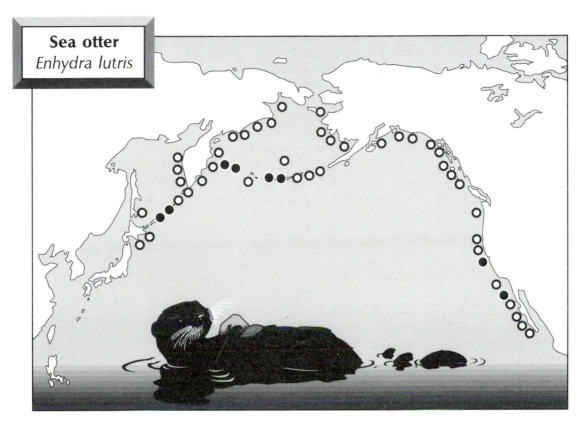

Sea otter
Enhydra lutris

Figure 6.12 Sea otter (*Enhydra lutris*) distribution. Circles are its mid-eighteenth-century distribution. Black dots are its distribution around 1910.
Source: Partly after Ziswiler (1967)

African elephants (*Loxodonta africana*) are relatively unspecialized herbivores. They have a diet of browse with a grass supplement. In feeding, they push over shrubs and small trees, thus helping to convert woodland habitats into grassland. They sometimes destroy large mature trees by eating their bark. As more grasses invade the woodland, so the frequency of fires increases, pushing the conversion to grassland even further. Grazing pressure from white rhinoceroses (*Ceratotherium simum*), hippopotamuses (*Hippopotamus amphibius*), and eland (*Taurotragus oryx*) then transforms medium-tall grassland into a mosaic of short and tall grass patches. The change from woodland to grassland is detrimental to the elephants, which begin to starve as woody species disappear, but beneficial to the many ungulates that are grassland grazers. Over millions of years, browsing and grazing by the megaherbivores of sub-Saharan Africa has created a mosaic of habitats and maintained a rich diversity of wildlife.

Some early human groups were keystone omnivores. Their invasion of North America may explain the mass extinction of megaherbivores (Owen-Smith 1987, 1988, 1989). Hunters may have liquidated the largest mammalian herbivores. These giant species had browsed and grazed, and maintained the grassland habitat. The reduction of the megaherbivores, possibly in conjunction with climatic change, led to forest expansion and open-habitat contraction. These habitat changes caused many mammal and bird populations to become reduced and fragmented. Coupled with additional hunting, loss of prey and carrion for predators and scavengers, and other changes, habitat fragmentation led to the extinction of other species. The human colonization of North American may have triggered a cascade of events that ultimately caused the extinction of the megafauna and other species in Pleistocene times.

Removing keystone species

What happens if a trophic level should be removed from an ecosystem? This is an enormously difficult question, to which there are at least four contradictory answers (Pimm 1991) (Box 6.1).

The effect of removing species depends upon the complexity of the food web (Figure 6.14). In a complex food web, removal of a plant at the bottom has little effect through the rest of the ecosystem (Figure 6.14a). This is borne out by the limited impact of American chestnut (*Castanea dentata*) removal from the eastern forests owing to chestnut blight (p. 80). Seven species of butterfly that feed exclusively on the chestnut are probably extinct but forty-nine other species of butterfly that also fed on the chestnuts found alternative food sources, as did the insect predators that fed on all fifty-six butterflies. On the other hand, removal of keystone predators or herbivores is not so innocuous an event – a major shock cascades all the way down the food web and shakes to lowermost level (Figure 6.14b). Kangaroo rats (*Dipodomys* spp.) are a keystone herbivores in the desert–grassland ecotone in North America. They have a major impact on seed predation and soil disturbance. Twelve years after their removal from plots of Chihuahuan Desert scrub, tall perennial and annual grasses increased threefold and rodent species typical of arid grassland had colonized (Brown and Heske 1990). Similarly, the cassowary is thought to be the sole disperser for over a hundred species of woody tropical rain-forest plants in Queensland, Australia (Crome and Moore 1990). It usually inhabits large forests. Logging and habitat fragmentation have removed the bird from several areas, in which only small remnants of forest remain. A progressive and massive loss of trees is likely to follow, unless the cassowary adapts or adjusts its behaviour.

For simple food chains, the situation is reversed (Figure 6.14c). Highly specialized herbivores or carnivores are extremely vulnerable to the loss of their sole food source. The koala (*Phascolarctos cinereus*) is an arboreal folivore. It feeds almost exclusively on the foliage of gum trees (*Eucalyptus*). Although still widespread, the koala population is controlled in some areas where overpopulation would otherwise lead to defoliation, the death of scarce food trees, and an endangerment of the koala population (Strahan 1995: 198). The sabre-toothed cats were one of the main branches of the cat family (Felidae) throughout much of the Tertiary. They had enormous upper canine teeth and probably specialized in preying on large, slow-moving, thick-hided herbivores, such as mastodons and giant sloths. They became extinct during the Pleistocene, most likely because their prey was at first thinned by extinctions and finally vanished.

The examples suggest that introducing generalists should have a greater impact on an ecosystem than introducing specialists. Correspondingly, introductions into ecosystems containing generalist predators (which can limit the number of intruders) should have less of an impact than introductions of specialist predators.

Contaminating food webs

Humans are part of food webs. Their skills at hunting, and later farming, enabled them to exploit plants and animals in a manner quite unlike any other organism. They are super-omnivores. Hunting skills, 10,000 years ago, were equal to the task of driving large herbivores to extinction. This dubious ability has lasted and evolved into a new, and perhaps more subtle, form – farming has transformed the global land cover. In large tracts of continents, naturally diverse plant communities have been replaced by vast expanses of monoculture crops – wheat, maize, and rice. Ecological efficiency runs at about 10 per cent, so it makes more sense for humans to eat plants than to let herbivores eat plants and then eat the herbivores. It makes little sense to let carnivores eat the herbivores and then eat the carnivores. Most cultures do eat

Plate 9 (a) Canadian beaver (*Castor canadensis*)
Photograph by Pat Morris

Plate 9 (b) Beaver dam and lodge, Rocky Mountains National Park, Colorado, United States
Photograph by Pat Morris

**Plate 10 Competing squirrels in the United Kingdom
(a) Red Squirrel
(Sciurus vulgaris)
(b) Grey squirrel
*(Sciurus carolinensis)***
Photographs by Pat Morris

Plate 12 (a) Reticulated velvet gecko *(Oedura reticulata)*
Photograph by Stephen Sarre

Plate 12 (b) A remnant patch of gimlet gum *(Eucalyptus salubris)* in a wheat crop, near Kellerberrin, Western Australia
Photograph by Stephen Sarre

Plate 11 A skipper butterfly *(Hesperia comma)*, **on the North Downs, Surrey, England**
Photograph by Owen Lewis

Plate 13 Golden lion tamarin *(Leontideus rosalia)*
Photograph by Pat Morris

Box 6.1

REMOVING TROPHIC LEVELS – THREE IDEAS

What happens if a trophic level is removed from a community? To answer this question it is helpful to ask what holds communities together. There are three competing ideas (Pimm 1991) (Figure 6.13).

'The world is green' hypothesis

Carnivores, which have no predators above them to keep them in check, should keep herbivore populations low. In turn, the herbivores should have little impact on plants, which thus thrive. Consequently, removing herbivores should have a minor effect upon an ecosystem, whereas removing carnivores should have major effect because it would allow herbivores to boom. In this model, competition is intense between plants and between predators, but not between herbivores. Removal of a predator will therefore have major consequences for other predators.

'The world is prickly and tastes bad' hypothesis

Most plants possess defence systems, such as toxins and sharp spines, that limit herbivore numbers. The scarcity of herbivores in turn limits the number of carnivores. If either carnivores or herbivores should be removed, the effect upon the community as a whole would be limited because they are already kept in check by the plants at the base of the food chain. In this model, competition is between herbivores and between predators, and if a species of either is removed, it should have consequences for the other species at the same level of the food chain.

'The world is white, yellow, and green' hypothesis

This considers the likely effects of ecosystem productivity. 'White' ecosystems with low productivity, such as tundra, contain scant plants that compete among themselves for limited resources. They do not produce enough food to support more than a few herbivores and even fewer carnivores. Removing herbivores and carnivores, which compete only weakly with their peers, from 'white' ecosystems will have little impact. 'Yellow' ecosystems, such as temperate forests and grasslands, have medium productivity. Enough plant material is produced to support a modest number of herbivores, certainly enough to keep plant numbers in check, but not support a large number of carnivores. In 'yellow' ecosystems, carnivore removals will have little

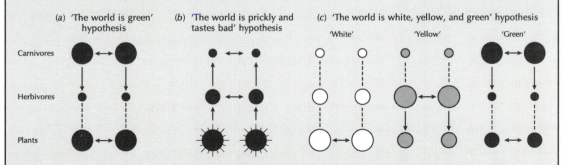

Figure 6.13 What holds communities together? Three possible answers. (a) 'The world is green' hypothesis. (b) 'The world is prickly and tastes bad' hypothesis. (c) 'The world is white, yellow, and green' hypothesis.
Source: After Pimm (1991)

impact, but herbivore removals will have a major impact on plants. 'Green' ecosystems, such as tropical forests, are highly productive. There are enough carnivores to keep herbivores in check. In 'green' ecosystems, removing carnivores will affect herbivore numbers, which will in turn affect plants.

These hypotheses are simplistic, but the field evidence tends to favour the 'world is white, yellow, and green' hypothesis, which offers plausible expla-nations for certain features of ecosystems. It explains why carnivores are sometimes important in eco-systems, and sometimes they are not important. It explains why controlling carnivore numbers some-times increases the population of a prey species, and sometimes it does not do so. 'White' ecosystems, such as the Arctic tundra, do seem resilient to major perturbations. 'Green' ecosystems, such as Yellow-stone National Park, do suffer from the removal of carnivores – herbivore numbers increase greatly with an accompanying impact on vegetation.

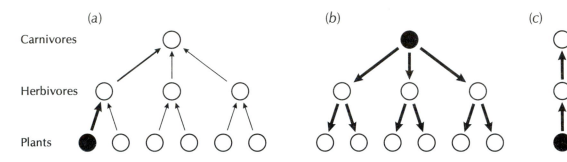

Figure 6.14 Simple and complex food webs – repercussions of removing trophic levels. (a) Removing a single plant from a food web has little impact on a predator, which draws upon a range of food sources. (b) Removing a predator from a food web creates a 'shock wave' that cascades down the trophic levels. (c) In a simple food chain, removing a single plant species may materially affect the predator.
Source: After Budiansky (1995)

some meat and herbivores are the main source of animal protein. Nonetheless, carnivores do appear in some diets.

The human exploitation of food chains is an enormous topic. Three issues will be considered here because they impinge on biogeography – biological magnification, the long-range transport of pesticides, and the long-range transport of radioactive isotopes.

Biological magnification

Some substances, including pesticides, may be applied in concentrations that are harmless to all but the pests they are deigned to eradicate. However, concentrations build up (or are magnified) as one organism eats another and the pesticide is fed along a food chain. This process is called **biomagnification** or **food-chain concentration**. Its inimical effects were brought to public notice by Rachel Carson in her book *Silent Spring* (1962). This book drew attention to the alarming build up of long-lasting pesticides, mainly DDT, in the environment and the damage that they were causing to wildlife and humans near the top of food chains.

DDT is a chlorinated hydrocarbon. It was first prepared by Othman Zeidler in 1874. Paul Müller discovered its insecticidal properties in 1939 and, for doing so, received the Nobel Prize for Physiology. It was thought to be a panacea, a complete solution to pest control. By the early 1960s, its persistence in the environment and accumulation in the food chain were becoming apparent. In 1972, after years of forceful lobbying and petitioning in the United States, DDT was banned for all but emergency use by the Environmental Protection Agency. The bio-magnification of DDT in the Long Island estuary, New York, clearly shows the increasing concentrations in moving up the trophic levels (Figure 6.15).

One reason that DDT accumulates along food chains is that it is firmly held in fatty animal tissues. Some heavy metals are also stored in body tissues and are subject to biomagnification. In the Alto Paraguay River Basin, Brazil, large quantities of mercury, used in gold mining, are dispersed directly into the air and the rivers running into Pantanal, a wildlife reserve (Hylander *et al*. 1994). Local, commercially important catfish (*Pseudoplatystoma coruscans*) had a mercury content above the limit for human consumption, and significantly above the natural background level. Mercury content in bird feathers also indicated biomagnification. No statistically significant accumulation of mercury was found in soil and sediment samples. Evidently, mercury originating from the gold-mining process is more readily absorbed by organisms than mercury naturally present in soil minerals.

Ditches along busy roads are liable to pollution by heavy metals. In Louisiana, United States, animals and plants living in ditches have accumulated cadmium and lead (Naqvi *et al*. 1993). In the red swamp crayfish (*Procambarus clarkii*), the cadmium level was 32 times that in the water, and the lead level 12 times that in the water, giving bioaccumulation factors of 5.1 and 1.7, respectively.

Long-range transport of radioactive isotopes

Although southern and temperate biological systems have largely cleansed themselves of radioactive fall-out deposited during the 1950s and 1960s, Arctic environments have not (D. J. Thomas *et al*. 1992). Lichens accumulate radioactivity more than many other plants because of their large surface area and long life-span; the presence and persistence of **radioisotopes** in the Arctic are of concern because of the lichen–reindeer (*Rangifer tarandus*)–human ecosystem.

Long-range transport of pesticides

Organochlorines (chlorinated hydrocarbons) include the best known of all the synthetic poisons – endrin, dieldrin, lindane, DDT, and others. They are widely used as biocides. **Polychlorinated biphenyls** (PCBs) are used in plastics manufacture and as flame retardants and insulating materials. A very worrying development is that high levels of organochlorines are recorded in Arctic ecosystems, where they are biomagnified and have inimical effects on consumers. The organochlorines are carried northwards from agricultural and industrial regions in the air. They then enter Arctic ecosystems through plants and start an upward journey through the trophic levels.

PCBs and DDT are the most abundant residues in peregrine falcons (*Falco peregrinus*) (D. J. Thomas *et al*. 1992). They reach average levels of 9.2 and 10.4 μg/g, respectively. These concentrations are more than ten times higher than other organochlorines. They are also found in polar bears (*Ursus maritimus*) (Polischuk *et al*. 1995). A wide range of organochlorine pesticides and PCBs were measured in muscle tissue and livers of lake trout (*Salvelinus namaycush*) and Arctic grayling (*Thymallus arcticus*) from Schrader Lake, Alaska (R. Wilson *et al*. 1995). PCBs are recorded in marine Arctic ecosystems, too. Samples were collected near Cambridge Bay, at Hall Beach, and at Wellington Bay, all in Northwest Territories, Canada (Bright *et al*. 1995). Organisms studied included clams (*Mya truncata*), mussels (*Mytilus edulis*), sea urchins (*Strongylocentrotus droebachiensis*), and four-horned sculpins (*Myoxocephalus quadricornis*), which are fish. Some of the high concentrations in these samples were attributable to local organochlorine sources, but long-distance sources were also implicated.

FROM BACTERIA TO BLUE WHALES: BIODIVERSITY

Biological diversity, or **biodiversity** for short, incorporates three types of diversity – genetic, habitat, and species. **Genetic diversity** is the variety of

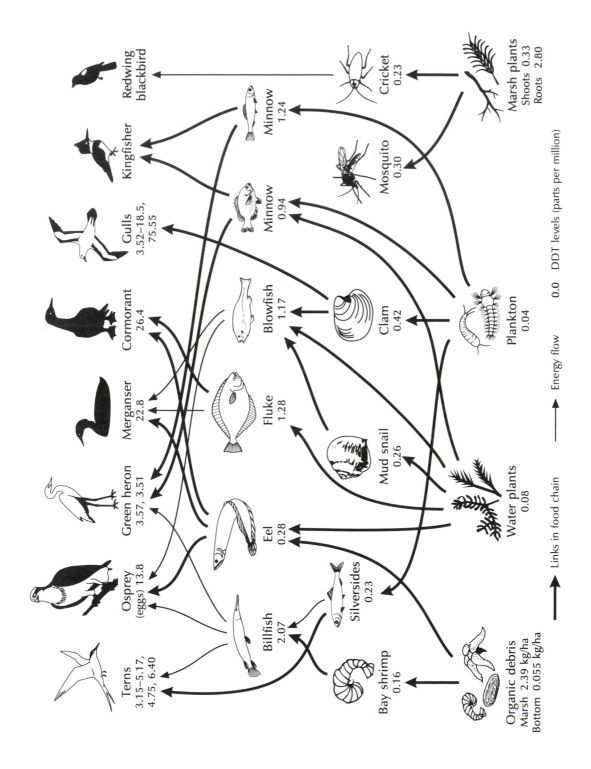

Figure 6.15 DDT biomagnification in the Long Island estuary food web, New York.
Source: After Woodwell (1967)

information (genetic characteristics) stored in a gene pool. **Habitat diversity** is the number of different habitats in a given area. **Species diversity** is the number of species in a given area. It is also called species richness and species number.

Guestimates of the total number of species currently living on the Earth vary enormously. They range from 4.4 million to 80 million. The number of known species is about 1,413,000. Therefore, the low estimate of 4.4 million would mean that 32 per cent of all species are known to date; the figure drops to a mere 2 per cent for the high estimate of 80 million.

The total species diversity disguises enormous differences between group of organisms (Table 6.3). Insects are by far the most numerous group on the planet. Total diversity also disguises three important geographical diversity patterns – species–area relationships, altitudinal and latitudinal diversity gradients, and diversity hot-spots.

Species–area relationships

Count the species in increasingly large areas, and the species diversity will increase. This, the **species–area effect**, is a fundamental biogeographical pattern. It applies to mainland species and to island species.

Table 6.3 The diversity of living things

Group	Number of species
Insects	751,000
Other animals	281,000
Higher plants	248,400
Fungi	69,000
Protozoa	30,800
Algae	26,900
Bacteria and similar forms	4,800
Viruses	1,000

Source: From data in E. O. Wilson (1992)

Species–area curves

The increasing number of species with increasing area is described by a **species–area curve**. Figure 6.16 is a species–area curve for plants in Hertfordshire, England. To construct the curve, plant species records from 10-km grid squares were grouped into successively larger contiguous areas, until the entire county was covered. Figure 6.16a shows the data plotted on arithmetic co-ordinates. The result is a curvilinear relationship. The same data plotted on logarithmic co-ordinates (Figure 6.16b) produces

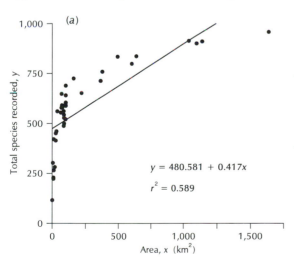

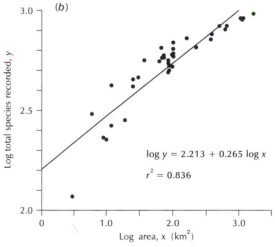

Figure 6.16 Species–area curve for Hertfordshire plants.
Source: Data from Dony (1963)

a linear relationship between species recorded and area. The line is described by the equation

$$\log S = \log c + z \log A$$

where S is the number of species, A is area, c is the intercept value (the number of species recorded when the area is zero), and z is the slope of the line. This equation may also be written as

$$S = cA^z$$

For the Hertfordshire plants, the equation is

$$S = 2.21A^{0.27}$$

The z-value (0.27) indicates that, for every 1 km² increase in area, an extra 0.27 plant species will be found.

Species also increase with area within island groups – large islands house more species than small islands. For amphibians and reptiles (the herpetofauna) living on the West Indian islands, the relationship, as depicted on Figure 6.17a, is

$$S = 2.16A^{0.37}$$

The line described by this equation fits the data for Sombrero, Redonda, Saba, Montserrat, Puerto Rico,

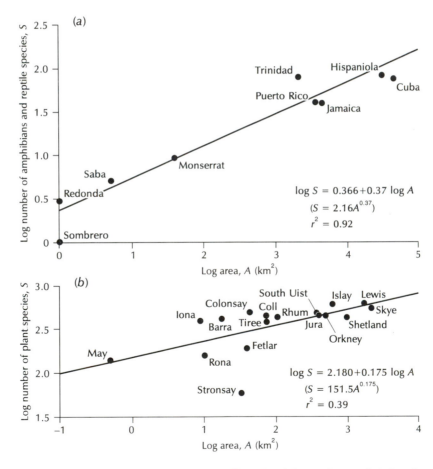

Figure 6.17 Species–area curve for island species. (a) Herpetofauna (amphibians plus reptiles) diversity on some West Indian islands. The equation for this classic best-fit line differs from that presented in many books and papers because Trinidad and the tiny island of Sombrero were included in the computations. (b) Bird species on Scottish islands.
Sources: (a) from data in Darlington (1957: 483) and (b) from data in M. P. Johnson and Simberloff (1974)

Jamaica, Hispaniola, and Cuba very well. Trinidad lies well above the line. It probably carries more species than would be expected because it was joined to South America 10,000 years ago. For plant species on a selection of Scottish islands, the relationship (Figure 6.17b) is

$$S = 151.51A^{0.18}$$

A large number of studies have established the validity of this kind of relationship.

As a rule, the number of species living on islands doubles when habitat area increases by a factor of ten. For islands, values of the exponent z normally range from 0.24 to 0.34. In mainland areas, z-values normally fall within the range 0.12 to 0.17 (the value of 0.27 for Hertfordshire plants may result from the high geodiversity of that county). This means that small areas contain almost as many species as large areas, but small islands contain fewer species than large islands. The difference may be partly attributable to relative isolation of islands, which makes colonization more difficult than on mainlands.

The habitat diversity and area-alone hypotheses

A crucial question is why there is such a good relationship between area and species diversity. The answer is not simple. Two rival hypotheses have emerged (Gorman 1979: 24). The **habitat diversity hypothesis** suggests that larger areas have more habitats and therefore more species. The **area-alone hypothesis** proposes that larger areas should carry more species, regardless of habitat diversity. A study of dicotyledonous plant species on forty-five un-inhabited, unimproved, small islands off Shetland Mainland, Scotland, plus two similar headlands treated as islands, sought to test the two hypotheses (Kohn and Walsh 1994). Species lists were complied during a systematic transect search covering each island. A second species list was derived by placing 50-cm × 50-cm square quadrats randomly on twenty-two islands sufficiently vegetated for a reasonable number of samples. Habitats were classified according to physical characteristics assumed to be

important for plants. Fourteen habitat types were identified (Table 6.4). The results show strong positive correlations between species per island, island area, and habitat diversity (Figure 6.18). Habitat diversity itself correlated with island size. A technique called path analysis was used to distinguish the effect of island size on species diversity from the direct effect of habitat diversity. The sum of the direct effects (area acting through area alone) and indirect effects (area acting through habitat diversity) of area were almost twice the overall effect of habitat diversity on species diversity. This finding strongly suggests a direct relationship between island area and species diversity, independent of habitat diversity.

Diversity gradients and hot-spots

The latitudinal diversity gradient

Many more species live in the tropics than live in the temperate regions, and more species live in temperate regions than live in polar regions. In consequence, a **latitudinal species diversity gradient** slopes steeply away from a tropical 'high diversity plateau'. The diversity gradient is seen in virtually all groups of organism. An example is the species richness gradient of American mammals (Figure 6.19). The diversity falls from a tropical high of about 450 mammal species to a polar low of about 50 mammal species.

Why are there so many species (and genera, and families, and orders) in the tropics? Or, conversely, why are there so few species in temperate and polar regions? These are fundamental questions in biogeography and ecology and have exercised the minds of researchers for over a century. There is no shortage of suggested answers (Table 6.5). A survey of twenty hypotheses suggests that each one contains an element of circularity or else is not supported by sufficient evidence (Rohde 1992). Hypotheses flawed by circular reasoning include biotic habitat diversity, competition, niche width, and predation. Hypotheses that lack an empirical base include: climatic stability, climatic variability, latitudinal ranges, area, geodiversity, and primary production. It is possible that none of the

Table 6.4 Habitat classification used in Shetland study

Class	Type		Number
Rocks	Sea cliff		1
	Scree		2
	Boulder field		3
Pasture	General, dry		4
	Rocky pasture		5
	Wet pasture:	Damp depressions, runnels	6
		Moss-dominated wet ground	7
		Mud	8
		Bog	9
		Open standing water	10
		Stream	11
	Grassy cliff		12
Shingle	Shingle		13
Sand	Sand		14
Total number of habitats			14

Source: After Kohn and Walsh (1994)

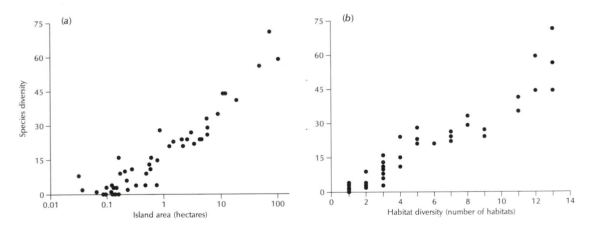

Figure 6.18 Results of the Shetland study. (a) Species diversity versus island area. (b) Species diversity versus habitat diversity.
Source: After Kohn and Walsh (1994)

proposed causes can alone account for the latitudinal species gradient. A universal explanation may lie in the changing impact of abiotic and biotic factors in moving from equator to poles (Kaufman 1995).

The relative importance of abiotic and biotic factors in determining species distributions is shown in Figure 6.20. In polar regions, productivity is low and extreme fluctuations in abiotic conditions occur. These extreme environmental conditions result in physiological stress in organisms. Such adverse conditions must be countered by adaptations in structure, physiology, and behaviour, which all require energy

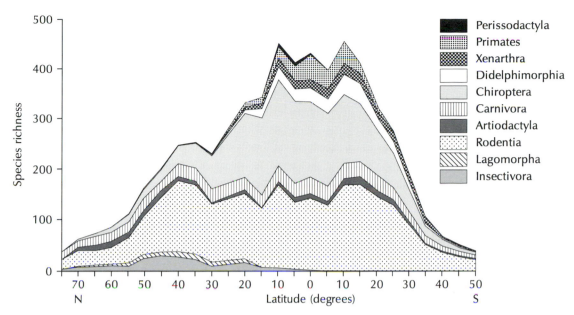

Figure 6.19 Latitudinal diversity gradient mammal species richness in the Americas. The species are grouped into Orders. Two Orders are omitted because they contain too few species. The Microbiotheria contains one species that ranges from 35°S to 45°S. The Paucituberculata contain three species found in bands from 15°N to 20°S and from 30°S to 45°S. Notice that species richness is fairly constant on the 'tropical species richness plateau' and declines steeply with increasing latitude outside the tropics.
Source: After Kaufman (1995)

expenditure. Abiotic factors are most likely to be severely limiting in polar regions and relax towards the equator. Contrariwise, biotic interactions are less limiting in polar regions because species diversity is low. They become severer towards the equator. As it becomes easier to cope with abiotic conditions, species can devote more resources to interacting with other species, that is to competing. By these arguments, in passing along the gradient from the abiotically stressful poles to the more abiotically congenial tropics, biotic interactions should become increasingly important in limiting species distributions and influencing species diversity.

A general explanation for latitudinal gradients emerges from these ideas (Kaufman 1995). Abiotic conditions limit diversity by setting the higher latitude boundaries of species distributions, and permit only a few species to live near the poles. Biotic interactions become limiting where abiotic conditions are more favourable, set the lower latitude boundaries of

species distributions, and allow many species to live in the tropics. Ultimately, the interplay of abiotic and biotic factors generates the latitudinal gradient in species diversity. Latitudinal diversity gradients occur in genera, families, and orders, as well as in species (Box 6.2).

Diversity hot-spots

Superimposed on latitudinal and altitudinal trends, are **diversity hot-spots**. These are areas where large numbers of endemic species occur. Eighteen hot-spots have been identified globally. Fourteen occur in tropical forests and four in Mediterranean biomes. These hot-spots contain 20 per cent of the world's plant species on 0.5 per cent of the land area. All are under intense development pressure.

Table 6.5 Some factors thought to influence species diversity gradients

Factor	Rationale
History	More time allows more colonization and the evolution of new species. In polar and temperate regions, diversity was greatly reduced during the ice ages and is now building up again
Climate	The warm, wet, and equable tropical conditions encourage a smaller niche breath, and therefore more species, than the colder and highly seasonal conditions elsewhere
Climatic stability	Tropical climatic stability is conducive to species specialization (smaller niche width) and therefore more species
Habitat heterogeneity	The greater the habitat diversity, the greater the species diversity. Forests contain more niches than grasslands. Tropical forests contain more niches than any other biome
Primary production	In food-deficient habitats, animals cannot be too choosy about their prey; in food-rich habitats, they can be selective. So food-rich environments allow greater dietary specialization and smaller niche width
Primary production stability	Habitats with stable and predictable primary production should allow greater dietary specialization (smaller niche width) than habitats with more variable and erratic primary production
Competition	Competition, which is most intense in the tropics, favours reduced niche width
Predation	Predation reduces competitive exclusion. Predators are therefore 'rarefying agents', reducing the level of competition between their prey species
Disturbance	Moderate disturbance mitigates against competitive exclusion
Energy	Species richness is limited by the partitioning of energy among species. In the energy-rich tropics, there is more energy to 'dish out' and it can be spread around a larger number of species than in temperate and polar regions
Latitudinal range size	Range size tends to increase towards the poles (Rapoport's rule), so fewer species can be packed into a given area
Area	The tropics cover more area than any other zone, which stimulates speciation and inhibits extinction (Rosenzweig 1992, 1995)

Diversity change

Change in species diversity depends upon gains and losses from the global diversity pool. Species are added by speciation and lost by extinction. Plainly, global biodiversity will rise whenever the speciation rate exceeds the extinction rate; and it will fall when the extinction rate exceeds the speciation rate. A point often overlooked in discussions of mass extinctions is that they could result from a reduction in the speciation rate while the extinction rate stays the same – they may not simply result from an elevated extinction rate.

Complications arise when the biodiversity of a

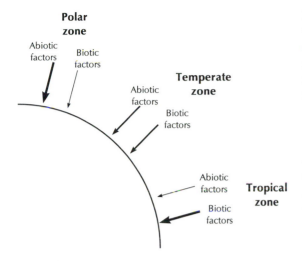

Figure 6.20 Controls on biotic diversity in polar, temperate, and tropical zones.
Source: After Kaufman (1995)

region is considered. In this case, immigration and emigration of species enter the equation. The biodiversity of the British Isles will depend upon the balance between the speciation rate and immigration rate on the one hand, and the extinction rate and the emigration rate on the other hand. These relationships are analogous to change in populations (Chapter 4). Speciation is equivalent to births, extinction to deaths, and immigration and emigration apply to species rather than to individuals.

Why all the fuss?

It is the fashion to bandy figures about extinction rates in the present wildlife holocaust. Extinction is said to be running at 25,000 times the natural rate. By the year 2000, one million animals and plants will have been driven to extinction, and the extinction rate will have soared to several species per hour! By 2050, half the species alive today could have gone. And, as they say, extinction is forever.

Although some ecologists are stepping back from these hyperbolic estimates, there is little doubt that extinction is running very fast at present and will continue to do so. The chief culprit is humans.

Pollution, overhunting, overfishing, the wildlife trade, and above all, habitat destruction (land cover change – the conversion of wild habitats for farming, fuel extraction, industry, and residential land) are the main problems. The main habitats lost are tropical rain forests, wetlands, and coral reefs.

Does biodiversity matter?

Although it is rather environmentally hostile to do so, it is worth asking if loss of biodiversity matters. Is the b-word simply a buzzword, or is there something important hiding behind the biodiversity rhetoric? This question may be answered by asking two questions. First, is the current biotic crisis unique? Second, what stands to be lost?

The only yardstick for comparing the current extinction rate is the rate of species loss during the mass extinctions that have occurred through geological time. Several past mass extinctions, though described as 'catastrophic', took hundreds of thousands of years. The Pleistocene mass extinctions were somewhat speedier. Some evidence suggests that the current loss of biodiversity is 'the biggest mass-extinction of them all' (Leggett 1989).

A loss of biodiversity should perhaps be deplored on moral grounds – most people would accept that all creatures have a right to exist. In addition, and this seems to carry far more weight with decision-makers, a loss of diversity is accompanied by a loss of genetic resources (genes, genomes, and gene pools), medicinal substances, and potential foodstuffs. It is surprising how dependent the pharmaceutical industry is upon the natural world. In the United States, of all prescriptions dispensed by pharmacies, 25 per cent are derived from plants, 13 per cent from microorganisms, and 3 per cent from animals. As for foodstuffs, the world currently depends on just twenty species to provide 90 per cent of its food. Three species – wheat, maize, and rice – account for over half the world's food production. The production of these foodstuffs is biased towards cooler climates and the crops are normally sown as monocultures that bring attendant problems of sensitivity to disease and attacks from insects and nematode worms. Many

Box 6.2

LATITUDINAL DIVERSITY GRADIENTS, IN GENERA, FAMILIES, AND ORDERS

Animals within genera, families, and orders share a basic body plan (bauplan) and many other characteristics that constrain their distribution and diversity (Kaufman 1995). Members of the same genus are alike in many particulars of their morphology, physiology, behaviour, and ecology. The grass voles, genus *Microtus*, are the dominant voles in European grasslands. They feed on grasses and sedges, and have ever-growing cheek teeth with a sharpy angular triangular pattern on the grinding surfaces. Members of the same Order share a basic suite of characteristics. All bats (Chiroptera), for instance, share traits associated with flight. Body plans are not limited by biotic interactions. Rather, they are constrained by abiotic factors – a body plan suitable for an aquatic existence is hardly suited to life in deserts (there are no desert-dwelling seals!). Abiotic factors have a top-down influence on the taxonomic hierarchy. They influence geographical distributions most strongly at the level of orders, followed

in decreasing strength by families, genera, and species. Just a few body plans can survive under the severe abiotic stresses of high latitudes, and those body plans tend to be generalized. So, abiotic factors act absolutely on diversity, limiting the number of body plans, and, because a species cannot exist where its body plan cannot exist, the number of species. The seventeen bat species found north of 45°N belong to one family body plan – the Vespertilionidae. Likewise, the eight xenarthran species found south of 35°S all belong to the family Dasypodidae. Conversely, biotic factors have a bottom-up influence on the taxonomic hierarchy, and their primary effects are felt at the species level. They do not limit the number of species per se, but they do limit each species individually by influencing population dynamics and niche width. Biotic factors lead to specialization and species packing. Biotic factors influence macrotaxonomic distribution only if the body has a feature that affects the interactions among species.

Each species and body plan trades off a resistance to abiotic pressures at its polar-most edge and to biotic pressures at its equatorial-most edge. The

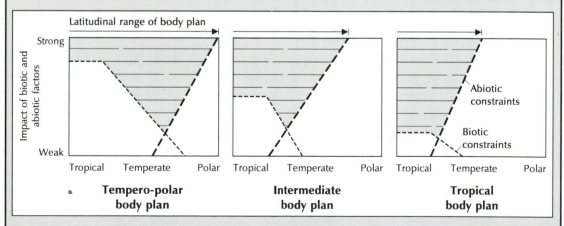

Figure 6.21 Latitudinal variations in abiotic pressure and biotic pressure acting on species of three hypothetical body plans – a tempero-polar body plan, an intermediate body plan, and a tropical body plan. The horizontal lines in the shaded areas represent latitudinal species ranges within a body plan.
Source: After Kaufman (1995)

trade-off strategies of species and body plans produce different latitudinal distributions (Figure 6.21). Abiotic limitations operate to a significant effect only outside the tropics. Biotic limitations have a nearly constant effect in the tropics and then tail off towards either pole. Combined, abiotic and biotic factors limit species distributions within a body plan, and influence the body plan itself. Body plans occupy larger latitudinal ranges than species, because most body plans consists of many species. Body plans tend to be limited only by abiotic factors. Normally, they straddle the tropics and are limited by abiotic factors at each polar-most edge of their distribution. On the other hand, species are potentially limited by abiotic and biotic pressures. Therefore, they have reduced distributions, compared with their body plans, owing to the trade-offs in adapting both these pressures.

alternative food sources are available, though the menu will shorten as biodiversity drops.

What can be done to safeguard biodiversity?

Four things, principally – take conservation measures, counter the wildlife trade, study diversity, and sell a greener environmental ethic. Conservation is a vast topic and beyond the scope of this book. A large amount of time and effort goes into saving species (e.g. elephants and whales), saving habitats (e.g. wetlands in Irian Jaya and Zambia), saving forests (e.g. in Ecuador, Costa Rica, and the Philippines), and saving islands (Madagascar and the Galápagos). Three main conservation measures are currently being taken. First, the establishing of protected areas and national parks. At present, 5 per cent of the land area is occupied by nature reserves, national parks, wildlife sanctuaries, and protected landscapes. Second, the setting up of IUCN/WWF (International Union for Conservation of Nature and Natural Resources/World Wildlife Fund) Centres of Plant Diversity. Some 250 sites and regions of high plant diversity would protect 90 per cent of world plant species. Third, the designating of more UNESCO Biosphere Reserves.

SUMMARY

Communities are collections of interacting populations living in a particular place. Ecosystems are communities and their physical environment. Communities and ecosystems range in size from a few cubic centimetres to the entire biosphere and ecosphere. Organisms have roles within a community. Some are producers, making biomass from energy and mineral resources. Gross primary production is the amount of biomass produced in a unit time. Net primary production is gross primary production less the energy burned by primary producers. The primary production is eaten by consumers. Macroconsumers are herbivores, carnivores, and top carnivores. Microconsumers are decomposers and detritivores. Energy flows through ecosystems. Biogeochemicals cycle around ecosystems, mineralization of dead organic remains providing food for renewed plant growth. Feeding relationships within communities produce food chains or food webs. Grazing food webs involve plant → herbivore → carnivore → top carnivore sequences. Decomposer food webs involve comminution spirals. Food chains have characteristic 'pyramids' of biomass, energy, and numbers. Keystone species are crucial to certain communities. Take them away and the community collapses or changes drastically. Humans have contaminated many food webs, largely because toxic substances become increasingly more concentrated as they move up the trophic levels. Biodiversity is a pressing concern. It has a consistent relationship with increasing area, as seen in species–area curves. There is a latitudinal diversity gradient – diversity decreases towards the poles from a tropical high-diversity plateau. There are also several diversity hot-spots. Biodiversity is currently falling, owing largely to habitat fragmentation. The loss of biodiversity appears to be real, despite all the hype, and measures are in hand to minimize losses during the next century.

ESSAY QUESTIONS

1 What are the similarities and differences between grazing food chains and detrital food chains?

2 How global are global biogeochemical cycles?

3 Why does it matter if global biodiversity falls?

FURTHER READING

Archibold, O. W. (1994) *Ecology of World Vegetation*, New York: Chapman & Hall.

Chameides, W. L. and Perdue, E. M. (1997) *Biogeochemical Cycles: A Computer-Interactive Study of Earth System Science and Global Change*, New York: Oxford University Press.

Hochberg, M. E., Colbert, J., and Barbault, R. (eds) (1996) *Aspects of the Genesis and Maintenance of Biological Diversity*, Oxford: Oxford University Press.

Jeffries, M. L. (1997) *Biodiversity and Conservation*, London and New York: Routledge.

Lawton, J. H. and May, R. M. (eds) (1995) *Extinction Rates*, Oxford: Oxford University Press.

Morgan, S. (1995) *Ecology and Environment: The Cycles of Life*, Oxford: Oxford University Press.

Polis, G. A. and Winemiller, K. O. (1995) *Food Webs: Integration of Patterns and Dynamics*, New York: Chapman & Hall.

Quammen, D. (1996) *The Song of the Dodo: Island Biogeography in an Age of Extinctions*, London: Hutchinson.

Reaka-Kudlka, M. L., Wilson, D. E., and Wilson, E. O. (1997) *Biodiversity II: Understanding and Protecting Our Biological Resources*, Washington, DC: National Academy Press.

Rosenzweig, M. L. (1995) *Species Diversity in Space and Time*, Cambridge: Cambridge University Press.

Schultz, J. (1995) *The Ecozones of the World: The Ecological Divisions of the Geosphere*, Hamburg: Springer.

Szaro, R. C. and Johnston, D. W. (1996) *Biodiversity in Managed Landscapes: Theory and Practice*, New York: Oxford University Press.

Wilson, E. O. (1992) *The Diversity of Life*, Cambridge, Massachusetts: Belknap Press of Harvard University Press.

7

COMMUNITY CHANGE

Communities seldom stay the same for long – community change is the rule. This chapter covers:

- communities and ecosystems in balance
- communities and ecosystems out of balance
- human-induced community change
- communities and global warming

As Nature abhors a vacuum, so the biosphere abhors bare ground. Land stripped clean by fire, flood, plough, or any other agency is soon colonized by life. What happens next is the subject of considerable debate. Proponents of two main schools of thought have their own views on the matter. These schools may be dubbed the 'climatic climax and balanced ecosystem' school, and the 'disequilibrium communities' school.

THE BALANCE OF NATURE: EQUILIBRIUM COMMUNITIES

Climatic climax

This influential school of thought had its heyday in the 1920s and 1930s. It stressed harmony, balance, order, stability, steady-state, and predictability within plant and animal communities. Its chief exponent was Frederic E. Clements, an American botanist. Clements (1916) reasoned that the first plant colonists are good at eking out a living under difficult conditions. By altering the local environment, they pave the way for further waves of colonization by species with a less pioneering spirit. And the process continues, each new group of species changing the environment in such a way as to entice further colonists, until a stable community evolves whose equilibrium endures. The full sequence of change is called **vegetation succession**.

Clements believed in the idea of **monoclimax**. He argued that a prisere was a developmental sequence that, for given climatic conditions, would always end in a reasonably permanent stage of succession – **climatic climax vegetation**. However, he realized that climatic climax formations contain patent exceptions to this rule. He termed these aberrant communities proclimax states. There are four forms of proclimax state (Figure 7.1): subclimax, disclimax, preclimax, and postclimax (anticlimax is something very different):

1 **Subclimax** is the penultimate stage of succession in all complete primary and secondary seres. It persists for a long time but is eventually replaced by the climax community. In eastern North America, an Early Holocene coniferous forest eventually gave way to a deciduous forest. Clementsians would therefore conclude that the coniferous forest was a subclimax community.

2 **Disclimax (plagioclimax)** partially or wholly replaces or modifies the true climax after an environmental disturbance. Most of the so-called

natural grasslands in the British Isles lying below 500 m are maintained by cattle and sheep grazing. If grazing were to cease, the grasslands would revert to a deciduous forest. In the arid and semi-arid United States, vegetation succession around some ghost towns is permanently altered by an annual grass – cheatgrass (*Bromus tectorum*) – introduced from the Mediterranean (Knapp 1992).

3   **Preclimax** and **postclimax** are caused by local conditions, often topography, producing climatic deviations from the regional norm. Cooler or wetter conditions promote postclimax vegetation; drier or hotter conditions promote preclimax vegetation. In the Glacier Bay area, south-east Alaska, the climax vegetation is spruce (*Sitka sitchensis*)–hemlock (*Tsuga* spp.) forest, but at wetter sites bog mosses (*Sphagnum* spp.) dominate with occasional shore pine (*Pinus contorta*) trees to form a postclimax vegetation (see pp. 174–7).

Clements's monoclimax hypothesis was superseded by a **polyclimax hypothesis** (e.g. Tansley 1939). According to the polyclimax hypothesis, several different climax communities could exist in an area with the same regional climate owing to differences in soil moisture, nutrient levels, fire frequency, and so on. The **climax-pattern hypothesis** (R. H. Whittaker 1953) was a variation on the polyclimax theme. As in the polyclimax hypothesis, it argued that natural communities are adapted to all environmental factors. But it suggested that there is a continuity of climax types that grade into one another along environmental gradients, rather than forming discrete communities that change through very sharp ecotones.

Balanced ecosystems

The idea of **balanced ecosystems** emerged in the 1940s and persisted until the late 1960s. It did not

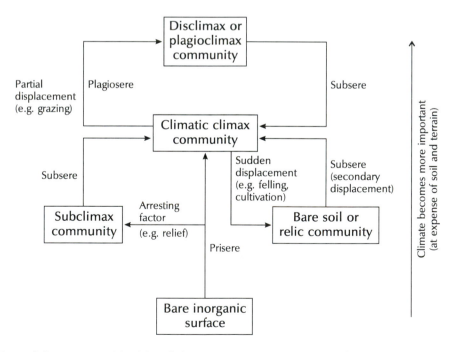

Figure 7.1 Types of climax communities. The subclimax community is also called 'deflected succession'.
Source: After Eyre (1963)

differ radically from the idea of Clements's enduring climaxes. But it did switch the emphasis from plant formations to ecosystems and related ideas of energy flow, trophic levels, homeostasis, and r- and K-selection. An ecosystem was defined as a biotic community together with its immediate life-support system (soil, water, and air) (Tansley 1935). During the next four decades, the ecosystem idea was promoted and refined by Raymond L. Lindeman, G. Evelyn Hutchinson, and, most persuasively, Eugene P. Odum. To Odum, each ecosystem has a strategy of development that leads to mutualism and co-operation between individual species; in other words, that leads to a balanced ecosystem. The **Gaia hypothesis** is, in part, an extension of this idea. It sees an overall **homeostasis** in the ecosphere, suggesting that life and its supporting environment are a globally balanced ecosystem (e.g. Lovelock 1979).

From bare soil to climax

Three mechanisms are thought to drive succession – facilitation, tolerance, and inhibition (Horn 1981). Researchers are divided over the relative importance of these mechanisms, so there are three main models of succession.

Facilitation model

The pioneer species make the habitat less suitable for themselves and more suitable for a new round of colonists. The process continues, each group of species facilitating the colonization of the next group. This is the classical model of succession, as expounded by Henry C. Cowles and Clements. According to Clements, vegetation succession involves a predetermined sequence of developmental stages, or **sere**, that ultimately leads to a self-perpetuating, stable community called climatic climax vegetation. He recognized six stages in any successional sequence:

1 **Nudation** – an area is left bare after a major disturbance.
2 **Migration** – species arrive as seeds, spores, and so on.

3 **Ecesis** – the plant seeds establish themselves.
4 **Competition** – the established plants complete with one another for resources.
5 **Reaction** – the established plants alter their environment and so enable other new species to arrive and establish themselves.
6 **Stabilization** – after several waves of colonization, an enduring equilibrium is achieved.

Reaction helps to drive successional changes. On a sandy British beach, the first plant to colonize is marram-grass (*Ammophila arenaria*). After a rhizome fragment takes hold, it produces aerial shoots. The shoots impede wind flow and sand tends to accumulate around them. The plant is gradually buried by the sand and, to avoid 'suffocation', grows longer shoots. The shoots keep growing and the mound of sand keeps growing. Eventually, a sand dune is produced. This is then colonized by other species, including sand fescue (*Festuca rubra* var. *arenaria*), sand sedge (*Carex arenaria*), sea convolvulus (*Calystegia soldanella*), and two sea spurges (*Euphorbia paralias* and *Euphorbia portlandica*), that help to stabilize the sand surface.

Tolerance model

According to this model, late successional plants, as well as early successional plants, may invade in the initial stages of colonization. In northern temperate forests, for example, late successional species appear almost as soon as early successional species in vacant fields. The early successional plants grow faster and soon become dominant. But late successional maintain a foothold and come to dominate later, crowding out the early successional species. In this model, succession is a thinning out of species originally present, rather than an invasion by later species on ground prepared by specific pioneers. Any species may colonize at the outset, but some species are able to outcompete others and come to dominate the mature community.

Inhibition model

This model takes account of chronic and patchy disturbance, a process that occurs when, for example, strong winds topple trees and create forest gaps. Any species may invade the gap opened up by the toppling of any other species. Succession in this case is a race for uncontested dominance in recent gaps, rather than direct competitive interference. No species is competitively superior to any other. Succession works on a 'first come, first served' basis – the species that happen to arrive first become established. It is a disorderly process, in which any directional changes are due to short-lived species replacing long-lived species.

Allogenic and autogenic succession

Autogenic succession is propelled by the members of a community. In the facilitation model, autogenic succession is a unidirectional sequence of community, and related ecosystem, changes that follow the opening up of a new habitat. The sequence of events takes place even where the physical environment is unchanging. An example is the heath cycle in Scotland (Watt 1947). Heather (*Calluna vulgaris*) is the dominant heathland plant. As a heather plant ages, it loses its vigour and is invaded by lichens (*Cladonia* spp.). In time, the lichen mat dies, leaving bare ground. The bare ground is invaded by bearberry (*Arctostaphylos uva-ursi*), which in turn is invaded by heather. The cycle takes about twenty to thirty years.

Allogenic succession is driven by fluctuations and directional changes in the physical environment. A host of environmental factors may disturb communities and ecosystems by disrupting the interactions between individuals and species. When a stream carries silt into lake, deposition occurs. Slowly, the lake may change into a marsh or bog, and in time the marsh may become dry land.

Primary succession

Clements distinguished between primary succession and secondary succession. **Primary succession** occurs on newly uncovered bare ground that has not supported vegetation before. New oceanic islands, ablation zones in front of glaciers, developing sand dunes, fresh river alluvium, newly exposed rock produced by faulting or volcanic activity, and such human-made features as spoil heaps are all open to first-time colonization. The full sequence of communities form a **primary sere** or **prisere**. Different priseres occur on different substrates: a **hydrosere** is the colonization of open water; a **halosere** is the colonization of salt marshes; a **psammosere** is the colonization on sand dunes; and a **lithosere** is the colonization of bare rock.

A hydrosere is recognized in the fenlands of England (e.g. Tansley 1939). Open water is colonized by aquatic macrophytes, and then by reeds and bulrushes. Decaying organic matter from these plants accumulates to create a reed swamp in which water level is shallower. Marsh and fen plants become established in the shallower water. Further accumulation of soil leads to even shallower water. Such shrubs as alder then invade to produce 'carr' (a scrub or woodland vegetation), and ultimately, so the classic interpretation claimed, carr would change into mesic oak woodland.

Glacier Bay, Alaska

A classic study of primary succession was made in Glacier Bay National Park, south-east Alaska (Cooper 1923, 1931, 1939; Crocker and Major 1955). The glacier has retreated considerably over the last 240 years (Figure 7.2; Plate 7.1). The pioneer stage is characterized by roadside rock moss (*Racomitrium canescens*) and hoary rock moss (*Racomitrium lanuginosum*), broad-leaved willow herb or river beauty (*Epilobium latifolium*), northern scouring rush (*Equisetum variegatum*), yellow mountain avens (*Dryas drummondii*), and the Arctic willow (*Salix arctica*). In the next stage, the Barclay willow (*Salix barclayi*), Sitka willow (*Salix sitchensis*), and feltleaf willow (*Salix alaxensis*), undergreen willow (*Salix commutata*), and other willows appear. They start as prostrate forms but eventually develop an erect habit, forming dense scrub. Later stages of succession vary from place

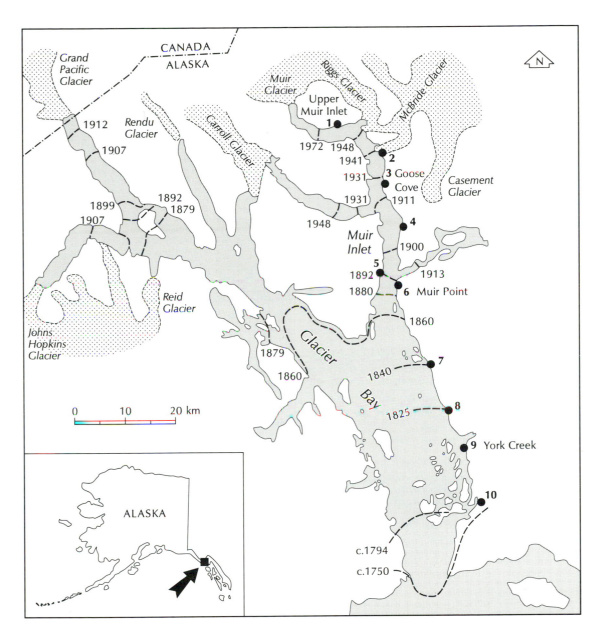

Figure 7.2 Glacier Bay, Alaska, showing the positions of the glacier termini and Fastie's (1995) study sites (referred to on p. 176).
Source: After Crocker and Major (1955) and Fastie (1995)

(a)

(b)

(c)

(d)

Plate 7.1 Plant succession at Glacier Bay, Alaska (a) Upper Muir inlet (Site 1 in Figure 7.2), 20 years old: glacial till surface with young plants of yellow mountain avens (*Dryas drummondii*), black cottonwood (*Populus trichocarpa*) and Sitka spruce (*Salix sitchensis*) (b) Goose Cove (Site 3 in Figure 7.2), 60 years old: alder thicket overtopped by black cottonwood (c) Muir Point (Site 6 in Figure 7.2), 105–110 years old: alder and willow thicket overtopped by scattered Sitka spruce (d) York Creek (Site 9 in Figure 7.2), 165 years old: Sitka spruce forest (with western hemlock) that was not preceded by a dense alder thicket or cottonwood forest
Photographs by Christopher L. Fastie

to place. They involve three main changes in vegetation. First, green or mountain alder (*Alnus crispa*) establishes itself in some areas. On the east side of Glacier Bay (Muir Inlet) and within 50 years, it forms almost pure thickets, some 10 m tall, with scattered individuals of black cottonwood (*Populus trichocarpa*). Second, Sitka spruce (*Picea sitchensis*) invades and after some 120 years forms dense, pure stands. The spruce finds it difficult to establish itself in those areas dom-

inated by alder thickets. Third, western hemlock (*Tsuga heterophylla*) enters the community, arriving soon after the spruce. Eventually, after a further 80 years or so, spruce–hemlock forest is the climax vegetation, at least on well-drained slopes. In wetter sites, where the ground is gently sloping or flat, *Sphagnum* species invade the forest floor. The more *Sphagnum* there is, the wetter the conditions become. Trees start to die and wetland forms. The wetland is dominated by *Sphagnum*, with occasional shore pine individuals (*Pinus contorta*), which are tolerant of the wet habitat.

Krakatau Islands, Indonesia

The eruptions of 20 May to 27 August 1883 largely or completely sterilized and greatly reshaped the islands in the Krakatau group. There are four islands in the group – Rakata, Sertung, Panjang, and Anak Krakatau. They lie in the Sunda Strait roughly equidistant 44 km from 'mainland' Java and Sumatra (Figure 7.3). Two large 'stepping stone' islands – Sebesi and Sebuku – connect them with Sumatra.

Rakata, Sertung, and Panjang were sterilized in 1883. Anak Krakatau emerged in 1930 and has a disturbed history of colonization.

Community succession on each of the three main islands follows a similar pattern for the first 50 years. The coastal communities were established rapidly. Typical dominant strand-line species were among the first to colonize, producing a one- or two-phase succession. In the interior lowlands and upland Rakata, an early phase of highly dispersive ferns, grasses (carried by winds), and a few Compositae slowly diversified. By 1897 they had gained such species as the heliophilous (sun-loving) terrestrial orchids

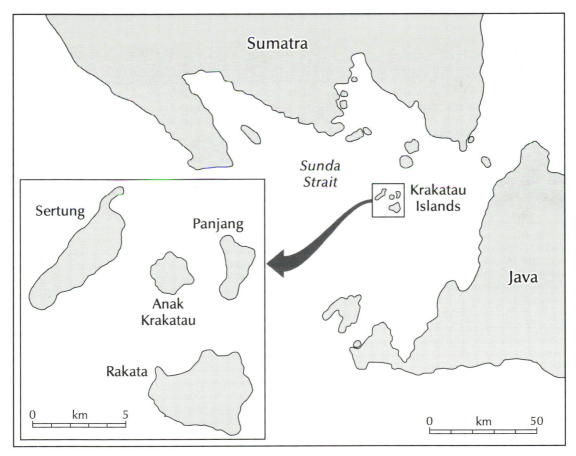

Figure 7.3 The location of the Krakatau island group. Rakata, Panjang, and Sertung are remnants of the pre-1883 volcano; Anak Krakatau, which appeared in 1930, is the only volcanically active island.
Source: After Whittaker and Bush (1993)

Arundina graminifolia, *Phaius tankervilliae*, and *Spathoglottis plicata*. A scattering of shrubs and trees, most of which were animal-dispersed colonists that were interspersed among the savannah vegetation, soon spread. Forest covered Rakata by the end of the 1920s. The forest closure led to the loss of a number of early colonists that were open-habitat species. Additionally, open areas were reduced to relatively small gaps and fauna dependent on open areas suffered. Losses included birds such as the red-vented bulbul (*Pycnonotus cafer*) and the long-tailed shrike (*Lanius schach bentet*).

Secondary succession

Secondary succession occurs on severely disturbed ground that previously supported vegetation. Fire, flood, forest clearance, the removal of grazing animals, hurricanes, and many other factors may inaugurate secondary succession.

Abandoned fields in Minnesota

Before the 1880s, upland habitats in the Cedar Creek Natural History Area, Minnesota, were a mosaic of oak savannah (open oak woodland), prairie openings, and scattered stands of oak forest, pine forest, and maple forest (Tilman 1988). Farming converted some of the land to fields. Some fields were abandoned at different times and secondary succession started. A set of twenty-two abandoned fields, placed in chronological order, provides a picture of the successional process (Figure 7.4). The initial dominants are annuals and short-lived perennials, many of which, including ragweed (*Ambrosia artemisiifolia*), are agricultural weeds. These are replaced by a sequence of perennial grasses. After fifty years, the dominant grasses are the little bluestem (*Schizachyrium scoparium*) and big bluestem (*Andropogon gerardi*), both native prairie species. Woody plants – mainly shrubs, vines, and seedlings or saplings of white oak (*Quercus alba*), red oak (*Quercus rubra*), and white pine (*Pinus strobus*) – slowly increase in abundance. After sixty years, they account for about 12 per cent of the cover. Many of the successional changes are explained by an increase of soil nitrogen over the sixty-year period.

Much of the pattern reflects the differing life histories and time to maturity of the component species.

Ghost towns in the western Great Basin

Terrill and Wonder, which are in central-western Nevada, United States, are abandoned gold- and silver-mining camps. Terrill was abandoned in about 1915 and Wonder in about 1925. Secondary succession following abandonment was studied for two levels of disturbance – greatly disturbed (abandoned roads) and moderately disturbed (within 5 m of building foundation edges) – and at 'undisturbed' control plots (Knapp 1992). Twenty-seven species were found at Terrill. Cheatgrass (called drooping brome-grass in the United Kingdom) (*Bromus tectorum*), Shockley's desert thorn (*Lycium shockleyi*), indian ricegrass (*Oryzopsis hymenoides*), and Bailey's greasewood (*Sarcobatus baileyi*) dominated the abandoned road (Plate 7.2a). Shadscale (*Atriplex confertifolia*), cheatgrass, and *Halogeton glomeratus* (an introduced Asiatic annual) dominated the foundation edges (Plate 7.2b). The control plot was dominated by greasewood, shadscale, and cheatgrass. At Wonder, just seven species were found. Big sagebrush (*Artemisia tridentata*) was the dominant species in the abandoned road and control site (Plate 7.2c). Big sagebrush and cheatgrass comprised 90 per cent of the cover around foundation edges (Plate 7.2d).

In both towns, a striking feature of succession is the higher percentages of therophytes around foundation peripheries. This is seen in the high percentage of cheatgrass at Terrill and Wonder, and in the high percentages of *Halogeton* and tumbleweed (*Salsola kali*) at Terrill (Plate 7.2). They are not so common on highly disturbed abandoned roads because soil bulk density is higher there, which hinders the growth of tumbleweed's and cheatgrass's extensive root systems, and because phosphorus and nitrogen levels are low, which hampers the establishment of cheatgrass. Vegetation has reverted to something approaching its original state (salt desert shrub in the case of Terrill and sagebrush–grass in the case of Wonder), but complete recovery to a 'climax' state is unlikely. Therophytes, especially cheatgrass, are likely to persist.

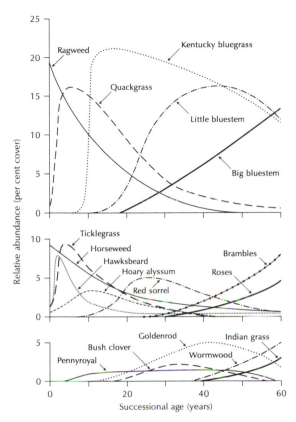

Figure 7.4 Secondary succession in Cedar Creek Natural History Area, Minnesota, United States. The diagram is based on an old field chronosequence and observations within some permanent plots. The species are quackgrass (*Agropyron repens*), ragweed (*Ambrosia artemisiifolia*), big bluestem (*Andropogon gerardi*), Kentucky bluegrass (*Poa pratensis*), little bluestem (*Schizachyrium scoparium*); ticklegrass (*Agrostis scabra*), hoary alyssum (*Berteroa incana*), hawksbeard (*Crepis tectorum*), horseweed (*Erigeron canadensis*), red sorrel (*Rumex acetosella*), brambles (*Rubus* spp.), roses (*Rosa* spp.); wormwood (*Artemisia ludoviciana*), pennyroyal (*Hedeoma hispida*), bush clover (*Lespedeza capitata*), goldenrod (*Solidago nemoralis*), Indian grass (*Sorghastrum nutans*). *Source*: After Tilman (1988)

Cheatgrass excels in disturbed areas. It establishes itself easily on disturbed sites and creates a self-enhancing cycle by promoting the occurrence of additional disturbances – fire intensity and frequency and grazing by small mammals. So, cheatgrass, which is a Mediterranean introduction (p. 172), will probably stay dominant at the expense of other, native,

species. The secondary succession has produced a new 'climax' vegetation, a sort of permanent disclimax.

EGOCENTRIC NATURE: DISEQUILIBRIUM COMMUNITIES

The idea that communities and ecosystems might not be in equilibrium, at least not in Clements's sense of stable climax communities or Odum's sense of balanced ecosystems, was first raised by Henry Allan Gleason and Alexander Stuart Watt in the 1920s. The **disequilibrium view** rose to stardom in the early 1970s, when some ecologists dared to suggest that succession leads nowhere in particular, that there are no long-lasting climatic climaxes (e.g. Drury and Nisbet 1973). Instead, each species 'does its own thing', communities are ever-changing, temporary alliances of individuals, and succession runs in several directions. This disequilibrium view emphasizes the individualistic behaviour of species and the evolutionary nature of communities. It stresses imbalance, disharmony, disturbance, and unpredictability in Nature. And it focuses on the geography of ecosystems – landscape patches, corridors, and matrixes replace the climax formation and ecosystem, and landscape mosaics replace the assumed homogeneous climaxes and ecosystems.

The individualistic comings and goings of species influence community change in a profound way. Communities change because new species arrive and old species are lost. New species appear in speciation events and through immigration. Old species vanish through local extinction (extirpation) and through emigration. Some species increase in abundance and others decrease in abundance, thus tipping the competitive balance within a community. Each species has its own propensity for dispersal, invasion, and population expansion. **Community assembly** is an unceasing process of species arrivals, persistence, increase, decrease, and extinctions played out in an individualistic way. Evidence that communities assemble (and disassemble) in this manner is seen in multidirectional succession, as revealed in recent vegetation chronosequences, and in community

(a)

(b)

(c)

(d)

Plate 7.2 Secondary succession in Great Basin ghost towns, United States (a) Terrill – view looking west. The 'main street' on the right is barely visible. It leads to the sole remaining structure. In the distance is the Terrill water tower. The small, pale, bushy plant is indian ricegrass (*Oryzopsis hymenoides*). The dominant shrubs are Bailey's greasewood (*Sarcobatus baileyi*) and Shockley's desert thorn (*Lycium shockleyi*) (b) A view of the sole remaining building. Species in the foreground are *Halogeton glomeratus* (mainly in the lower left corner) and tumbleweed (*Salsola kali*) (next to the scrap wood) (c) Wonder – view looking north-west towards tailings pile. Almost the entire area in the foreground was part of the town. The shrubs are nearly all big sagebrush (*Artemisia tridentata*); the grass is cheatgrass (*Bromus tectorum*) (d) View of the foundation at Wonder. Big sagebrush, cheatgrass and piñon pine (*Pinus edulis*) are in the foreground
Photographs by Paul A. Knapp

impermanence, as revealed in palaeobotanical and palaeozoological studies.

Multidirectional succession

A result of individualistic community assembly is that succession may continue along many pathways, and is not necessarily fenced into a single predetermined path. Some field studies support this idea.

Hawaiian montane rain forest

Montane rain forest on a windward slope of Mauna Loa, Hawaii, displays a primary successional sequence on lava flows with ages ranging from 8 years to 9,000 years (Kitayama *et al.* 1995). Both downy (pubescent) and smooth (glabrous) varieties of the tree *Metrosideros polymorpha* (Myrtaceae) dominated the upper canopy layers on all lava flows in the age range 50 to 1,400 years. The downy variety was replaced by the smooth variety on the flows more than 3,000 years old. Lower forest layers are dominated a matted fern, *Dicranopteris linearis*, for the first 300 years, and then by tree ferns (*Cibotium* spp.). The *Cibotium* cover declined slightly after 3,000 years, while other native herb and shrub species increased. A 'climax' vegetation state was not reached – biomass and species composition changed continuously during succession. Such divergent succession may be unique to Hawaii, where the flora is naturally impoverished and disharmonic due to its geographical isolation (Kitayama *et al.* 1995). Against this view, it could be argued that Hawaii is the sort of place where succession should have the best chance of following the classical model (R. J. Whittaker pers. comm.). It is very interesting that montane forest on Hawaii does not follow the classical model. Rather, it may be an example of a general pattern of non-equilibrium dynamics of the kind revealed by new work at Glacier Bay and the Krakatau Islands.

Glacier Bay revisited

A re-evaluation of the succession in Glacier Bay, Alaska, used reconstructions of stand development based on tree-ring records from 850 trees at ten sites of different age (Fastie 1995). The findings suggested that succession there is multidirectional. The three oldest sites were deglaciated before 1840. They differ from all younger sites in three ways. First, they were all invaded early by Sitka spruce (*Picea sitchensis*). Second, they all support western hemlock (*Tsuga heterophylla*). Third, they all appear to have had early shrub thickets. Sitka alder (*Alnus sinuata*) is a nitrogen-fixing shrub. Only at sites deglaciated since 1840 has it been an important and long-lived species. Black cottonwood (*Populus trichocarpa*) has dominated the overstorey only at sites deglaciated since 1900. The new reconstruction of vegetation succession in the Glacier Bay area suggests additions or replacements of single species. These single-species changes distinguish three successional pathways that occur in different places and at different times. This re-evaluation of the evidence dispels the idea that communities of different age at Glacier Bay form a single chronosequence describing unidirectional succession.

The multiple successional pathways seem to result from the number and timing of woody species arriving on the deglaciated surfaces. They do not appear to stem from spatial differences in substrate texture and lithology. For example, site proximity to a seed source at the time of deglaciation accounts for up to 58 per cent of the variance in early Sitka spruce recruitment. Nitrogen-fixing is another factor influencing successional pathways. Long-lived alder thickets produce soil nitrogen pools that tend to reduce conifer recruitment and prevent the rapid development of spruce–hemlock forest.

Krakatau Islands revisited

The latest studies of primary succession on Krakatau favour a disequilibrium interpretation (e.g. R. J. Whittaker *et al.* 1989, 1992; Bush *et al.* 1992; R. J. Whittaker and Jones 1994). With the exception of the strand-line species, all components of the Krakatau fauna and flora are still changing. A lasting equilibrium does not prevail because the faunal and floral diversity is continually being altered by dispersal opportunities, and by disturbance events ranging from continuing volcanism to the fall of individual trees. The patchy nature of these processes helps to explain why the forest-canopy architecture on Panjang and Sertung changed materially from 1983 to 1989, while the pace of change on Rakata over the same period was fairly slow. It also accounts for significant differences within and between forests on the four islands. Overall, the forest dynamics of Krakatau have been highly episodic, with stands experiencing times of relatively minor change punctuated by pulses of rapid turnover and change; and all the while, new colonists species have arrived and have spread (R. J. Whittaker pers. comm.).

On Rakata, for example, there are four common successional pathways that have produced distinct forest communities (R. J. Whittaker and Bush 1993). The first occurs on coastal communities, where either *Terminalia catappa–Barringtonia asiatica* woodland is rapidly established, or else the same woodland is established after a stage of *Casuarina equisetifolia* woodland (Plate 7.3a). Path two is followed inland (but not in uplands) and runs from ferns, to grass savannah, to *Macaranga tanarius–Ficus* spp. forest, to *Neonauclea calycina* forest (Plate 7.3b). Path three is the same as path two, but the current stage is *Ficus pubinervis–Neonauclea calycina* forest (Plate 7.3c). Path four occurs in the upland (Plate 7.3d). It runs from ferns, to grass savannah, to *Cyrtandra sulcata* scrub, to *Schefflera polybotrya–Saurauia nudiflora–Ficus ribes* submontane forest.

The vegetation of Rakata is quite different from the vegetation on Panjang and Sergung (Tagawa 1992). *Neonauclea calycina* seeds successfully formed forest on Rakata, but failed to do so on Panjang and Sertung. Factors in seed dispersal, soil conditions, and disturbance by volcanic activity on Anak Krakatau all help to explain these differences. The *Timonius compressicaulis* and *Dysoxylum gaudichaudianum* forests found on Panjang and Sertung developed on both islands after the appearance of Anak Krakatau. Both *Timonius compressicaulis* and particularly *Dysoxylum gaudichaudianum* have larger seeds than *Neonauclea* and they were able to germinate and grow under the regenerated mixed forest canopy, but it was not so

easy for them to colonize the *Neonauclea forest*. *Dysoxylum gaudichaudianum* invasion progresses only gradually on Rakata.

Fleeting communities

Communities, like individuals, are impermanent. Species abundances and distributions constantly change, each according to its own life-history characteristics, largely in response to an ever-changing environment. This **community impermanence** is seen in the changing distributions of three small North American mammals since the Late Quaternary (Figure 7.5). The northern plains pocket gopher (*Thomomys talpoides*) lived in south-western Wisconsin around 17,000 years ago and persisted in western Iowa until at least 14,800 years ago. Climatic change associated with deglaciation then caused it to move west. The same climatic change prompted the least shrew (*Cryptotis parva*) to shift eastwards. The collared lemming (*Dicrostonyx* spp.), which lived in a broad band south of the Laurentide ice sheet, went 1,600 km to the north.

With each species acting individually, it follows that communities, both local ones and biomes, will come and go in answer to environmental changes (e.g. Graham and Grimm 1990). This argument leads to a momentous conclusion: there is nothing special about present-day communities and biomes. However, this bold assertion should be tempered with a cautionary note – insect species and communities have shown remarkable constancy in the face of Quaternary climatic fluctuations (Coope 1994).

Community impermanence is evidenced in three aspects of communities – in communities with no modern analogues, in communities that are disharmonious, and in communities that behave chaotically.

No-modern-analogue communities

Some modern communities and biomes are similar to past ones, but most have no exact fossil counterparts. Contrariwise, many fossil communities and biomes have no precise modern analogues. This finding

(a)

(b)

(c)

(d)

Plate 7.3 Successional pathways on Rakata (a) Path one: *Casuarina equisetifolia* woodland (at north end of Anak Krakatau) (b) Path two: lowland *Neonauclea calycina* forest (c) Path three: lowland *Ficus pubinervis–Neonauclea calycina* forest (d) Path four: upland forest
Photographs by R. J. Whittaker

supports the disequilibrium view of communities. The list of past communities that lack modern analogues is growing fast. In the Missouri–Arkansas border region, United States, from 13,000 to 8,000 years ago, the eastern hornbeam (*Ostyra virginiana*) and the American hornbeam (*Carpinus caroliniana*) were sig-

nificant plant community components (Delcourt and Delcourt 1994). These communities, which were found between the Appalachian Mountains and the Ozark Highlands, bore little resemblance to any modern communities in eastern North America. They appear to have evolved in a climate characterized by

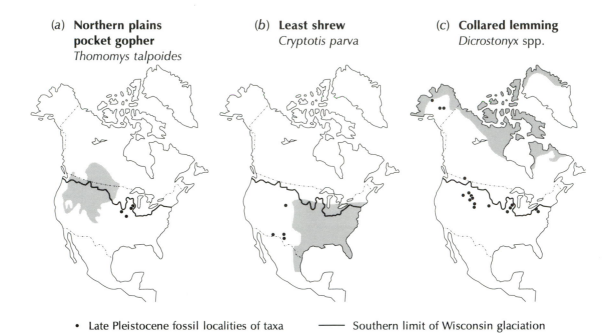

(a) Northern plains pocket gopher
Thomomys talpoides

(b) Least shrew
Cryptotis parva

(c) Collared lemming
Dicrostonyx spp.

• Late Pleistocene fossil localities of taxa   —— Southern limit of Wisconsin glaciation

Figure 7.5 Individualistic response of some small North American mammals since the Late Quaternary. (a) Northern plains pocket gopher (*Thomomys talpoides*). (b) Least shrew (*Cryptotis parva*). (c) Collared lemming (*Dicrostonyx* spp.).
Source: After Graham (1992)

heightened seasonality and springtime peaks in solar radiation. Farther north, in the north-central United States, a community rich in spruce and sedges existed from about 18,000 to 12,000 years ago. This community was a boreal grassland biome (Rhodes 1984). It occupied a broad swath of land south of the ice sheet and has no modern counterpart, though it bore some resemblance to the vegetation now found in the southern part of the Ungava Peninsula, in northern Quebec, Canada.

Disharmonious communities

The fauna and flora of communities with no modern analogues are commonly described as **disharmonious communities**. This inapt name inadvertently conjures an image of animal and plants struggling for survival in an alien environment. It is meant to convey the idea that these communities had evolved in, and flourished under, climatic types that no longer exist anywhere in the world (Graham and Mead

1987). In the southern Great Plains and Texas, United States, present-day grassland or deciduous forest species – including the least shrew (*Cryptotis parva*), the bog lemming (*Synaptomys cooperi*), the prairie vole (*Microtus ochrogaster*), and short-tailed shrews (*Blarina* spp.) – lived cheek-by-jowl with present-day boreal species – including the long-tailed shrew (*Sorex cinereus*), white-tailed jack rabbit (*Lepus townsendii*), ermine or stoat (*Mustela erminea*), and meadow vole (*Microtus pennsylvanicus*) (Lundelius *et al.* 1983; also see Figure 7.5). During the Late Pleistocene epoch, disharmonious animal communities were found over all the United States, except for the far west where vertebrate faunas bore a strong resemblance to their modern day equivalents, and date from at least 400,000 years ago to the Holocene epoch. These disharmonious communities evolved from species responding individually to changing environmental conditions during Late Pleistocene times (Graham 1979). At the end of the Pleistocene, new environmental changes led to the disassembly of the communities. The climate became more seasonal and individual species had to readjust their distributions. Communities of a distinctly modern mark emerged during the Holocene epoch.

North America does not have a monopoly in disharmonious communities. In Australia, an Early Pliocene fauna from Victoria – the Hamilton local fauna – contains several extant genera whose living species live almost exclusively in rain forest or rain-forest fringes (Flannery *et al.* 1992). The indication is, therefore, that the Pliocene fauna lived in rain-forest environment, but a more complex rain forest than exists today. Modern representatives of four genera (*Hypsiprymnodon*, *Dorcopsis*, *Dendrolagus*, and *Strigocuscus*) are almost entirely rain-forest dwellers, but they live in different kinds of rain forest. Living species of *Dorcopsis* (kinds of kangaroo) live in high mountain forests, lowland rain forests, mossy montane forests, and mid-montane forests. Living species of *Dendrolagus* (tree kangaroos) live mainly in montane rain forests. *Hypsiprymnodon moschatus* (the musky rat-kangaroo) is restricted to rain forest where it prefers wetter areas. The modern New Guinea species *Strigocuscus gymnotus* is chiefly a rain-forest dweller, though it also lives in areas of regrowth, mangrove swamps, and woodland savannah. Living species of *Thylogale* (pademelons) live in an array of environments, including rain forests, wet sclerophyll forests, and high montane forests. Other modern relatives of the fossils in the Hamilton fauna, which includes pseudocheirids, petaurids, and kangaroos and wallabies, live in a wide range of habitats. It contains two species, *Trichosurus* (brushtail possums) and *Strigocuscus* (cuscuses), whose ranges do not overlap at present. It is thus a disharmonious assemblage. Taken as a whole, the Hamilton mammalian assemblage suggests a diversity of habitats in the Early Pliocene. The environmental mosaic consisted of patches of rain forest, patches of other wet forests, and open area patches. Nothing like this environment is known today.

Chaotic communities

In the 1990s, the notion of non-equilibrium was formalized in the **theory of chaotic dynamics**.

The main idea is that all Nature, including the communities and ecosystems, is fundamentally erratic, discontinuous, and inherently unpredictable. This view of communities and ecosystems arose from mathematical models. The models confirmed what ecologists had felt for a long time but had been unable to prove – emergent community properties, such as food webs and a resistance to invasion by alien species, arise from the host of individual interactions in an assembling community. In turn, the emergent community properties influence the local interactions.

All evolving ecosystems, from the smallest pond to the entire ecosphere, possess emergent properties and appear to behave like superorganisms. But this superorganic behaviour is the result of a continuing two-way feedback between local interactions and global properties. It is not the outcome of some mystical global property determining the local interactions of system components, as vitalists would contend. Nor is it the cumulative result of local interactions in the system, as mechanists would hold. No, the whole system is an integrated, dynamic structure powered by energy and involving two-way interaction between all levels.

Experiments with 'computer communities' showed that species-poor communities were easy to invade (Pimm 1991). Communities of up to about twelve species offered essentially open access to intruding species. Beyond that number, in species-rich communities, there were two results. First, newly established species-rich communities were more difficult to invade than species-poor communities. Second, long-established communities were even harder to invade than newly established species-rich communities. Other mathematical experiments started with a 125-species pool of plants, herbivores, carnivores, and omnivores (Drake 1990). Species were selected one at a time to join an assembling community. Second chances were allowed for first-time failed entrants. An extremely persistent community emerged comprising about fifteen species. When the model was rerun with the same species pool, an extremely persistent community again emerged, but this time with different component species than in the first community. There was nothing special about the species in the communities: most species could become a member of a persistent community under the right circumstances; the actual species present depended on happenstance. It was the dynamics of the persistent communities that was special: a persistent community of fifteen species could not be reassembled from scratch using only those fifteen species. This finding suggests that communities cannot be artificially manufactured from a particular set of species – they have to evolve and to create themselves out a large number of possible species interactions.

UNDER THE AXE AND PLOUGH: LAND COVER TRANSFORMATION

In almost all parts of the world, biomes are being changed on a massive scale. Much of this **land-cover transformation** has taken place in the past two centuries (Buringh and Dudal 1987) (Figure 7.6). It resulted from the western European core region pushing out successive waves of exploitation into peripheral regions of the globe. As the frontiers expanded, so isolated subsistence economies were drawn irresistibly into a single world market. Basic commodities – crops, minerals, wood, water, and wild animals – were traded globally. Combined with steam power, medicine, and advances in agricultural technology, these economic changes set in motion the 'great transformation' – a world-wide alteration of land cover.

The land-cover transformation has changed biomes. From 1860 to 1978, 8,517,000 km^2 of land were converted to cropping (Revelle 1984). The converted land was originally forest (28.5 per cent), woodland (18.3 per cent), savannah (13.8 per cent), other grassland (34.9 per cent), wetland (2.5 per cent), and desert (2.0 per cent). A major problem with this ecosystem conversion is its patchy nature – a field here, a clearing there – that has broken the original habitat into increasingly isolated habitat fragments. Land-cover change is caused by five processes – perforation, dissection, fragmentation, shrinkage, and attrition (Forman 1995: 406–15). Forests, for example, are being perforated by clearings,

dissected by roads, broken into discrete patches by felling, and the newly created patches are shrinking and some of them disappearing through attrition. The overall effect of land-cover change is habitat fragmentation.

Habitat fragmentation

This is the breaking up of large habitats or areas into smaller parcels. It is enormously significant for wildlife and poses a global environmental problem. As habitat patches become smaller and more isolated, so several community changes occur. The numbers of generalist species, species that can live in more than one habitat, edge species, and exotic species all rise. The nest predation rate increases, populations fall, and extinctions become more common. The numbers of species specialized for life within a habitat interior, and species with a large home range, both decrease, as does the diversity of habitat interior species. Habitat fragmentation also changes ecosystem processes. It disturbs the integrity of drainage networks, affects water quality in aquifers, alters the disturbance regime to which species have adapted, and influences other ecosystem processes.

The effects of fragmenting habitats on wildlife are evident in the examples of the skipper butterfly in Britain, the malleefowl in Australia, and the reticulated velvet gecko in Australia.

The skipper butterfly in Britain

In Britain, the skipper butterfly (*Hesperia comma*) lives in heavily grazed calcareous grasslands (Colour plate 11) (C. D. Thomas and Jones 1993). The grazing is carried out largely by rabbits (*Oryctolagus cuniculus*). During the mid-1950s, the rabbits were devastated by myxomatosis and the habitat became overgrown. The skipper butterfly population shrank and became a metapopulation, occupying forty-six or fewer localities in ten refuges (Figure 7.7). The rabbits had recovered by 1982, and many former grassland sites, now neatly cropped, appeared suitable for recolonization by the skipper butterfly. Indeed, from 1982 to 1991, the number of populated habitat patches

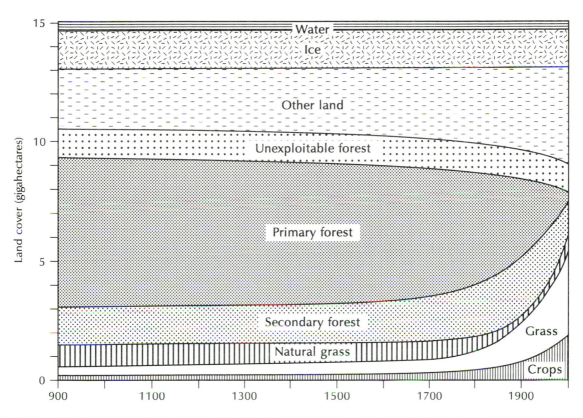

Figure 7.6 Land-cover transformation, AD 900–1977. The greatest transformation took place in the last two centuries.
Source: After M. Williams (1996)

increased by 30 per cent in the South and North Downs, with most of the increase taking place in East Sussex. Most of the recolonized habitat patches were fairly large and close to other populated patches. But even by 1991, many suitable habitat patches had not been reoccupied. Were the habitat not fragmented, the skipper butterfly might be expected to fill its previous range in south-east England within fifty to seventy-five years. In the fragmented landscape, it is likely that little or no further spread will take place in the twenty-first century, except in East Sussex. In most places, bands of unsuitable habitat, which are more than 10 km wide, halt the spread. Conservation of the skipper butterfly requires the protection of metapopulations in networks of habitat patches.

The malleefowl in Australia

Mallee is an Australian sclerophyllous shrub formation (equivalent to the maquis in the Mediterranean region) characterized by high bushes of shrubs and small trees. Loss, fragmentation, and degradation of mallee habitat within the New South Wales wheatbelt have caused a marked decline in the range and local abundance of malleefowl (*Leipoa ocellata*) (Priddel and Wheeler 1994). The malleefowl is a large, ground-dwelling bird that makes a mound for its nest and keeps the temperature tightly controlled. Small, disjunct populations of malleefowl now occupy small and isolated remnants of mallee habitat. Several of these populations have recently become locally extinct. As an experiment, young malleefowl

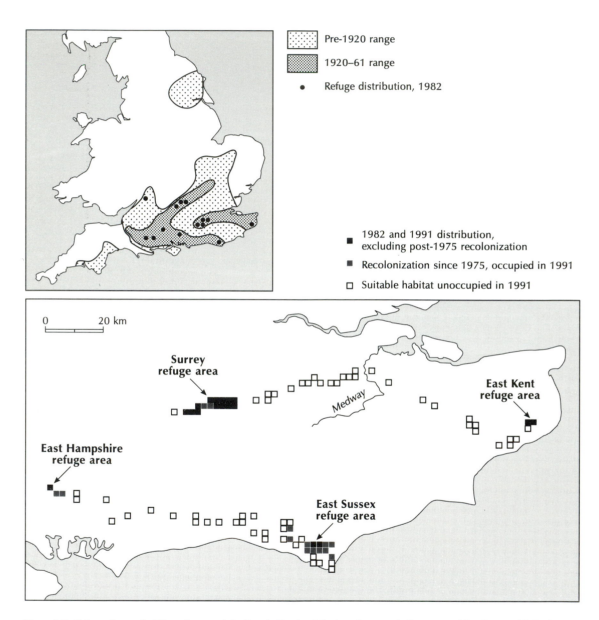

Figure 7.7 Skipper butterfly (*Hesperia comma*) decline in England during the twentieth century. The three published maps show the skipper butterfly beyond the range shown, but each of the five records occurs on just one of the maps and they are excluded. The bottom map shows the results of a recent survey.
Source: After C. D. Thomas and Jones (1993)

(8–184 days old) reared in captivity were released in March and June 1988 into a 55-ha remnant of mallee vegetation that contained a small but declining population of malleefowl. From the first day after release, malleefowl were found dead. Deaths continued until, within a relatively short time, no malleefowl remained alive. The main cause of their demise was predation, which accounted for 94 per cent of the deaths. Raptors accounted for 26–39 per cent of the loss, and introduced predators, chiefly the red fox (*Vulpes vulpes*), accounted for 55–68 per cent. Helpings of supplementary food made no difference to survival. Young malleefowl rely principally on camouflage for safety. They have no effective defence or escape behaviour to evade ground-dwelling predators. Foxes are imposing severe predation pressure on young malleefowl, and are probably curtailing recruitment into the breeding population. Foxes are thus a major threat to the continuance of malleefowl remnant populations in the New South Wales wheat-belt.

An Australian gecko

The reticulated velvet gecko (*Oedura reticulata*) is restricted in distribution to south-west Western Australia (Sarre 1995) (Colour plate 12a). Within this region, it occurs mainly in smooth-barked eucalypt woodland habitat, mainly consisting of gimlet gum (*Eucalyptus salubris*) and salmon gum (*Eucalyptus salmonophloia*). Much of the woodland has become fragmented by clearing and remains as islands within a sea of wheat (Colour plate 12b). For example, before land clearance, about 40 per cent of the Kellerberrin region contained smooth-barked eucalypt woodland suitable for reticulated velvet gecko populations. The same habitat now occupies just 2 per cent of the region. Only four remnants, including three nature reserves, contain stands of more than 570 gimlet gum and salmon gum trees. Demographic characteristics of nine *Oedura* populations that currently survive in Kellerberrin eucalypt woodland remnants were assessed and compared with *Oedura* populations in three nature reserves. The woodland remnants vary in area from 0.37 to 5.40 ha, but are similar in age,

tree species composition, and vegetation structure. *Oedura* population sizes vary considerably among remnants. They are poorly correlated with remnant area or the number of smooth-barked eucalypts. Population size does not appear to be limited by habitat availability in most remnants. The number of adults of breeding age is small in most populations suggesting that they may be susceptible to stochastic extinction pressures. The species is a poor disperser – interaction between local populations is small or zero, even over distances of 700 m or less. This poor dispersal ability means that the possibility of remnant recolonization after an extinction event is unlikely. The occupancy rate of *Oedura reticulata* in remnant woodland is thus likely to decline, and the species is likely to become restricted to a few, large remnants. It may therefore be futile to direct conservation effort into protecting small woodland remnants of a few hundred hectares.

Loss of wetlands

The world's **wetlands** are the swamps, marshes, bogs, fens, estuaries, salt marshes, and tidal flats. They cover about 6 per cent of the land surface and include some very productive ecosystems. But this figure is falling as wetlands are reclaimed for agricultural or residential land, drowned by dam and barrage schemes, or used as rubbish dumps. Sea-level rise during the twenty-first century also poses a major threat to coastal wetlands. The following two cases illustrate the complexity of changes in wetland habitats.

Bottomland forests in the Santee River floodplain

The coastal plain drainage of the Santee River, South Carolina, United States, was drastically modified in 1941. Almost 90 per cent of the Santee River discharge was diverted into the Cooper River as part of a hydroelectric power project (Figure 7.8). However, this diversion of water has led to silting in Charleston Harbor, the main navigation channels in which require year-round dredging. To alleviate this

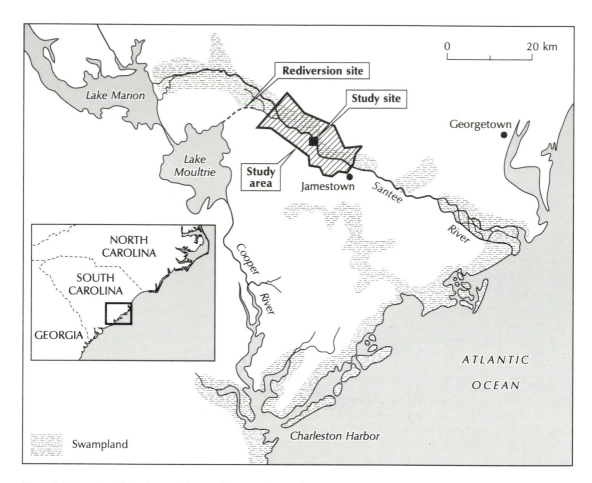

Figure 7.8 Location of the Santee River and proposed river diversion.
Source: After Pearlstine *et al.* (1985)

costly problem, authorization was granted in 1968 to redivert most of the water back into the Santee River through an 18.5 km long canal feeding off Lake Moultrie (Figure 7.8). An additional hydroelectric power plant would be constructed on this canal. Fears have been voiced that this rediversion would inundate much of the Santee floodplain and cause a substantial decline in the bottomland forest. After the diversion, the Santee River would return to 80 per cent of its pre-diversion flow rate with seasonal spates of floods and stages of low flows. A crucial change would be a reduction of flow early in the growing season.

A forest growth model, called FORFLO, was developed to quantify the effects of the changed hydrological regime on the bottomland forests (Pearlstine *et al.* 1985) (Box 7.1). It was used to simulate the effect of the proposed river rediversion, and a modified version of it, along a 25 km reach of the Santee forested floodplain from the rediversion site downstream to Jamestown (Figure 7.8). In the proposed rediversion, flow from Lake Marion to the Santee River stays the same, flow to the Cooper River is reduced to 85 m³/s (the level which prevents silting in Charleston Harbor), increasing the annual flow to the Santee River via the rediversion canal to

Box 7.1

KEY FEATURES OF FORFLO

The FORFLO model allows hydrological variables to influence tree species composition through seed germination, tree growth, and tree mortality. Key features of the model are the assumed relationships between flooding and various aspects of forest growth succession. First, for all tree species, save black willow (*Salix nigra*) and eastern cottonwood (*Populus deltoides*), seeds will not germinate when the ground is flooded. If the plot should be continuously flooded during that period of the year when a species would germinate, then the germination of that species fails. Black willow and eastern cottonwood can germinate whether the land is flooded or not. Second, after having germinated, the survival of seedlings depends on environmental conditions. A notable determinant of the seedling survival rate is the duration of the annual flood. Each species has a tolerance to flooding, and will survive if its range of tolerance should fall within the flood duration for the plot. Third, the optimum growth of trees was reduced by, among other things, a water-table function that models floodplain conditions. The water-table function modifies the tree-growth equation to account for the tolerance of species to the level of water on the plot during the growing season. The model computes the height of water for each half month during the growing season. It is assumed that all trees will fail to grow during the half months when they are more than three-quarters submerged by flood water. At lower levels of submergence, tree growth was related to water level by a curvilinear function in which the optimum water-table depth for each species is taken into account.

413 m³/s. The modified rediversion was the same as the proposed rediversion, except that during the early growing season (April to July), flow through the rediversion canal would be kept to a level which could be handled by just one of the three turbines at the power station. Although this would mean that the Cooper River would exceed the critical flow of 85 m³/s for the four months of the growing season, and thus cause some silting at the coast, it might promote the preservation of the bottomland forest.

The simulated responses of the forests to the rediversion are shown in Figures 7.9a and b. The annual duration of flooding is a crucial factor in determining the course of vegetational change. With an annual flood duration of more than 30–35 per cent (Figure 7.9a), bottomland hardwood forest is replaced by cypress–tupelo forest, bald cypress (*Taxodium distichum*) and water tupelo (*Nyssa aquatica*) being the only species that would manage to regenerate. When annual flood duration was above 65–70 per cent, no species was able to survive and the forest was replaced by a non-forest habitat; this happened more rapidly in the subcanopy.

The effects of the proposed rediversion and the modified rediversion version on the frequency of habitat types are indicated in Figure 7.10. In both cases there is a large loss of bottomland forest: a 97 per cent loss in the case of the proposed rediversion and a 94 per cent loss in the modified rediversion. The saving grace of the modified rediversion plan is that a forest cover is maintained: bottomland forest changes to cypress–tupelo forest, rather than to open water. If these predictions of sweeping changes in the bottomland forest along the Santee River should be trustworthy, then plainly it would be advisable to rethink the plans for rediverting the flow from the Cooper River.

A coastal ecosystem in southern Louisiana

Ecosystems occupying coastal locations are under threat from a variety of human activities: gas and oil exploration, urban growth, sediment diversion, and greenhouse-induced sea-level rise, to name but a few. To protect and preserve these ecosystems, it is valuable to know what the effects of proposed human

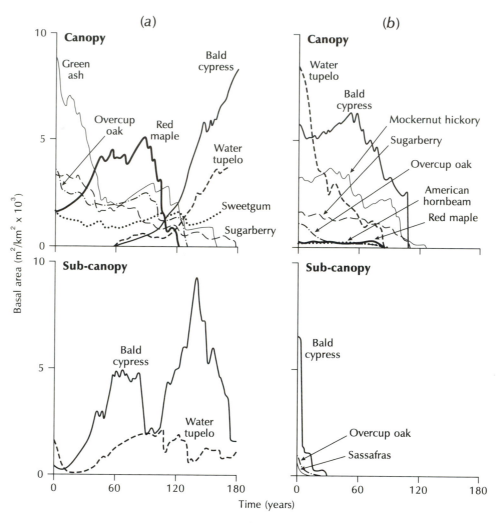

Figure 7.9 Results of simulations. (a) Bottomland forest community subjected to an annual flood duration of 45 ± 4 per cent. (b) Bottomland forest community subjected to an annual flood duration of 72 ± 5 per cent. The flood duration is the percentage of a year during which a plot is flooded. The plant species are bald cypress (*Taxodium distichium*), water tupelo (*Nyssa aquatica*), and sassafras (*Sassafras albidum*).
Source: After Pearlstine *et al.* (1985)

activities are likely to be, and how these effects differ from natural changes. These questions were addressed in the Atchafalaya Delta and adjacent Terrebonne Parish marshes in southern Louisiana, United States, using a mathematical model to simulate likely changes (Figure 7.11) (Costanza *et al.* 1990). This landscape, part of the Mississippi River distributary system, is one of the most rapidly changing land-

scapes in the world. The Atchafalaya River is one of the two principal distributaries of the Mississippi River. It carries about 30 per cent of the Mississippi discharge to the Gulf of Mexico. Since the mid-1940s, sediments transported by the Atchafalaya River have been laid down in the bay area. In consequence, the bay area has gradually become filled in, and in 1973 a new subaerial delta appeared that has

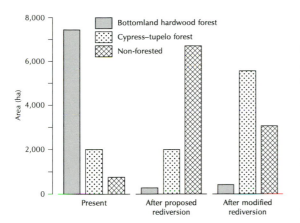

Figure 7.10 Habitat changes in the Santee River study area.
Source: After Pearlstine *et al.* (1985)

since grown to about 50 km². Other changes are taking place in the area. The western Terrebonne marshes are becoming less salty, while the eastern part of the area is becoming more salty: the boundary between fresh and brackish marshes has shifted closer to the Gulf in the western marshes, and farther inland in the east (Figure 7.12). As a whole, the study region is losing wetland, but rate of loss in the Terrebonne marshes has slowed and reversed since the mid-1970s owing to river deposition. The hydrology of the area has been greatly altered by dredging of waterways and the digging of access canals for petroleum exploration.

A dynamic spatial model was developed and christened the Coastal Ecological Landscape Spatial Simulation model (CELSS model for short). It divided the marsh–estuary complex into 2,479 square grid-cells, each with an area of 1 km². The model was used to predict changes of habitats under a range of climatic, management, historical, and boundary scenarios. The results of several scenario analyses, which predicted changes to the year 2033, are summarized in Table 7.1. Climate scenarios take 'climate' to mean all the driving variables including rainfall, Atchafalaya River flow, wind, and sea-level. They address the impact of climatic changes on the study area. Management scenarios look at the effect of specific human manipulations of the system. Historical

scenarios consider what the system would have been like had not the environment been altered by human action, and if climatic conditions had been different. Boundary scenarios delve into the potential impacts of natural and human-induced variations in the boundary conditions of the system, such as sea-level rise. The management and boundary scenarios were run by restarting the model with the actual habitat map for 1983, rather than the predicted habitat map for that year, to add more realism. Climate and historical scenarios were run starting in 1956, so that the full impacts of climate and historical variations could be assessed. After 1978, the rate of canal and levee construction for oil and gas exploration slowed appreciably compared with the 1956–78 period, so all scenarios assumed that no canals nor levees were built after 1978, except those specifically mentioned in the scenarios.

It is clear from Table 7.1 that the assumptions made about climatic change have a big influence on habitat distribution by the year 2033. The mean climate scenario, wherein the long-term average for each variable was used for all weeks, produced a modest loss in land area. On the other hand, the weekly average climate scenario, wherein each weekly value of each climatic variable for the entire run from 1956 to 2033 was set to the average value for that week in the 1956–83 data, produced a drastic loss of land area. This finding indicated that the annual flood cycle and other annual cycles in climatic variables are important to the land building process, but that chance events, such as major storms and floods, tend to have a net erosional effect on marshland. If the global climate should become less predictable in the future as a result of global warming, then the stability of coastal marshes may be in jeopardy.

Several management scenarios were run. As the data in Table 7.1 indicate, the largest loss of land would arise from the full six-reach levee extension scheme that had been considered at one time. With this scheme, 48 km² of brackish marsh and 6 km² of fresh marsh would be lost by 2033, largely because the extended levees prevent sediment-laden water reaching the brackish marsh bordering Four League Bay, where most of the loss occurs. Boundary scenarios

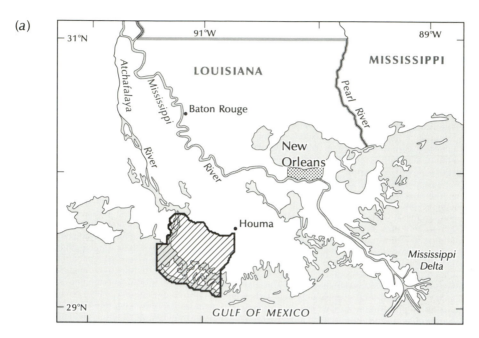

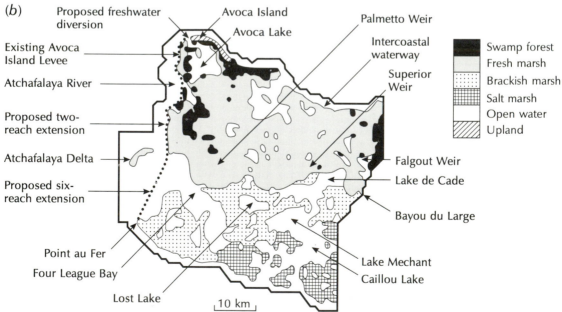

Figure 7.11 The Atchafalaya Delta and Terrebonne Parish marshes study area in southern Louisiana, United States. (a) General location map. (b) Major geographical features, types of habitat in 1983, and management options considered in the simulations.

Source: After Costanza *et al.* (1990)

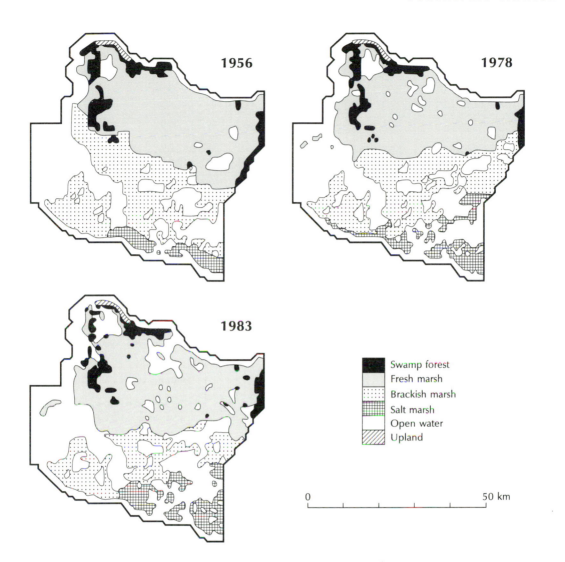

Figure 7.12 Observed distribution of habitats in the Atchafalaya–Terrebonne study area.
Source: After Costanza *et al.* (1990)

considered the effects of projected rates of sea-level rise, both high and low projections, on the area. The results were unexpected. Surprisingly, doubling the rate of eustatic sea-level rise from 0.23 to 0.46 cm/yr caused a net gain in land area of 10 km^2 relative to the base case. This was probably because, so long as sediment loads are high, healthy marshes can keep pace with moderate rates of sea-level rise.

Two historical scenarios were tested. The first probed the changes in the system that might have taken place had not the original Avoca Island levee been built. The second considered the changes that might have ensued had not the Avoca levee nor any of the post-1956 canals been constructed. The results shown in Table 7.1 suggest that the original levee and the canals had a major influence on the development of the system, causing a far greater loss of land than would have occurred in their absence.

Table 7.1 Area (km²) occupied by each habitat type for three years for which data are available, and for 2033, under various scenarios

	Swamp	Fresh marsh	Brackish marsh	Saline marsh	Upland	Total land	Open water
Survey data							
1956	130	864	632	98	13	1,737	742
1978	113	766	554	150	18	1,601	878
1983	116	845	347	155	18	1,481	998
Climate scenarios							
Base case	84	871	338	120	10	1,423	1,056
Mean climate	94 (+10)[a]	974 (+103)	402 (+64)	136 (+16)	11 (+1)	1,617 (+194)	862 (−194)
Weekly average climate	128 (+44)	961 (+90)	813 (+475)	300 (+180)	11 (+1)	2,213 (+790)	266 (−790)
Management scenarios							
No levee extension[b]	100	796	410	123	15	1,444	1,035
Full six-reach levee extension	103 (+3)	790 (−6)	362 (−48)	122 (−1)	15 (0)	1,393 (−52)	1,087 (+52)
Freshwater diversion	103 (+3)	803 (+7)	404 (−6)	123 (0)	15 (0)	1,448 (+4)	1,031 (−4)
Boundary scenarios							
Low sea-level rise[c]	104 (+4)	800 (+4)	411 (+1)	124 (+1)	15 (0)	1,454 (+10)	1,025 (−10)
High sea-level rise[d]	89 (−11)	794 (−2)	396 (−14)	131 (+8)	15 (0)	1,425 (−19)	1,054 (+19)
Historical scenarios[e]							
No original Avoca levee	84	951	350	126	13	1,524	955
No effects	130	863	401	144	12	1,550	929

Notes: [a] Parentheses indicate changes from the base case
[b] This is the base case for the management scenarios
[c] 50-cm rise by the year 2100
[d] 200-cm rise by 2100
[e] No comparisons with a base case are given for the historical scenarios because these runs started in 1956 rather than 1983

Source: After Costanza et al. (1990)

GETTING WARMER: COMMUNITIES IN THE TWENTY-FIRST CENTURY

Individuals and communities, by processes of natural selection, become adapted to the environment in which they live. If that environment should change, life systems must adapt, move elsewhere, or perish.

Species under pressure

The fate of many species in the twenty-first century, as the world warms and habitats shrink or vanish, is worrying (Peters 1992a, b). It seems that extinction will continue apace and global biodiversity will drop. But which species are the most vulnerable to global warming? The safest species are mobile birds, insects, and mammals who can track their preferred climatic zone. In the British Isles, the white admiral butterfly (*Ladoga camilla*) and the comma butterfly (*Polygonia c-album*) expanded their ranges over the past century as temperatures rose by about 0.5°C (Ford 1982). However, even mobile species would need a food source, and their favourite menu item might be extinct or have wandered elsewhere.

Five kinds of species appear to be most at risk:

1 **Peripheral species**. Populations of animals or plants that are at the contracting edge of a species range are vulnerable to extinction.
2 **Geographically localized species**. Many currently endangered species live in alarmingly limited habitats. The golden lion tamarin (*Leontideus rosalia*) is a small primate that lives in lowland Atlantic forest, Brazil (Colour plate 13). Lowland Atlantic forest is one of the most endangered rain forests in the world. It once covered 1 million km² along the Brazilian coastline. Now, a mere 7 per cent of the original forest remains, putting the golden lion tamarin under enormous threat (but see p. 212).
3 **Highly specialized species**. Many species have a close association with only one other species. An example is the Everglade (or snail) kite (*Rostrhamus sociabilis*) that feeds exclusively on the large and colourful *Pomacea* snail in Florida wetland.

Swamp drainage has already robbed the kite of its prey in some areas.
4 **Poor dispersers**. Many trees have heavy seeds that are not broadcast far. Plants with limited dispersal abilities have knock-on effects along the food chain. Some birds, mammals, and insects are closely associated with specific forest trees, which cannot migrate rapidly. These species would be hard pressed to survive. The endangered Kirtland's warbler (*Dendroica kirtlandii*), for example, breeds in and nests only upon the ground under jack-pine (*Pinus banksiana*) forest on well-drained sandy soil in north-central Michigan, United States. But not any jack-pine forest will do – it has to be young, secondary growth forest that emerges in the aftermath of a forest fire. As the global thermometer rises, the jack pines may move northwards onto less well-drained soils, and the Kirtland's warbler may find itself without any suitable nesting sites (Botkin *et al*. 1991). This species could be the first casualty of global warming.
5 **Climatically sensitive species**. Species in climatically sensitive communities are vulnerable to global warming. Communities in this class include wetlands, montane and alpine biomes, Arctic biomes, and coastal biomes. Wetlands will dry out, with grave consequences for amphibians (Box 7.2), mountains will become warmer towards their tops, tundra regions will warm up, and coastal biomes will be flooded. Climatic warming is predicted to be greatest at high latitudes. Animals and plants in these regions will have to cope with the most rapid changes. Tundra vegetation may be pushed northwards as much as 4 degrees of latitude. This could mean that, for a climatic warming of 3°C, 37 per cent of present tundra will become forest. The same degree of climatic warming would fuel local temperature increases on mountains. Animals and plants would need to shift their ranges upwards by about 500 m. Animal and plant populations on mountains may retreat upwards as climate warms, but eventually some of them will have nowhere to go. This would probably happen to the pikas (*Ochotona* spp.) living on

alpine meadows on mountains of the western United States. Pikas feed on grasses and do not range into boreal forests. As the boreal forest habitat advances up the mountainsides, so the pikas will retreat into an ever-decreasing area that may one day vanish. This process is already happening to Edith's checkerspot butterfly (*Euphydryas editha*), which lives western North America (Parmesan 1996). Records of this species suggest that its range is shifting upwards and northwards, and extinctions have occurred at some sites.

Communities under pressure

Community composition and structure change in the face of environmental perturbations. During the twenty-first century, many communities will be perturbed by global warming. They will affect the geographical location of biomes and the species composition of communities.

Prairie wetlands in North America

Prairie wetland habitat occupies a broad swath of land across the American Midwest and southern Canadian prairie provinces (Figure 7.13) (Poiani and Johnson 1991). It is the main breeding area for waterfowl on the North American continent. The habitat consists of relatively shallow-water temporary ponds (holding water for a few weeks in spring), seasonal ponds (holding water from spring until early summer), semi-permanent ponds (holding water throughout much of the growing season in most years), and large permanent lakes (Plate 7.4). For breeding purposes, the waterfowl require a mix of open water and emergent vegetation. The temporary and seasonal ponds provide a rich food source in the spring and are used by dabbling ducks. The semi-permanent ponds provide food and nesting areas for birds as seasonal wetlands dry. As temperatures rise over the next decades, water depth and the numbers of seasonal ponds in these habitats should decrease. The lower water levels would favour the growth of emergent vegetation and reduce the amount of open water. Waterfowl may respond by migrating to different geographical locations, relying more upon semi-permanent wetlands but not breeding, or failing to re-nest as they do at present during droughts.

Box 7.2

AMPHIBIANS AND GLOBAL WARMING

Amphibians need water. They must live in it, near it, or else in very humid conditions. Warmer and drier conditions could be disastrous for them. In places where global warming is associated with increasing aridity, they are likely to suffer. Evidence of this comes from endemic golden toad (*Bufo periglenes*) and harlequin frog (*Atelopus varius*) populations in the Monteverde Cloud Forest Preserve, Costa Rica (Pounds 1990; Pounds and Crump 1994). In May and April 1987, there were thousands of golden toads thriving along streams in the reserve. The males live in underground burrows near the water table; they emerge only to mate. There was also a considerable number of harlequin frogs living in wet stream-bank patches. These amphibians retreat into crevices or gather in damp pockets during the dry season. Four years later, the golden toad and the harlequin frog had vanished. Rainfall during May 1987 was 64 per cent less than average, the lowest ever recorded for that month. Normally, early rains in May and June help the amphibians to recover from the November and December dry season. The 1987 dry season was extra dry. During the drought, water tables fell, springs all but dried up, and streams fed by aquifers, which would usually give enough water to keep the mossy riverbanks wet, dwindled. As the amphibians disappeared suddenly, it may be that the extreme drought killed the adults and their tadpoles or eggs. This catastrophe augurs ill for amphibians during the twenty-first century.

Figure 7.13 Location of North American prairie wetlands. The prairie wetlands are relatively shallow, water-holding depressions of glacial origin.
Source: After Poiani and Johnson (1991)

Mires in the Prince Edward Islands

Temperatures in the Prince Edward Islands have increased by approximately 1°C since the early 1950s, while precipitation has decreased (Chown and Smith 1993). The changing water balance has led to a reduction in the peat-moisture content of mires and higher growing season 'warmth'. In consequence, the temperature-sensitive and moisture-sensitive sedge, *Uncinia compacta*, has increased its aerial cover on Prince Edward Island. Harvesting of seeds by feral mice (*Mus musculus*), which can strip areas bare, has prevented an increase in sedge cover on Marion Island. Such extensive use of resources suggests that prey switching may be taking place at Marion Island. Mice not only are eating ectemnorhinine weevils to a greater extent than found in previous studies of populations at Marion Island, but also prefer larger weevils. A decrease in body size of preferred weevil prey species (*Bothrometopus randi* and *Ectemnorhinus similis*) has taken place on Marion Island (1986–1992), but not on Prince Edward Island. This appears to be a result of increased predation on weevils. Adults of the prey species, *Ectemnorhinus similis*, are relatively more abundant on Prince Edward Island than adults of the smaller congener *Ectemnorhinus marioni*, and could not be found on Marion Island in the late southern summer of 1991. Results not only provide support for previous hypotheses of the effect of global warming on mouse–plant–invertebrate interactions on the Prince Edward Islands, but also provide limited evidence for the first recorded case of predator-mediated speciation.

Plate 7.4 Semi-permanent prairie wetland, Stutsman County, North Dakota. This wetland is part of the Cottonwood Lake site where long-term hydrological and vegetation data have been collected by the United States Geological Survey (Water Resources and Biological Resources Division). The plants in the foreground are cattails (*Typha* spp.)
Photograph by Karen A. Poiani

Biome and ecotone shifts

Global warming fuelled by a doubling of carbon dioxide levels could produce large shifts in the distribution of biomes. One study, which explored five different scenarios for vegetation redistribution with a doubling of atmospheric carbon dioxide levels, predicted large spatial shifts, especially in extratropical regions (Neilson 1993a). Large spatial shifts in temperate and boreal vegetation were predicted (Figure 7.14). Boreal biomes (taiga and tundra) retreat northwards and decrease in size by 62 per cent (the range is 51 to 71 per cent, depending on the scenario used). Boreal and temperate grassland increase in area under two scenarios. The increase is 36 per cent one case

and 82 per cent in the other. Temperate forest area alters little. Tropical forests show a slight reduction in area in all but one scenario, which predicts a slight increase. Maps of predicted change in the leaf–area index (which reflects the maximum rate of transpiration) imply drought-related biomass losses in most forested regions, even in the tropics. The areas most sensitive to drought-induced vegetation decline are eastern North America and eastern Europe to western Russia.

In detail, spatial shifts of biomes should produce discrete 'change' zones and 'no change' zones (Figure 7.15) (Neilson 1993b). Such changes alter the pattern of vegetation. Large, relatively uniform biomes are reduced in size, while new biomes, which combine

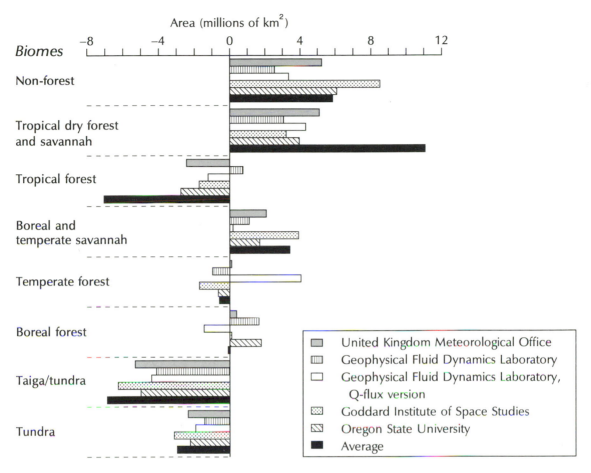

Figure 7.14 Changes in biome area predicted by the MAPPS (Mapped Atmosphere–Plant–Soil System) model under various scenarios.
Source: After Neilson (1993a)

the original biome and the encroaching biomes, emerge. Biomes and ecotones alike would probably be affected by drought followed by infestations and fire if climate were to change rapidly. Ecotones would be especially sensitive to climatic change, whether it were slow or fast. The pattern of ecotone change would be sensitive to water stress (Figure 7.16). With global warming unaccompanied by water stress, habitats should not fragment as under water stress, but should display a wave of high habitat variability as the ecotone gradually tracks the climatic shifts. Under extreme drought stress, the entire landscape

should become fragmented as every variation in topography and soil become important to site water-balances and the survivorship of different organisms. The ecotone disappears for a while and reappears at a new location. It does not visibly shift geographically, as it would with no water stress, but disassembles and then re-establishes itself later.

Forest change in eastern North America

It is probably common for species to move in the same general direction. However, that does not mean

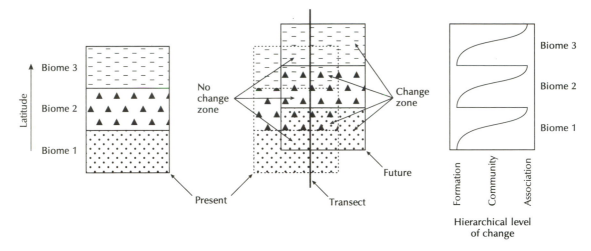

Figure 7.15 Geographical shift in biomes induced by global warming. Discrete 'change' and 'no change' zones are produced. Large, uniform biomes shrink. New biomes, which combine characteristics of original biomes and encroaching biomes, emerge.
Source: After Neilson (1993b)

that entire communities move together. Quite the contrary, species move at different rates in response to climatic change. The result is that communities disassemble, splitting into their component species and losing some species that fail to move fast enough or cannot adapt. Mathematical models are helpful in understanding the possible changes in communities as the world warms up. Clearly, profound changes in the composition and geography of communities would occur. This is evident in the following case studies.

An early model simulated forest growth at twenty-one locations in eastern North America, as far west as a line joining Arkansas in the south to Baker Lake, North West Territories, in the north (Solomon 1986). All forests started growing on a clear plot and grew undisturbed for 400 years under a modern climate. After the year 400, climate was changed to allow for a warmer atmosphere. A linear change of climate between the years 400 and 500 was assumed, the new climate at the year 500 corresponding to a doubling of atmospheric carbon dioxide levels. Climatic change continued to change linearly after the year 500, so that, by the year 700, the new climate corresponded to a quadrupling of atmospheric carbon dioxide

levels. At the year 700, climate stabilized and the simulation went on for another 300 years. Simulation runs at each site were repeated ten times and the results averaged. Validation of the results was achieved by testing with independent forest composition data, as provided by pollen deposited over the last 10,000 years and during the last glacial stage, 16,000 years ago.

Results for some of the sites are set out in Figure 7.17. At the tundra–forest border, the vegetation responds to climatic change in a relatively simple way (Figure 7.17a). With four times the carbon dioxide in the atmosphere, the climate supports a much increased biomass and, at least at Shefferville in Quebec, some birches, balsam poplar, and aspens. The response of northern boreal forest to warming is more complicated (Figure 7.17b). With a four-fold increase in atmospheric carbon dioxide levels, the climate promotes the expansion of ashes, birches, northern oaks, maples, and other deciduous trees at the expense of birches, spruces, firs, balsam poplar, and aspens. In all northern boreal forest sites, the change in species composition is similar, but takes place at different times. This underscores the time-transgressive nature of vegetational response to climatic change. The

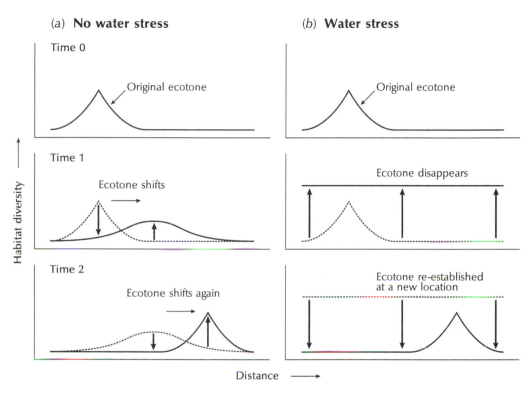

Figure 7.16 Ecotone changes induced by global warming. (a) With no water stress. (b) With water stress.
Source: After Neilson (1993b)

southern boreal forest and northern deciduous forest (Figures 7.17c and d) also have a complicated response to climatic change. In both communities, species die back twice, the first time around and just after 500 years, and the second time from about 600 to 650 years. In the southern boreal forest, the change is from a forest dominated by conifers (spruces, firs, and pines) to a forest dominated initially by maples and basswoods, and later by northern oaks and hickories. At some sites in the northern deciduous forest, species appear as climate starts to change then vanish once the stable climate associated with a quadrupled level of atmospheric carbon dioxide becomes established. Species that do this include the butternut (*Juglans cinerea*), black walnut (*Juglans nigra*), eastern hemlock (*Tsuga canadensis*), and several species of northern oak. In western and eastern deciduous forests, the response of trees to climatic warming is

remarkably uniform between sites (Figure 7.17e and f). Biomass declines everywhere, generally as soon as warming starts in the year 400. The drier sites in the west suffer the greatest losses of biomass, as well as a loss of species; this is to be expected as prairie vegetation would take over. The decline of biomass in the eastern deciduous forest probably results chiefly from the increased moisture stress in soils associated with the warmer climate. The increased moisture stress also gives a competitive edge to smaller and slower-growing species, such as the southern chinkipin oak (*Quercus muehlenbergii*), post oak (*Quercus stellata*), and live oak (*Quercus virginiata*), the black gum (*Nyssa sylvatica*), sugarberry (*Celtis laevigata*), and the American holly (*Ilex opaca*), which come to dominate the rapid-growing species such as the American chestnut (*Castanea dentata*) and various northern oak species.

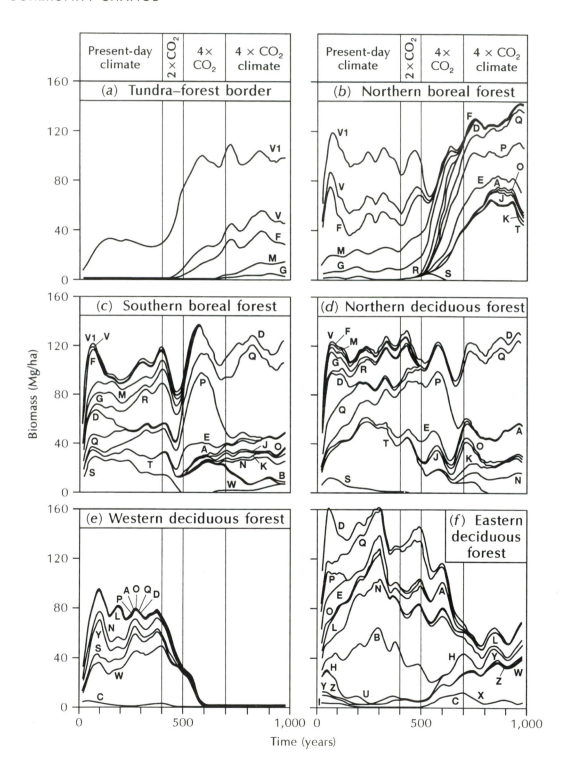

Soil water regimes and forest change

Forest growth is sensitive to changes in soil water regimes. Global warming is sure to alter water regimes and this will have an inevitable impact upon forests. A forest-growth model was used in conjunction with a soil model to assess the impact on forests of a doubling of atmospheric carbon dioxide levels (Pastor and Post 1988). The model was run for several sites in the north-eastern United States, including a site in north-eastern Minnesota. The climate of this region is predicted to become warmer and drier. The major changes will occur at the boundary between the boreal forest and cool temperate forest. Forest growth was modelled on two soil types: soils with a high water-holding capacity and soils with a low water-holding capacity. The simulations began in 1751. Bare plots were 'sown' with seeds of the tree species common in the area. The model forest was then allowed to grow for 200 years under present climatic conditions. Next, carbon dioxide concentrations were gradually increased until, after a century, their present value had been doubled. After having reached that value, carbon dioxide levels were held constant for the next 200 years.

The model predicted that on soils with a high water-holding capacity, where there will be enough water available to promote tree growth, productivity and biomass will increase. On soils with a low water-holding capacity, productivity and biomass will decrease in response to drier conditions. In turn, raised productivity and biomass will up levels of soil nitrogen, whereas lowered productivity and biomass will down levels of soil nitrogen. The vegetation changes are, therefore, self-reinforcing. On water-retentive soils, global warming favours the expansion of northern hardwoods (maples, birches, basswoods) at the expense of conifers (spruces and firs); on well-drained, sandy soils, it favours the expansion of a stunted oak–pine forest, a relatively unproductive vegetation low in nitrogen, at the expense of spruce and fir.

Figure 7.17 Simulations of forest biomass dynamics over one millennium in response to climatic change induced by increasing levels of carbon dioxide in the atmosphere at six sites in eastern North America. (a) Shefferville, Quebec (57°N, 67°W). (b) Kapuskasing, Ontario (49°N, 83°W). (c) West upper Michigan (47°N, 88°W). (d) North central Wisconsin (45°N, 90°W). (e) South central Arkansas (34°N, 93°W). (f) Central Tennessee (36°N, 85°W). The tree species are as follows: A. American beech (*Fagus grandifolia*); B. American chestnut (*Castanea dentata*); C. American holly (*Ilex opaca*); D. ashes: green ash (*Fraxinus pennsylvanica*), white ash (*Fraxinus americana*), black ash (*Fraxinus nigra*), blue ash (*Fraxinus quadrangulata*); E. basswoods: American basswood (*Tilia americana*) and white basswood (*Tilia heterophylla*); F. birches: sweet birch (*Betula lenta*), paper birch (*Betula papyrifera*), yellow birch (*Betula alleghaniensis*), and gray birch (*Betula populifolia*); G. balsam poplar (*Populus balsamifera*), bigleaf aspen (*Populus grandidentata*), trembling aspen (*Populus tremuloides*); H. black cherry (*Prunus serotina*); I. black gum (*Nyssa sylvatica*); J. butternut (*Juglans cinerea*) and black walnut (*Juglans nigra*); K. eastern hemlock (*Tsuga canadensis*); L. elms: American elm (*Ulmus americana*) and winged elm (*Ulmus alata*); M. firs: balsam fir (*Abies balsamea*) and Fraser fir (*Abies fraseri*); N. hickories: bitternut hickory (*Carya cordiformis*), mockernut hickory (*Carya tomentosa*), pignut hickory (*Carya glabra*), shagbark hickory (*Carya ovata*), shellbark hickory (*Carya laciniosa*), and black hickory (*Carya texana*); O. hornbeams: eastern hornbeam (*Ostrya virginiana*) and American hornbeam (*Carpinus caroliniana*); P. maples: sugar maple (*Acer saccharum*), red maple (*Acer rubra*), and silver maple (*Acer saccharinum*); Q. northern oaks: white oak (*Quercus alba*), scarlet oak (*Quercus coccinea*), chestnut oak (*Quercus prinus*), northern red oak (*Quercus rubra*), black oak (*Quercus velutina*), bur oak (*Quercus macrocarpa*), gray oak (*Quercus borealis*), and northern pin oak (*Quercus ellipsoidalis*); R. northern white cedar (*Thuja occidentalis*), red cedar (*Juniperus virginiana*), and tamarack (*Larix laricina*); S. pines: jack pine (*Pinus banksiana*), red pine (*Pinus resinosa*), shortleaf pine (*Pinus echinata*), loblolly pine (*Pinus taeda*), Virginia pine (*Pinus virginiana*), and pitch pine (*Pinus rigida*), and, T. white pine (*Pinus strobus*); U. yellow buckeye (*Aesculus octandra*); V spruces: black spruce (*Picea mariana*), and red spruce (*Picea rubens*); and V1 white spruce (*Picea glauca*); W. southern oaks: southern red oak (*Quercus falcata*), overcup oak (*Quercus lyrata*), blackjack oak (*Quercus marilandica*), chinkipin oak (*Quercus muehlenbergii*), Nuttall's oak (*Quercus nuttallii*), pin oak (*Quercus palustris*), Shumard's red oak (*Quercus shumardii*, post oak (*Quercus stellata*), and live oak (*Quercus virginiana*); X. sugarberry (*Celtis laevigata*); Y. sweetgum (*Liquidambar styraciflua*); Z. yellow poplar (*Liriodendron tulipifera*).
Source: After Solomon (1986)

Disturbance and forest dynamics

Global warming would favour a rise in the rate of forest **disturbance** owing to an increase in meteorological conditions likely to cause forest fires (drought, wind, and natural ignition sources), convective winds and thunderstorms, coastal flooding, and hurricanes. Simulations suggest that changes in forest composition associated with global warming would depend upon the disturbance regime (Overpeck *et al*. 1990). Two sets of simulations were run. In the first set ('step-function' experiments), simulated forest was grown from bare ground under present-day climate for 800 years. This enabled the natural variability of the simulated forest to be characterized. At year 800, a single climatic variable was changed in a single step to a new mean value, which perturbed the forest. The simulation was then continued for a further 400 years. In each perturbation experiment, the probability of a catastrophic disturbance was changed from 0.00 to 0.01 at year 800. This is a realistic frequency of about one plot-destroying fire every 115 years when a 20-year regeneration period (during which no further catastrophe takes place) of the trees in a plot is assumed. In each of the step-function simulations, three types of climatic change (perturbation) were modelled: a 1°C increase in temperature; a 2°C increase in temperature; and a 15 per cent decrease in precipitation. In the second set of simulation runs ('transient' experiments), forest growth was, as in the step-function experiments, started from bare ground and allowed to run for 800 years under present climatic conditions. Then, from the years 800 to 900, the mean climate, both temperature and precipitation, was changed linearly year by year to simulate a twofold increase in the level of atmospheric carbon dioxide, until the year 1600 when mean climate was again held constant. As in the step-function experiments, the probability of forest disturbance was changed from 0.00 to 0.01 at year 800. In all simulation runs, a relatively drought-resistant soil was assumed, and the results were averaged from 40 random plots into a single time series for each model run.

The model was calibrated for selected sites in the mixed coniferous–hardwood forest of Wisconsin and the southern boreal forest of Quebec. Selected summary results are presented in Figure 7.18. An increase in forest disturbance will probably create a climatically induced vegetation change that is equal to, or greater than, the same climatically induced change of vegetation without forest disturbance. In many cases, this enhanced change resulting from increased disturbance is created by rapid rises in the abundances of species associated with the early stages of forest succession, which appear because of the increased frequency of forest disturbance. In some cases, as in Figures 7.17a, d, and e, a step-function change of climate by itself does not promote a significant change in forest biomass, but the same change working hand-in-hand with increased forest disturbance does have a thoroughgoing effect on forest composition and biomass. Interestingly, the altered regimes of forest disturbance, as well as causing a change in the composition of the forests, also boost the rate at which forests respond to climatic change. For instance, in the transient climate change experiments, where forest disturbance is absent through the entire duration of the simulation period, vegetation change lags behind climatic change by about 50 to 100 years, and simulated vegetation takes at least 200 to 250 years to attain a new equilibrial state. In the simulation runs where forest disturbance occurred from year 800 onward, the vegetational change stays hard on the heels of climatic change and takes less than 180 years to reach a new equilibrium composition after the climatic perturbation at year 800.

Ecological lessons from simulation models

Species show a wide range of responses to climate. In consequence, the response of different animal and plant species to climatic change will be varied. This should mean that natural communities are likely to disassemble and that habitats restructure in a transient, non-equilibrium fashion as climatic change unfolds. Validated models that help forecast these events are needed to aid scientists in better

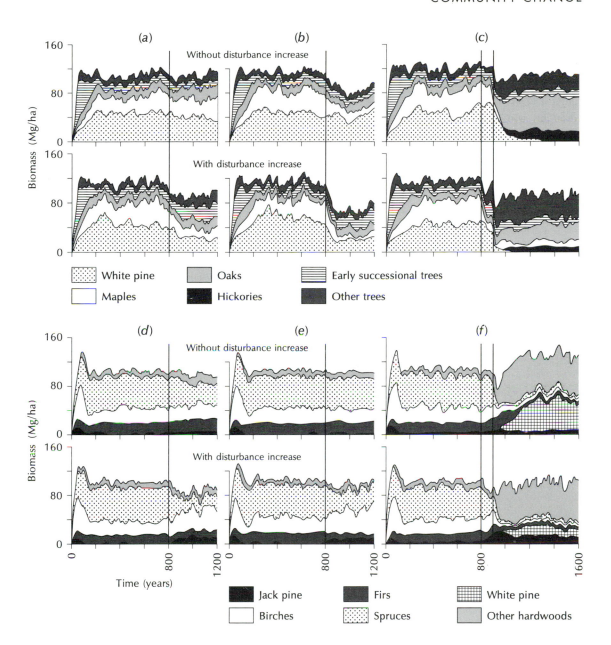

Figure 7.18 Simulated changes in species composition of forests at two sites investigated in eastern North America. (a) to (c) is a site in Wisconsin, and (d) to (f) is a site in southern Quebec. At both sites, experiments were run with an increase in disturbance at year 800 (top of figure for each site) and without an increase in disturbance at year 800 (bottom figure for each site). Additionally, three climatic change scenarios were simulated: a 1°C temperature increase at year 800 (left-hand figures, a and d); a 15 per cent decrease in precipitation (middle figures, b and e); and a transient change in which mean monthly precipitation and temperature were changed linearly from year 800 to year 900 and thereafter held constant (right-hand figures, c and f).

Source: After Overpeck *et al*. (1990)

understanding the ecological ramifications of global climatic change. Also, and perhaps more important for conservation biology, such validated models can help provide probabilities for the occurrence of these events, which will allow policy makers to make better, informed decisions (T. L. Root and Schneider 1993).

SUMMARY

Communities change. The nature of this change is debatable. The classic view saw climatic climaxes and balanced ecosystems resulting from unidirectional succession. Succession is a complex process and is explained by at least three models – the facilitation model, the tolerance model, and the inhibition model. It may also be driven by factors external to the community (allogenic factors). Primary succession is the colonization of land or submarine surfaces that have never existed before. Secondary succession is the invasion of newly created surfaces resulting from removal of pre-existing vegetation. A modern view of community change stresses the disequilibrium behaviour of communities. It sees succession going in many possible directions, and sees communities as temporary collections of species that assemble and disassemble as the environment changes. Much community change has been caused by land cover transformation over the past two hundred years. Two important aspects of this transformation are habitat fragmentation, and it attendant effects on wildlife, and the loss of wetlands. Global warming during the twenty-first century is likely to put peripheral, geographically restricted, highly specialized, poorly dispersive, and climatically sensitive species under intense pressure. It will also cause significant changes in many communities. Wetlands, tundra, and alpine meadows are especially vulnerable.

ESSAY QUESTIONS

1 Why has Clements's unidirectional view of succession been revised?

2 Why does habitat fragmentation pose a threat to many species?

3 What community changes are likely to occur as a result of global warming?

FURTHER READING

Committee on Characterization of Wetlands, National Research Council (1995) *Wetlands: Characteristics and Boundaries*, Washington, DC: National Academy Press.

Gates, D. M. (1993) *Climate Change and Its Biological Consequences*, Sunderland, Massachusetts: Sinauer Associates.

Huggett, R. J. (1993) *Modelling the Human Impact on Nature: Systems Analysis of Environmental Problems*, Oxford: Oxford University Press.

Huggett, R. J. (1997) *Environmental Change: The Evolving Ecosphere*, London: Routledge.

Jackson, A. R. W. and Jackson, J. M. (1996) *Environmental Science: The Natural Environment and Human Impact*, Harlow: Longman.

Matthews, J. A. (1992) *The Ecology of Recently-Deglaciated Terrain: A Geoecological Approach to Glacier Forelands and Primary Succession*, Cambridge: Cambridge University Press.

Peters, R. L. and Lovejoy, T. E. (eds) (1992) *Global Warming and Biological Diversity*, New Haven, Connecticut and London: Yale University Press.

Schwartz, M. (1997) *Conservation in Highly Fragmented Landscapes*, New York: Chapman & Hall.

8

LIFE, HUMANS, AND MORALITY

Human attitudes towards the living world are relevant to biogeographers involved with conservation and ecosystem management. This chapter covers:

- non-human rights
- kinds of environmentalism
- biogeographical theory and conservation practice

Biogeographical and ecological principles naturally inform questions of environmental management. However, environmental management does not just have a 'scientific' dimension. It has social and ethical dimensions, too. All dimensions must be considered in discussing the relations between humans and other living things. This chapter will explore the rights of animals and Nature, attitudes towards Nature, as seen in the different brands of environmentalism, and the relationships between biogeography, ecology, and conservation practice.

BIORIGHTS

Environmental ethics deals with value judgements about the biological and physical worlds. It was begun, at least in its modern form, by philosophers during the 1970s. Environmental ethics is a diverse and immense field of study. It arose from three considerations (Botkin and Keller 1995: 619). First, human actions, aided by technological developments, affect Nature deeply, so it seems necessary to examine the ethical consequences of these actions. Second, many human actions have world-wide consequences, and a global perspective raises its own moral questions. Third, moral concerns have expanded so that

animals, and even trees and landscapes, possess moral and legal rights – civilization includes all Nature within its ethical systems. This expansion of concerns demands a new environmental ethics.

Animal wrongs: do organisms have rights?

What are rights?

Rights are a social contrivance that help people to live together. They are not some quality that individuals possess. Rather, they are a quality that members of a society are willing to pretend that individuals possess. Such a pretence assures socially acceptable behaviour (most of the time). This idea is straightforward, though open to philosophical debate. The big question is whether the scope of 'rights' should be extended to include other species, and even rocks, rivers, and landscapes. There is no convincing reason why they should not be so extended. If rights should help humans to avoid conflict with one another, why should not animal rights, plant rights, and landscape rights help humans to protect their environment?

An all-encompassing definition of rights would help humans to respect Nature. But the matter is complicated by three preconditions, without which an object cannot be granted rights – commensurability, responsibility, and sensibility (Tudge 1997). **Commensurability** means that rights are a reciprocal agreement – you respect my rights and I'll respect yours. Unfortunately, this equitable arrangement is distorted by power structures. Some groups of people have less power to infringe the rights of others. More powerful groups (the richer for instance), therefore, tend to possess more rights than less powerful groups (the poorer for instance). As animals, plants, and landscapes have no effective power, why should they be granted rights? Stated so bluntly, this argument smacks of insensitivity. To soothe consciences, a rider is added – to earn rights, people must accept **responsibility**. Animals, plants, and landscapes cannot accept responsibility, therefore they can have no rights. However, there is no reason why rights and responsibilities should be inseparable. If animals are worthy of human respect, then they can be awarded rights, even though they should have no feelings toward humans. **Sensibility**, or lack of it, in all species save humans, arose from a mechanistic view of animal behaviour instigated by Ivan Pavlov and his dogs. To be brief, most scientists are now prepared to see other animals as something more than biochemical machines. It would seem that 'respectable science has caught up with the animal-lovers' and it is no longer perverse 'to perceive animals as sentient beings, who register their emotional responses with varying degrees of consciousness' (Tudge 1997: 41; see also Regan 1983).

So, if animals have rights, how should they be treated? This is a difficult issue with many aspects. All the aspects stem from a sea change in human attitudes toward animals, at the root of which is a new-found sympathy towards the natural world, and towards animals in particular (e.g. Fisher 1987). This sympathy has driven the forces of animal welfare over recent years. The **animal liberation movement**, through films and other media, has shown people how many animals are forced to live in pounds, on factory farms, and in laboratories. People have been stirred into various kinds of action – boycotting pet shops, breaking into laboratories to free animals, ceasing to eat meat, spurning the wildlife trade, and

so on. However, the change in human attitudes toward animals does not express itself solely in militant action. There is a quiet side to the change that is less evident but possibly more fundamental. Two examples from Australia and New Zealand (O'Brien 1990) should demonstrate this subtle and yet profound attitude change.

In Australia, the Victorian Acclimatization Society was formed in 1861. A central objective of this early wildlife management society was 'the introduction, acclimatization and domestication of all innoxious animals, birds, fishes, insects and vegetables whether useful or ornamental' (Rolls 1969). This objective was regarded as worthy and serving the public interest by scientists, natural historians, and landholders of the Society. In modern Australia, the acclimatization of exotic species would be regarded by many as irresponsible and bizarre, and is prevented by legislation. Attitudes towards 'pests', some of which are the ancestors of the 'innoxious' animals introduced in the nineteenth century, have also shifted. Take the example of the rabbit. The huge Australian rabbit population was controlled in the 1950s by introducing myxomatosis. But in 1987, when the same virus was considered as a control agent for the wild rabbit population in New Zealand, the welfare of wild rabbits was put first and the virus was not introduced.

Three moral issues arising from the notion of animal rights deserve closer scrutiny – culling animals, the wildlife trade, and keeping animals in captivity.

Culling

Controlling animal populations may mean culling. Few people raise objections to the eradication of viruses, bacteria, or pestilential insects. Attempts to limit populations of organisms higher up the evolutionary ladder cause eyebrows to be raised. Mammal **culling** causes public outcry. The culling of seal pups is met by a vociferous and appalled response.

The shooting of feral horses from helicopters began in Australia in late 1985. By 1988, the Commonwealth Department of Primary Industries was receiving more than 7,000 protest letters a year, mostly 'campaign' mail from correspondents overseas (O'Brien 1990). In the United States, feral horses are protected by law. The Wild Free-Roaming Horse and Burro Act (1971) decrees that feral horses and burros must be considered as an integral part of the natural systems in which they are found. The Act provides for the Bureau of Land Management to set suitable levels for herd size, and for surplus animals to be removed and offered for private maintenance by qualified individuals, or destroyed if they are old, sick, lame, or private care cannot be found. In 1987, 8,000 horses were held in captivity awaiting 'adoption'.

Many people accept that some humane culling of large mammals is a necessary evil to prevent over-exploitation of vegetation and environmental degradation. In Great Britain, for example, the culling of deer populations is probably necessary. With animal rights now to the fore, the method of culling is the subject of scientific study. On Exmoor and the Quantock Hills, in Somerset, red deer (*Cervus elaphas*) are culled by rifle (stalking) and by hunting with hounds. A study carried out for the National Trust (the Bateson Report) concluded that hunts with hounds cause deer great distress. The exertions of the chase change muscles and blood far beyond any changes that would be expected in normal life. Red deer are sedentary animals that escape natural predators (mainly the wolf, although there are none now in Britain) by brief sprints. Hunting deer with hounds is not 'natural', because wolves would never chase deer for long distances. As a result of the Bateson Report, members of the National Trust voted to ban deer hunting with hounds on all its lands.

The wildlife trade

The **wildlife trade** is an international business worth about $15 billion a year. Much of it is legal and controlled by national laws and an international treaty – CITES (Convention on International Trade in Endangered Species of Wild Fauna and Flora). CITES was drawn up in 1975 and 139 countries are now party to it. Collectively, these countries act to regulate and monitor trade in around 25,000 species

and to ban trade in a further 800 species threatened with extinction.

About a quarter of the wildlife trade is unlawful commerce in rare and endangered species. The unfortunate animals are poached in the wild and smuggled across frontiers. A typical year will see the sale of 50,000 primates, tusk ivory from 70,000 African elephants, 4 million live birds, 10 million reptile skins, 15 million pelts, 350 million tropical fish, and about 1 million orchids (Lean and Hinrichsen 1992: 145). People caught illegally trafficking in wildlife are arrested, fined, and even jailed. The growing willingness and determination to stop the wildlife trade is another instance of humans upholding the rights of animals.

Keeping animals in captivity

Zoological gardens are not what they were. As society's attitudes toward animals have changed, zoos have responded by rethinking the reasons for their existence. Zoos began in the heyday of natural history, when almost every educated Victorian was an amateur natural historian. Interests at the time centred around collecting, identifying, and naming. Zoos were places to display the ordinary and bizarre creatures found in far-flung parts of the globe. This role for zoos lingered well into the twentieth century. Even in 1970, the lion enclosure at Newquay Zoo, in Cornwall, England, was designed to keep lions on display to the public for as long as possible. This was achieved by keeping them in a relatively small space, and letting them return to their even smaller sleeping quarters only at night. In the 1990s at the zoo a large lion enclosure incorporates many features that resemble the lions' natural savannah environment. There is, for instance, a platform (simulating a small knoll) and trees. Food is hidden around the enclosure and the lions have to find it. And the lions are free to 'go indoors' when they wish.

A brochure for Newquay Zoo captures the spirit of the times when it says that good zoos are good for animals and good for people. They are good for animals for one reason only – conservation. Five aspects of conservation are paramount:

1 Breeding programmes save animals from extinction.
2 Wildlife research helps to preserve habitats and save animals.
3 Animal behaviour research assists their survival in the wild.
4 Providing sanctuaries for animals that have lost their habitat.
5 Providing hospitals for sick and injured animals.

Unless wild populations are supplemented by captive breeding, and culling populations that outgrow their reserves, then most of the world's largest terrestrial mammals will vanish. Zoos make significant contributions to conservation as genetic refuges and reservoirs, especially for large vertebrate species threatened with extinction (Rabb 1994). Przewalski's horses (*Equus przewalskii*) are believed to be extinct in the wild; the current known population of 797 animals exists in zoos (Boyd 1991). It is hoped to reintroduce this species to its former Mongolian habitat by the early years of the twenty-first century. Zoo resources for conservation and research are limited, so zoos are encouraging the development of criteria to help prioritize actions for conservation of biodiversity. North American, European, and Australian zoos are assisting the development of technical capacities among zoo counterparts, government agencies, and protected areas in both developing and developed countries of the world to further the conservation of biodiversity.

A remarkable success story (so far) is the golden lion tamarin (*Leontideus rosalia*) (p. 197). This squirrel-sized primate lives in the much-reduced lowland Atlantic forest, Brazil. Since the mid-1990s, a combination of relocating five tamarin families from vulnerable areas into the protection of the Poço das Antas Biological Reserve, and reintroducing 147 captive-bred tamarins into the wild, has put the little primate on the road to recovery. For long-term survival, the tamarin's habitat area will need to be doubled.

Does all Nature have rights?

A belated effect of the Darwinian revolution is that humans are now seen as part of Nature. Environmental philosophers now passionately prosecute this idea, but remain unclear as to its implications. What does it mean to be part of Nature? This philosophical issue is complex and beset by a terminological labyrinth (see Colwell 1987). But all commentators agree that 'being part of Nature' dispenses with the duality between humans and the natural world, provides a moral justification for treating Nature more humanely, and gives a philosophical justification for seeing intrinsic value in all things (e.g. Callicott 1985; McDaniel 1986; Zimmerman 1988).

An early and eloquent advocate of the intrinsic worth of the natural world was Aldo Leopold (1949), who advanced an all-embracing **land ethic**. Leopold affirmed the right of all resources, including plants, animals, and earth materials, to continued existence, and, at least in places, to continued existence in a natural state. This land ethic assumes that humans are ethically responsibility to other humans and human societies, and to the wider environment that includes, animals, plants, landscapes, seascapes, and the air. Granting rights of survival to animals does not necessarily mean that they cannot be eaten, but it does mean that endangered species should be cared for and helped to recuperate. A land ethic behoves humans to sustain Nature for the present generation and for future generations.

The question of Nature's rights arose as a legal issue in the 1970s. Mineral King Valley is a wilderness area in the Sierra Nevada, California. Disney Enterprises, Inc., wished to develop the valley as a ski resort within multimillion dollar recreational facilities. The Sierra Club (founded in 1892 to help preserve the Yosemite Valley and other natural wonders in California) objected that the development would destroy the aesthetic value and ecological balance of the area, and brought a suit against the government. The Sierra Club could not claim direct harm from the development, and the land was government owned and the government represented the people, so people in general were not wronged.

A lawyer, Christopher D. Stone (1972), suggested that, as there were precedents for inanimate objects such as ships having legal standing, trees should also have legal standing. So the suit was made on behalf of the non-human wilderness. The case was taken to the US Supreme Court but was turned down. However, in a famous dissenting statement, Justice William O. Douglas proposed the establishment of a new federal rule that would allow 'environmental issues to be litigated before federal agencies or federal courts in the name of the inanimate object about to be despoiled, defaced, or invaded by roads and bulldozers and where injury is the subject of public outrage'. Although trees were not given legal rights in this case, the ethical values and legal rights for wilderness were discussed.

ATTITUDES TOWARDS NATURE

The Age of Ecology opened on 16 July 1945 in the New Mexican desert 'with a dazzling fireball of light and a swelling mushroom cloud of radioactive gases' (Worster 1994: 342). Observing the scene, a phrase from the *Bhagavad-Gita* came into project leader J. Robert Oppenhiemer's mind: 'I am become Death, the shatterer of worlds'. For the first time in human history, a weapon of truly awesome power existed, a weapon capable of destroying life on a planetary scale. From under the chilling black cloud of the atomic age arose a new moral concern – environmentalism – that sought to temper the modern science-based power over Nature with ecological insights into the radiation threat to the planet. The publication of Rachel Carson's *Silent Spring* (1962), the first study to bring the insidious effects of DDT application to public notice, was more grist for the environmentalist's mill. The dual threats of radiation and pesticides set environmentalism moving.

Brands of environmentalism

There are several brands of **environmentalism**, each of which promotes a different attitude towards the environment (see O'Riordan 1988, 1996). These brands tend to polarize around, at one extreme, **ecocentrism**, with its decidedly motherly attitude towards the planet; and, at the other extreme, **technocentrism** with a more *laissez-faire*, exploitative attitude (Table 8.1). The 'green' end of this spectrum was initiated by the atomic bomb and galvanized into action by pesticide abuse. Each strand represents a system of beliefs, although no individual would necessarily believe in just one strand.

Technocentrics (pale greens and very pale greens)

Technocentrics tend to have a human-centred view of the Earth coupled with a managerial approach to resource development and environmental protection. They favour maintaining the status quo in existing government structures. Technocentrics tend to be conservative politicians of all political parties, leaders of industry, commerce and trades' unions, and skilled workers. In short, they are members of the productive classes. They aspire to improve wealth for themselves and for society at large. They enjoy material acquisition for own sake and for the status it endows. And they are politically and economically very powerful.

There are two brands of technocentric – the **cornucopians** (or **optimists**) and the **accommodationists** (or **environmental managers**). Conservative technocentrics are cornucopians or optimists. They are named after the legendary goat's horn that overflowed with fruit, flowers, and corn, and signified prosperity. They have faith in the application of science, market forces, and managerial ingenuity to sustain growth and survival.

Liberal technocentrics are accommodationists. They have faith in the adaptability of institutions and mechanisms of assessment and decision making to accommodate environmental demands. They believe that adjustments by those in positions of power and significance can meet the environmental challenge. The adjustments involve adapting to and moulding regulation (including environmental impact assessment) and modifying managerial and business practices to reduce resource wastage and economically

Table 8.1 Brands of environmentalism and their characteristics

Environmental inclination	Type and colour	Characteristics
Ecocentrism (Nature-centred)	Deep ecologists or Gaians (dark greens)	Stress the intrinsic importance of Nature for humanity Believe that ecological (and other) laws should dictate human morality Ardently promote biorights – the right of endangered species, communities, and landscapes to remain unmolested
	Communalists or self-reliance soft-technologists (medium greens)	Emphasize the smallness of scale ('small is beautiful') and hence community identity in settlement, work, and leisure Interpret concepts of work and leisure through a process of personal and communal improvement Stress the importance of participation in community affairs, and guarantee of the rights of minority interests
Technocentrism (human-centred)	Accommodationists or environmental managers (pale greens)	Believe that economic growth and resource exploitation can continue assuming: suitable economic adjustments to taxes, fees, etc.; improvements in the legal rights to a minimum level of environmental quality; compensation arrangements satisfactory to those who experience adverse environmental and social effects Accept new project appraisal techniques and decision reviews arrangements to allow for wider discussion or genuine search for consensus among representative groups of interested parties
	Cornucopians or optimists (very pale greens – dark blue under the surface)	Believe that humans can always find a way out of any difficulty, be it political, scientific, or technological Accept that pro-growth goals define the rationality of project appraisal and policy formulation Optimistic about the human ability to improve the lot of the world's people Faith that scientific and technological expertise provides a firm foundation for advice on matters pertaining to economic growth, public health, and public safety Suspicious of attempts to widen basis for participation and lengthy discussion in project appraisal and policy review Believe that all impediments can be overcome given a will, ingenuity, and sufficient resources arising out of growth

inconvenient pollution, without any fundamental shift in the distribution of political power. Pressure groups aligned to technocentrics tend to be NIMBY (not-in-my-back-yard) groups – they do not mind development so long as it does not take place near them – and private interest groups (consumer groups, civil liberties groups, and so on). They are very vocal and command much media attention.

Ecocentrics (medium greens and dark greens)

Ecocentrics have a Nature-centred view of the Earth in which social relations cannot be divorced from people–environment relations. They hold a radical vision of how future society should be organized. They would give far more real power to communities and to confederations of regional interests. Central control and national hegemony, so treasured by technocentrics, are the very antithesis of ecocentrism. Ecocentrics tend to be found among those on the fringe of modern economies, that is, the 'non-productive' classes – clerics, artists, teachers, students, and, to an ever increasing extent, women. They are characterized by a willingness to eschew material possessions for their own sake, and to concern themselves more with relationships; and by a lack of faith in large-scale modern technology and associated demands on elitist expertise and central state authority. Pressure groups aligned to ecocentrics tend to be green groups and alternative subcultures (e.g. feminists, the peace movement, religious movements).

There are two brands of ecocentrics. The more conservative ecocentrics are **communalists** or **self-reliance**, **soft technologists**. They believe in the co-operative capabilities of societies to be collectively self-reliant using appropriate science and technology. That is, they believe in the ability of people to organize their own economies if given the right incentives and freedoms. This view extends a 'bottom up' approach to the development of the Third World based on the application of indigenous customs and appropriate technical and economic assistance from western donors.

Liberal ecocentrics are **Gaians** or **deep ecologists** (Box 8.1). They believe in the rights of Nature and the essential co-evolution of humans and natural phenomena. Extreme ecocentrics believe that the moral basis for economic development must lie in the interconnections between natural and social rights: there is no purely anthropocentric ethic. Within this group might be included a powerful religious lobby (Box 8.2).

Alignments of ecologists

It would be wrong, if excusable, to imagine that all ecologists and environmental scientists hold ecocentric views. The relationship between ecocentric environmentalists and ecologists has not always been as cosy as might be supposed: ecologists do not always say what ecocentrics want to hear. Paul Colinvaux (1980: 105) wrote: 'If the planners really get hold of us so that they can stamp out all individual liberty and do what they like with our land, they might decide that whole counties full of inferior farms should be put back into forest'. His displeasure with land-use planning and environmentalism is evident. Later in the book (Colinvaux 1980: 119), his words smack of **social Darwinism** and 'Nature red in tooth and claw', when he talks of different species 'going about earning their livings as best they may, each in its own individual manner', and what 'look like community properties' being 'the summed results of all these bits of private enterprise', though elsewhere he sees peaceful coexistence, not struggle, as the outcome of natural selection, the peace being broken in upon by a deviant aggressor species such as *Homo sapiens* who seek to encroach on another's niche (Colinvaux 1980: 131). Today, few ecologists wholeheartedly accept Colinvaux's viewpoint, and many approach Nature green in head and heart. One of the many is Daniel B. Botkin (1990) who advocated using modern technology in a constructive and positive manner, a position that tries to bridge the middle ground between ecocentrism and technocentrism.

Box 8.1

DEEP ECOLOGY

The deep ecology movement began in the early 1970s. It was named by the Norwegian, Arne Naess, in 1973. Naess's original article has spawned an ecological cult. It set the deep ecologists against the shallow ecologists. The central objective of the shallow ecologists, in fighting against pollution and resource depletion, is the health and affluence of people in developed countries. The deep ecology manifesto, as originally penned by Naess, was embodied in a set of slogans (Table 8.2). The differing attitudes of deep ecologists and shallow ecologists are summarized in Table 8.3. However, some commentators believe that deep and shallow ecological perspectives are a somewhat crude division of an environmental-perspectives continuum that has grown from a long-standing critique of western development (e.g. Jacob 1994).

Table 8.2 Slogans for deep and shallow ecology

Shallow ecology	*Deep ecology*
Natural diversity is valuable as a resource	Natural diversity has its own intrinsic value
Nature is there for humans to exploit as they will, even if that should cause some damage	Nature does not belong to humans and humans have no right to destroy natural features of the planet
It is nonsense to talk about value except as value for humankind	Equating value with value for humans reveals a radical prejudice
Animals are valuable only as resources for people	Animals have their own intrinsic value and have a right to live, even if they should be no use to humans
Plant species should be saved because they are valuable genetic reserves for agriculture and medicine	Plant species should be saved because of their intrinsic value
Pollution should be decreased if it should threaten economic growth	Decrease of pollution has priority over economic growth
Third World population growth threatens ecological equilibrium	World population at the present level threatens ecosystems, the major threat being posed by the people and attitude of industrial states more than by those of many others. Human population is today excessive
'Resource' means resource for humans	'Resource' means resource for living beings
People will not tolerate a broad decrease in their standard of living	People should not tolerate a broad decrease in quality of life, but they should be ready to accept a reduction in the living standard in overdeveloped countries
Nature is cruel and necessarily so	Humans are cruel but not of necessity

Source: After Naess (1973)

Table 8.3 Different attitudes of deep and shallow ecologists

Environmental problem	Shallow ecology	Deep ecology
Pollution	Advocate a technology that seeks to purify the air and water and to spread pollution more evenly; laws limit possible pollution; polluting industries are preferably exported to developing countries	Evaluate pollution from an ecospheric point of view, not focusing on its effects on human health, but on the effects on life as a whole, including life conditions of every species and ecosystem
Resources	Emphasize resources for human beings, especially for the present generation in affluent societies. Resources belong to those who have the technology to exploit them. Resources will not be depleted because, as they get rarer, higher market prices conserve them, and substitutes will be found through technological progress. Plants, animals, and natural objects are valuable only as resources for humans; if they should have no use to humans, they can be destroyed	Deeply concerned with resources and habitats for all life-forms for their own sake; no natural object is regarded solely as a resource; this view leads to a critical examination of human modes of production and consumption
Population	See overpopulation as a problem of developing countries. The question of the optimum human population is discussed without reference to the optimum population of other life forms. The destruction of wild habitats is accepted as a necessary evil	Recognize that excessive pressures on planetary life stem from the human population explosion. The pressure stemming from industrial societies is a major factor, and population reduction in these societies, as well as in the developing countries, is imperative
Cultural diversity and appropriate technology	Regard industrialization of the kind manifested in the west as an example for developing nations	Concerned with maintaining cultural diversity, and limiting the impact of western society upon presently existing non-industrial societies, defending the fourth world against foreign domination, and applying local, soft technologies where appropriate

Source: After Naess (1986)

Box 8.2

RELIGION, NATURE, AND THE WORLD WILDLIFE FUND

In autumn 1986, a unique alliance was forged between conservation, as represented by the World Wildlife Fund, and five of the world's chief religions. The venue was Assisi, in central Italy, home of the patron saint of animals. A pilgrimage, conference, cultural festival, retreat, and interfaith ceremony in the Basilica of St Francis formed the basis of a permanent bond between conservation and religion.

For religions, ecological problems pose major challenges in relating theology and belief to the damage to Nature and human suffering caused by environmental degradation. All too often, religious teaching has been used or abused to justify the destruction of Nature and natural resources. In the declarations issued at Assisi, leading religious figures mapped out the way to re-examine and reverse this process. And in 1987, the Baha'i' Movement, which had its origins in Islam, joined the World Wildlife Fund's New Alliance. At the same time, starting with the conference and retreat held in Assisi, religious philosophers are helping inject some powerful moral perspectives into conservation's ill-defined ethical foundations.

A development of the Assisi event was the making of the Rainbow Covenant at Winchester Cathedral, England in October 1987. Covenants have a long history. For the ancient Hittites, covenants were made between rulers and subjects for the benefit of both. For Jews and Christians, 'The Covenant' is between God and his people, and was made after the Flood. In the Bible, Noah, at God's command, builds an ark to save species on Earth from extinction when God sends a grand Flood to punish humanity. After the Flood has finished, God makes a covenant – the first in the Bible. He makes his covenant not just with Noah, but with all creatures on Earth, in the seas, and in the skies. God promises never wantonly to destroy life again, and sets the rainbow in the sky to bear witness to his promise.

The Rainbow Covenant has been inspired by God's covenant with Noah, and was specifically designed by the International Consultancy on Religion, Education and Culture (ICOREC) to be usable by anyone, religious or not. So, unlike the Hittite and biblical covenants, it is unilateral – a contrite pledge by humans to other living beings and their ecosystems. The rainbow's arc embraces not only those who live, but those who will live; it encompasses all living creatures and the earth, air, and water that sustain them. This is the Rainbow Covenant:

Brothers and Sisters in Creation, we covenant this day with you and with all creation yet to be;
With every living creature and all that contains and sustains you.
With all that is on Earth and within the Earth itself;
With all that lives in the waters and with the waters themselves;
With all that flies in the skies and with the sky itself.
We establish this covenant, that all our powers will be used to prevent your destruction.
We confess that it is our own kind who put you at risk of death.
We ask for your trust and as a symbol of our intention we mark this covenant with you by the rainbow.
This is the sign of the covenant between ourselves and every living thing that is found on the Earth.

A balancing act: biorights, human rights, and ecosystem management

Do humans still have rights in the world of the deep ecologist? People are part of Nature. The philosophy behind a new brand of conservation – ecosystem management – accepts that simple fact.

Ecosystem management arose in the late 1980s and is advocated by many scientists and other people interested in the environment (e.g. Agee and Johnson 1988). Its ultimate aim is to enhance and to ensure the diversity of species, communities, ecosystems, and landscapes. It is a fresh and emerging model of resource management. It covers a spectrum of approaches. A mild, technocentric version simply extends multiple use, sustained yield policies, prosecuting a stewardship approach, in which the ecosystem is seen merely as a human life-support system (e.g. Kessler *et al.* 1992). In this view, public demands for habitat protection, recreation, and wildlife uses are simply seen as constraints to maximizing resource output (Cortner and Moote 1994). A more radical, ecocentric approach is to accept Nature on its own terms, even where doing so means controlling incompatible human uses (e.g. Keiter and Boyce 1991). This extreme, but eminently sensible, form of ecosystem management reflects a willingness to place environmental values, such as biodiversity and animal rights, and social and cultural values, such as the upholding of human rights, on an equal footing. It means accepting that environmental concerns are not purely factual; they have do with values and ethics.

People exert a profound influence on ecological patterns and processes, and in turn ecological patterns and processes affect people. The connection between technology and the environment is well studied; not so the connection between social system and the environment. However, policies are tending to move away from the administrator-as-neutral-expert approach to policies that engender public deliberation and the discovery of shared values. Naturally, such extension of ecological matters to the social and political arena presents difficulties, though these may not be insuperable (e.g. Irland 1994). Ecosystem management accepts that human values must play a leading role in policy decision-making. Conservation strategies must take account of human needs and aspirations; and they must integrate ecosystem, economic, and social needs. The key players in ecosystem management are scientists, policy makers, managers, and the public. The public, many of whom have a keen interest in environmental matters, are becoming more involved in ecosystem management as professionals recognize the legitimacy of claims that various groups make on natural resources. In Jervis Bay, Australia, the marine ecosystem is used by many existing and proposed conflicting interests (national park, tourism, urbanization, military training) (Ward and Jacoby

1992). Similarly, in the forests of the south-western United States, ecosystem, economic, and social needs are considered in policy decision-making concerning ecology-based, multiple-use forest management (Kaufmann *et al.* 1994) (Figure 8.1).

BIOGEOGRAPHY, ECOLOGY, AND CONSERVATION PRACTICE

Biogeographical and ecological ideas inform conservation practice. When ruling theories in ecology and biogeography change, conservation practice soon follows suit. This generalization, though valid, should be treated cautiously, as the relationships between ecological science and environmental policy are complicated (Shrader-Frechette and McCoy 1994). Nevertheless, it is fair to say that the major paradigm shifts in twentieth-century ecological thought have engendered quite distinct conservation philosophies. Four paradigms, which were mentioned in Chapter 7, are recognized: enduring equilibrium, balanced ecosystems, disequilibrium communities, and the edge of chaos.

Environmental exploitation and enduring equilibrium

Environmental problems in the offing

In his *Walden; or, Life in the Woods* (1893 edn), Henry David Thoreau expressed concern for wild Nature and wild environments. He urged that, when a conflict arises between Nature and human society, then the first thing to consider are the laws of Nature. George Perkins Marsh wrote *Man and Nature; or, Physical Geography as Modified by Human Action* in 1864. He examined 'the greater, more permanent, and more comprehensive mutations which man has produced, and is producing in earth, sea, and sky; sometimes, indeed, with conscious purpose, but for the most part, as unforeseen though natural consequences of acts performed for narrower and more immediate ends' (Marsh 1965 edn: 19). He believed that 'Man is everywhere a disturbing agent. Wherever he plants

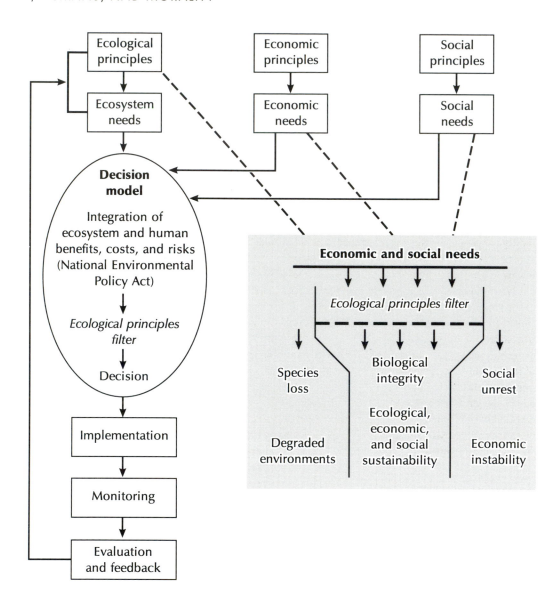

Figure 8.1 The integration of ecological, economic, and social needs in a decision-analysis model. Economic and social needs are tested against an 'ecological filter', which is shown in the shaded box. The aim is to determine economic and social actions that will produce the most desirable balance between biological integrity and ecological, economic, and social sustainability. Bowing fully to economic and social needs would lead to species loss and environmental degradation. Bowing fully to ecological needs would lead to social unrest and economic instability. A compromise position allows the maintenance of biological integrity while catering for economic and social needs. The resulting decision model leads to the implementation of an environmental policy. The effects of the policy are carefully monitored and evaluated. If the policy should fail to work as desired, then amendments can be made and the process started anew, until a satisfactory outcome is achieved.

Source: After Kaufmann *et al.* (1994)

his foot, the harmonies of nature are turned to discords' (Marsh 1965 edn: 36). Marsh was the first modern scholar to see humans as a factor, and not merely as a subject, in Nature.

In 1922, Robert Lionel Sherlock published an enormously important book entitled *Man as a Geological Agent: An Account of His Action on Inanimate Nature*. From extensive observations within the British Isles, Sherlock showed the effects that human activities were having on the landscape, the effects that forestry, grazing, agricultural management, and the technological and industrial activity of twentieth-century society were having on landscape processes. Specifically, he looked at the effects of road and rail-way construction, open-cast coal mining, and slate, stone, gravel, and sand quarrying; and the management of waterways and coasts. He gathered statistics giving the amounts of material extracted or removed, and furnished estimates of the rates of erosion and sedimentation.

During the late 1920s and the 1930s, the recognition of the human impact on the soil and the need for soil conservation was a pressing concern, largely because of the dust-bowl that devastated the Great Plains region in the United States. The main problems of environmental degradation were found mainly in western Europe and North America.

Broken equilibrium

Whatever the root cause of environmental degradation should be, Clements's notion of enduring equilibrium suggested that humans had erred. Its message for con-servation was clear: the climax state is Nature's design for biotic communities so humans should respect and preserve it. It was thought that the equilibrium between humans and Nature had been upset and should be restored, that Nature's design principles had been forsaken and needed immediate reinstatement. This response to environmental degradation was evident in Paul Sears's *Deserts on the March* (1949), the first edition of which appeared in 1935, in the after-math of the American dust-bowl calamity. Sears advocated that each county in the United States should appoint an ecologist to advise on matters of land use. This practice, he averred, should stop environmental degradation and place the entire nation on a biologically and economically sustainable foot-ing. In other words, he was proposing ecological principles to guide conservation practice and achieve a relationship with Nature that ensured an enduring equilibrium.

Balanced ecosystems

The balanced ecosystem view, and its extension in the Gaia hypothesis, supplanted 'enduring equilibrium' as a blueprint for Nature's design. It guided conservation strategy in two different ways. First, ecocentric envi-ronmentalists argued that unrestrained interference with Nature's development strategy leads to eco-system damage, and that the world's endangered ecosystems should be defended against the excesses of human actions. 'Spaceship Earth', a term first used by Adlai Stevenson, a United States Ambassador, in a speech given before the United Nations Economic and Social Council in Geneva on 9 July 1965, became a leading environmental idea.

The image of Spaceship Earth has now lost its potency. Nonetheless, it sparked off an environ-mentally friendly brand of economics. Free enterprise and Marxist economists placed humans at the stage centre and tended to disregard the natural environ-ment, seeing it as a free and inexhaustible reservoir and a bottomless rubbish dump (Cottrell 1978: 7). This exploitative attitude fostered traditional resource management practice that maximizes production of goods and services through sustained yield from balanced ecosystems; adopts utilitarian values that regard human consumption as the best use of resources; and holds a continuous supply of goods for human markets as the purpose of resource manage-ment (Cortner and Moote 1994). Such barefaced 'resourcism' is ecologically unsound. It fails to recog-nize limits to exploitation and, in consequence, a growing number of species, and even entire eco-systems, are currently endangered. But, unsound or not, it persists. Even now, ecosystems are viewed by some as long-lived, multi-product factories (Gottfried 1992), or, if you prefer, as Nature's superstores.

In the mid-1960s, a new 'environmental economics' emerged that was sensitive to the needs of the environment. Kenneth Ewart Boulding (1966, 1970), for instance, complained that humans have operated a 'cowboy economy', in which success is measured in the amount of material turned over and in which resources are exploited in a profligate and reckless manner. Instead, he insisted, life on a small and crowded planet demanded a 'spaceship economy', in which success is gauged by turnover and the watchwords are conservation, maintenance, and recycling. Eugene P. Odum employed the spaceship analogy as recently as 1989. He recalled the crisis on board the spacecraft Apollo 13. An explosion in the craft had caused damage to the vessel and threatened the lives of the crew. After several anxious hours, the crew managed to reach the safety of the lunar module and abandoned ship. Odum compared this event to the current environmental crisis on Earth. Our life-support systems are being destroyed, but, unlike the Apollo 13's crew, humans cannot abandon ship and find a safe haven. The only option is to stay on board Spaceship Earth and repair the damage. Unfortunately, he said, humans do not know how to repair the planet, although he hinted that a civilization capable of putting people on the Moon should have wit enough to comprehend the mechanics of ecospheric dynamics and keep the cogs of the ecological machine running freely. Luckily, he observed, the Earth has its own regeneration systems, its own means of healing wounds. The view that the Earth has its own powers of healing is supported by the Gaia hypothesis, the medical analogy being made clear in Lovelock's (1989) use of the term 'geophysiology'. (As an aside, I would point out that the impossibility of indefinitely sustaining life in a spacecraft was explored in Brian Aldiss's fictional 'Helliconia' trilogy, which also used an exotic setting to speculate on the possibilities of the Gaia hypothesis.)

Second, technocentric environmentalists used the balanced ecosystem view of Nature to promote the understanding of balanced ecosystem function as a basis for planetary management. Both ecocentrics and technocentrics thought they could determine what was safe for humans to do and what was not. Largely, they were mistaken (see Worster 1994: 416).

Many ideas of the equilibrium ecologists are used today. Equilibrium ecological thought also underpins the theory of island biogeography, as formulated by Robert H. MacArthur and Edward O. Wilson in the 1960s, which is still used to aid the designing of nature reserves. Some conservationists have espoused the steady-state Gaia hypothesis as a blueprint for planetary management (Myers 1987), and some very green environmentalists have used it as a design for living. The equilibrial ecological theme persists in the notion of sustainability. This is the well-meaning, but possibly misguided, idea that the ecosphere should be developed without compromising the ability of future generations to satisfy their needs. Or, put another way, sustainable development urges that diverse, functioning ecosystems should be preserved without damaging the economy; and that economies and human welfare need preserving as much as the integrity of ecosystems (Gerlach and Bengston 1992). It was first presented in 1980 at an international forum in the World Conservation Strategy, by the International Union for the Conservation of Nature (IUCN), and has become mainstay of much environmentalist philosophy. The IUCN identified three key areas for action: the maintenance of essential ecological processes and life-support systems, the preservation of genetic diversity, and the sustainable use of species and ecosystems (Allen 1980).

But the idea of sustainability is questionable. One problem lies in integrating human and biophysical factors. Humans interact with Nature through culture, often in symbolic ways that are not comprehended by biological or physical ecosystem models (Gerlach and Bengston 1992). In other words, humans can generate wants and capabilities of meeting these wants that lie outside the natural ecological order. This is often difficult for scientists to understand, and leads to their focusing on environmental protection and restoration, rather than facing the greater challenge of understanding social and cultural interactions with the biophysical world. The very idea of sustainability itself is a curious human construct. Laudable though the idea of sustainability might be, there are problems with it, not the least of which is the lack of a definition. It was originally taken to mean meeting the needs of the present generation without compromising the ability of future generations to satisfy their needs (World Commission on Environment and Development 1987), but at least eight interpretations are available (Gale and Cordray 1991). It is therefore a buzzword that lacks bite. Another problem is its inadvertent arrogance – should the present generation be so presumptuous and foolhardy as to anticipate the needs of future generations? These pithy questions should not be ducked in ecosystem management initiatives.

Evolutionary disequilibrium

Some time in the 1970s, the notions of homeostasis, balance, and stability in Nature collapsed. They were supplanted by the idea of **evolutionary disequilibrium**. The dramatic change of paradigms was, in part, a reflection of the human society at that time. It arose when it was fashionable to see neither Nature nor society as a stable entity, steadfastly withstanding all attempts to dislodge it from a global equilibrium state. Instead, a view emerged of all history as a record of disturbance springing from both 'cultural *and* natural agents, including droughts, earthquakes, pests, viruses, corporate takeovers, loss of markets, new technologies, increasing crime, new federal laws, and even the invasion of America by French literary theory' (Worster 1994: 424, emphasis in original).

Disequilibrium ecologists seemed divided among themselves on the advice they should give to society about how to act over the environment. One group, reflecting some of the new disequilibrium thinking, began to challenge the public perception that ecology and environmentalism were the same thing. Some ecologists became disenchanted with trying to conserve a healthy planet. Nature is characterized by highly individualistic associations, so why attempt to constrain it? This anarchic argument, if taken to the extreme, could have revived social Darwinism and stood in antithesis to the conservation ethic of

co-operation and collective action suggested by Odum's balanced ecosystems. In the event, a mildly anarchic view of Nature was taken by some ecologists (p. 215). Another group of ecologists, in stark contrast, drew different conclusions from the disequilibrium trends within their discipline. Daniel B. Botkin (1990) was one of the most articulate advocates of a new, chastened set of environmentalist policies. He favoured an environmentalism that was more friendly towards manipulating and dominating Nature, but that tolerated modern technology and progress, and human desires for greater wealth and power. This was not the preservation of a balanced nature, so much as a responsible attitude to technological developments and their effects on an environment that would inescapably change. Botkin's specific proposals were somewhat vague. He cautioned that we should not engineer Nature at an unnatural rate nor in novel ways. But in a world where disturbance is commonplace and change is the rule, how may the unnaturally rapid and the novel be identified and defined?

The edge of chaos

The implications of **chaotic ecosystem dynamics** for conservation are far from clear. How are conservationists to use a chaotic design? Answers to this question have to confront a paradox: a chaotic view of Nature is at once exhilarating and threatening. Chaotic Nature, so irregular and individualistic in character, appears almost impossibly difficult to admire or to respect, to understand or to predict. It seems to be a world in which the security of stable, permanent rules is gone forever, a dangerous and uncertain world that inspires no confidence (Prigogine and Stengers 1984: 212–13). This dark aspect of chaos might promote a feeling of alienation from the natural world, and cause people to withdraw into doubt and self-absorption (cf. Worster 1994: 413). It might also set people wondering how they should behave in a world where chaos reigns. With natural disturbance found everywhere, why should humans worry about doing their own bit of disturbing? Why not join individuals of all other species and do their own thing, free of guilt that in doing so they would cause any specific damage? As Worster (1994: 413) put it, what does the phrase 'environmental damage' mean in a world so full of natural upheaval and unpredictability?

However, chaos does not have to be portrayed in such gloomy and doom-laden terms. Chaotic Nature has a bright and edifying aspect, too. In a chaotic world, communities, ecosystems, and societies are sensitive to disturbance. Small disturbances can, sometimes, grow and cause the communities, ecosystems, or societies to change. Consequently, it is a world in which individual activity may have major significance (cf. Prigogine and Stengers 1984: 313). It is a post-modern world in which increased individuality and diversity encourages a great overall harmony (a notion akin to Taoist beliefs). Moreover, the newfangled theory of chaotic dynamics is leading to the discovery of hidden regularities in natural processes, the application of which is proving most salutary (Stewart 1995). Satellites can now be sent to new destinations, ferried by gravitational forces through a fractal Solar System, with far less fuel than was once thought possible. A dishwasher, involving two rotating arms that spin chaotically, has been invented in Japan that removes dirt from dishes using less energy than in a conventional dishwasher. Some forests and fisheries are being better managed through an understanding of their chaotic behaviour. And, chaos theory is giving us a breathtaking view of our Universe, and persuading us to look again at our place within it. It offers us a new design with which to reconsider our world, and a new framework within which to review our fears and our hopes.

Forward from the past: the biodiversity bandwagon

By the last decade of twentieth century, ecologists were undecided about the implications of their subject for modern technological civilization. Nonetheless, they found themselves regrouping around a new conservation ideal – **biodiversity**. It mattered not that Nature was chaotic and unpredictable. It was also gloriously diverse, and, for all its disturbing and perverse ways, and its abiding ability to evade our understanding, it still needed our love, our respect, and our help (cf. Worster 1994: 420). This view has supplied a conservation ethic, and a source of inspiration (and funding) for biogeographers, for the closing years of the twentieth century, and perhaps beyond.

SUMMARY

There are moral grounds for granting rights to animals, plants, and even landscapes. In the late 1990s attitudes towards animals are more sympathetic than they were in the past. This sympathy is seen in the public outcry against cruel methods of culling, in the measures taken to stop the wildlife trade, and in the new role for zoos in conservation. Attitudes towards Nature are varied and reflected in different brands of environmentalism. The two main opposing brands of environmentalism are ecocentrism and technocentrism. Ecosystem management accommodates human needs and aspirations within an ecocentric perspective. Biogeographical and ecological theory informs conservation practice. During the twentieth century the leading ideas in biogeography and ecology have changed. In consequence, conservation guidelines have changed. As the twenty-first century draws near, the chaotic world of the theoretical ecologist stands parallel with the biodiverse world of the practical ecologist.

ESSAY QUESTIONS

1 Should animals, plants, and landscapes be granted rights?

2 To what extent does biogeographical and ecological theory inform conservation practice?

3 What is so special about ecosystem management?

FURTHER READING

Anderson, E. N. (1996) *Ecologies of the Heart: Emotion, Belief, and the Environment*, New York: Oxford University Press.

Botkin, D. B. (1990) *Discordant Harmonies: A New Ecology for the Twenty-First Century*, New York: Oxford University Press.

Committee on Scientific Issues in the Endangered Species Act, National Research Council (1995) *Science and the Endangered Species Act*, Washington DC: National Academy Press.

Cooper, N. and Carling, R. C. J. (eds) (1996) *Ecologists and Ethical Judgements*, London: Chapman & Hall.

Norton, B. G. (1991) *Toward Unity Among Environmentalists*, New York: Oxford University Press.

Ryder, R. D. (ed.) (1993) *Animal Welfare and the Environment*, London: Duckworth and RSPCA.

Shafer, C. L. (1990) *Nature Reserves: Island Theory and Conservation Practice*, Washington and London: Smithsonian Institution Press.

Sylvan, R. and Bennett, D. (1994) *The Greening of Ethics: From Human Chauvinism to Deep-Green Theory*, Cambridge: White Horse Press; Tucson, University of Arizona Press.

GLOSSARY

Table G1 The geological time-scale for the last 570 million years

Era and sub-era	Period	Epoch	Age range (millions of year ago)	
			Start	Finish
Cenozoic era				
Quaternary sub-era	Pleistogene	Holocene	0.01	0
		Pleistocene	1.64	0.01
Tertiary sub-era	Neogene	Pliocene	5.2	1.64
		Miocene	23.3	5.2
	Palaeogene	Oligocene	35.4	23.3
		Eocene	56.5	35.4
		Palaeocene	65	56.5
Mesozoic era				
	Cretaceous	Late	97	65
		Early	145.6	97
	Jurassic	Late	157	145.6
		Middle	178	157
		Early	208	178
	Triassic	Late	235	208
		Middle	241	235
		Early	245	241
Palaeozoic era				
	Permian		290	245
	Carboniferous		362.5	290
	Devonian		408.5	362.5
	Silurian		439	408.5
	Ordovician		510	439
	Cambrian		570	510

abiotic Characterized by the absence of life; inanimate.

acidophiles Organisms that love acidic conditions.

adaptation The adjustment or the process of adjustment by which characteristics of an entire organism, or some of its structures and functions, become better suited for life in a particular environment.

aerobic Depending on, or characterized by, the presence of oxygen.

algae Simple, unicellular or filamentous plants.

alkaliphiles Organisms that flourish in alkaline environments.

allelopathic Pertaining to the influence of plants upon each other through toxic products of metabolism – a sort of phytochemical warfare.

allogenic Originating from outside a community or ecosystem.

anaerobic Depending on, or characterized by, the absence of oxygen.

anthrax An infectious and often fatal disease caused by the bacterium *Bacillus anthracis*; common in cattle and sheep.

apical buds Buds located at the tips of shoots.

aquifer A subsurface geological formation containing water.

archipelago A group of islands, typically a large group.

aridity The state or degree of dryness.

atmosphere The gaseous envelope of the Earth, retained by the Earth's gravitational field.

autogenic Originating from inside a community or ecosystem.

autotrophs Organisms – green plants and some micro-organisms – capable of making their own food from inorganic materials.

available moisture Precipitation less evaporation.

azonal soils Soils in which erosion and deposition dominate soil processes, as in soils formed in river alluvium and sand dunes.

bacteria Micro-organisms, usually single-celled, that exist as free-living decomposers or parasites.

barrier Any terrain that hinders or prevents the dispersal of organisms.

basal metabolic rate The rate of energy consumption by an organism at rest.

behaviour The reaction of organisms to given circumstances.

benthic Pertaining to the bottom of a water body.

biocide A poison or other substance used to kill pests (and inadvertently other organisms).

bioclimates The climatic conditions that affect living things.

biodiversity The diversity of species, genetic information, and habitats.

biogeochemical cycles The cycling of a mineral or organic chemical constituent through the biosphere; for example, the carbon cycle.

biological control The use of natural ecological interrelationships to control pest organisms.

biomagnification The concentration of certain substances up a food chain.

biomantle The upper portion of soil that is worked by organisms.

biomass The mass of living material in a specified group of animals, or plants, or in a community, or in a unit area; usually expressed as a dry weight.

biome A community of animals and plants occupying a climatically uniform area on a continental scale.

biosphere Life and life-supporting systems – all living beings, atmosphere, hydrosphere, and pedosphere.

biota All the animals (fauna) and plants (flora) living in an area or region.

biotic Pertaining to life; animate.

biotic potential The maximum rate of population growth that would occur in the absence of environmental constraints.

boreal forest A plant formation-type associated with cold-temperate climates (cool summers and long winters); also called taiga and coniferous evergreen forest. Spruces, firs, larches, and pines are the dominant plants.

browsers Organisms that feed mainly on leaves and young shoots, especially those of shrubs and trees.

bryophytes Simple land plants with stems and leaves but no true roots or vascular tissue: mosses, hornworts, and liverworts.

cacti New World plants belonging to the family Cactaceae with thick, fleshy, and often prickly stems.

calcicoles (calciphiles) Plants that love soils rich in calcium.

calcifuges (calciphobes) Plants that hate soils rich in calcium.

capillary pressure The force of water in a capillary tube.

carrion Dead and decaying flesh.

carrying capacity The maximum population that an environment can support without environmental degradation occurring.

caviomorph rodents A suborder of the Rodentia.

Cenozoic era A slice of geological time spanning 65 million years ago to the present; the youngest unit of the geological eras.

chasmophytes Plants that live in crevices.

chomophytes Plants that live on ledges.

chronosequence A time sequence of vegetation or soils constructed by using sites of different age.

circulatory system The system of vessels through which blood is pumped by the heart.

cohort Individuals in a population born during the same period of time, e.g. during a particular year.

commensalism An interaction between two species in which one species (the commensal) benefits and the other species (the host) is unharmed.

community assembly The coming together of species to form a community.

competition An interaction between two species trying to share the same resource.

competitive exclusion principle The rule that two species with identical ecological requirements cannot coexist at the same place.

congener An organism belonging to the same genus as another or others.

conifers Trees of the Order Coniferales, commonly evergreen and bearing cones.

consumption Community respiration.

continental drift The differential movement of continents caused by plate tectonic processes.

corolla The inner envelope of a flower; it is made of fused or separate petals.

corridors Dispersal routes offering little or no resistance to migrating organisms.

Cretaceous period A slice of geological time spanning 145.6 million to 65 million years ago; the youngest unit of the Mesozoic era.

culling The act of removing or killing 'surplus' animals in a herd or flock.

cursorial Adapted for running.

cuticle A protective layer of cutin (a wax-like, water-repellent material) covering the epidermis (outermost layer of cells) of plants.

cycad Any gymnosperm of the family Cycadaceae looking like a palm tree but topped with compound, fern-like leaves.

decomposer An organism that helps to break down organic matter.

deme A group of interbreeding indivduals; a local population.

density-dependent factors Causes of fecundity and mortality that become more effective as population density rises.

density-independent factors Causes of fecundity and mortality that act independently of population density; natural disasters are an example.

detritivore An organism that comminutes dead organic matter.

detritus Disintegrated matter.

diatom A microscopic, unicellular, marine (planktonic) alga with a skeleton composed of hydrous opaline silica.

dicotyledonous plant species Plants belonging to the Dicotyledoneae, one of the two major divisions of flowering plants. They are characterized by a pair of embryonic seed leaves that appear at germination.

disharmonious community A community of animals and plants adapted to a climate that has no modern counterpart.

dispersal The spread of organisms into new areas.

dispersal route The path followed by dispersing organisms.

distribution The geographical area occupied by a group of organisms (species, genus, family, etc.); the same as the geographical range.

drought A prolonged period with no rain.

ecological equivalents Species of different ancestry but with the same characteristics living in different biogeographical regions. Also called vicars.

ecosphere The global ecosystem – all life plus its life-support systems (air, water, and soil).

ecosystem Short for ecological system – a group of organisms together with the physical environment with which they interact.

ecotone A transitional zone between two plant communities.

El Niño The appearance of warm water in the usually cold water regions off the coasts of Peru, Ecuador, and Chile.

endangered species A species facing extinction.

endemic species A species native to a particular place.

environment The biological (biotic) and physical (abiotic) surroundings that influence individuals, populations, and communities.

environmental ethics A philosophy dealing with the ethics of the environment, including animal rights.

environmental factor Any of the biotic or abiotic components in the environment, such as heat, moisture, and nutrient levels.

environmental gradient A continuous change in an environmental factor from one place to another; for example, a change from dry to wet conditions.

environmentalism A movement that strives to protect the environment against human depredations.

Eocene epoch A slice of geological time spanning 56.5 million to 35.4 million years ago.

epiphyte A plant that grows on another plant or an object, using it for support but not nourishment.

estuarine Of, pertaining to, or found in an estuary.

evaporation The diffusion of water vapour into the atmosphere from sources of water exposed to the air (cf. evapotranspiration).

evapotranspiration Evaporation plus the water discharged into the atmosphere by plant transpiration.

evergreen A plant with foliage that stays green all year round.

extinction The demise of a species or any other taxon.

extirpation The local extinction of a species or any other taxon.

extremophiles Organisms that relish ultra-extreme conditions.

fauna All the animals living in an area or region.

fecundity Reproductive potential as measured by the number of eggs, sperms, or asexual structures produced.

filter route A path followed by dispersing organisms that, owing to the environmental or geographical conditions, does not permit all species to pass.

flora All the plants living in an area or region.

food chain A simple expression of feeding relationships in a community, starting with plants and ending with top carnivores.

food web A network of feeding relationships within a community.

frugivorous Pertaining to fruit-eating.

Gaia hypothesis The idea that the chemical and physical conditions of the surface of the Earth, the atmosphere, and the oceans are actively controlled by life, for life.

genetic variability A measure of the number of genetic diversity within a gene pool.

genus (plural **genera**) A taxonomic group of lower rank than a family that consists of closely related species or, in extreme cases, only one species.

geodiversity The diversity of the physical environment.

geographical range The area occupied by a group of organisms (species, genus, family, etc.); the same as the distribution.

geosphere The solid Earth – core, mantle, and crust.

geothermal springs Springs of water made hot by the Earth's internal heat.

germination The time of sprouting in plants.

Gondwana A hypothetical, Late Palaeozoic supercontinent lying chiefly in the Southern Hemisphere and comprising large parts of South America, Africa, India, Antarctica, and Australia.

grazer An organism that feeds on growing grasses and herbs.

gross primary production Production before respiration losses are accounted.

habitat The place where an organism lives.

habitat selection The selection of a particular habitat by a dispersing individual.

heath A tract of open and uncultivated land supporting shrubby plants, and especially the heaths (*Erica* spp).

heliotropism The growth of a plant towards or away from sunlight.

helophytes Marsh plants.

heterogeneous An environment comprising a mosaic of dissimilar spatial elements; non-uniform environment.

heterotrophs Organisms that obtain nourishment from the tissues of other organisms.

homeostasis A steady-state with inputs counter-balancing outputs.

homeotherm An animal which regulates its body temperature by mechanisms within its own body; an endotherm or 'warm-blooded' animal (cf. poikilotherm).

homogeneous Referring to a mosaic of similar spatial elements; uniform.

humification The process of humus formation.

hummock–hollow cycle A peatland cycle involving the infilling of a wet hollow by bog mosses until a hummock is formed and colonized by heather. The heather eventually degenerates and the process starts again.

humus An amorphous, colloidal substance produced by soil micro-organisms transforming plant litter.

hurricane A violent and sometimes devastating tropical cyclone.

hydrophyte A plant that grows wholly or partly submerged in water.

hydrosphere All the waters of the Earth.

hyperparasite A parasite that lives on another parasite.

hyperthermophile An organism that thrives in ultra-hot conditions.

Ice Age An old term for the Quaternary glacial–interglacial sequence.

ice age A time when ice forms broad sheets in middle and high latitudes (often in conjunction with the widespread occurrence of sea ice and permafrost) and mountain glaciers form at all latitudes.

interspecific competition Competition among species.

intestinal endosymbionts Symbiotic organisms living in another organism's gut.

intraspecific competition Competition within a species.

intrazonal soils Soils whose formation is dominated by local factors of relief or substrate; an example is soils formed on limestone.

introduced species or **introduction** A species released outside its natural geographical range.

isolines A line on a map joining points of equal value.

Jurassic period A slice of geological time spanning 208 million to 145.6 million years ago; the middle unit of the Mesozoic era.

keystone species A species that plays a central role in a community or ecosystem, its removal having far-reaching effects.

land bridge A dry land connection between two land masses that were previously separated by the sea.

land ethic A set of ethical principles declaring the rights of all natural objects (animals, plants, landscapes) to continued existence and, in some places at least, continued existence in a natural state.

Laurasia The ancient, Northern Hemisphere land mass that broke away from Gondwana about 180 million years ago and subsequently split into North America, Greenland, Europe, and Northern Asia.

leaf–area index The ratio of total leaf surface to ground surface. For example, a leaf–area index of 2 would mean that if you were to clip all the leaves hanging over 1 m^2 of ground, you would have 2 m^2 of leaf surface.

legume A plant of the family Leguminosae, characteristically bearing pods that split in two to reveal seeds attached to one of the halves.

lichen A plant consisting of a fungus and an alga.

life table A tabulation of the complete mortality schedule of a population.

life-form The characteristics of a mature animal or plant.

littoral Shallow water zone.

local extinction (extirpation) The loss of a species from a particular area.

macronutrients Chemical elements needed in large quantities by living things.

maquis A shrubby vegetation of evergreen small trees and bushes.

mass extinctions Extinction episodes in which large numbers of species disappear.

meadow Grassland mown for hay.

mechanists Proponents of the view that all biological and ecological phenomena may be explained by the interaction of physical entities.

megafauna Large mammals, such as mammoths and sloths, contrasted with small mammals, such as rodents and insectivores.

megaherbivores Large browsers and grazers, such as elephants and rhinoceroses.

mesophyte A plant flourishing under mesic (not too dry and not too wet) conditions.

Mesozoic era A slice of geological time spanning 245 million to 65 million years ago comprising the Triassic, Jurassic, and Cretaceous periods.

metabolic processes The many physical and chemical process involved in the maintenance of life.

micronutrients Chemical elements required in small quantities by at least some living things.

micro-organism An organism not visible to the unaided naked eye.

mineralization The release of nutrients from dead organic matter.

Miocene epoch A slice of geological time spanning 23.3 million to 5.2 million years ago.

mutualism An obligatory interaction between two species where both benefit.

natural selection The process by which the environmental factors sort and sift genetic variability and drive evolution.

Neogene period A slice of geological time between 23.3 million and 1.64 million years ago.

net primary production Gross primary production less respiration.

neutralism An interaction in which neither species is harmed or benefited.

neutrophiles Plants that like neutral conditions, not too acidic and not too alkaline.

nitrogen-fixing The conversion of atmospheric nitrogen into ammonium compounds by some bacteria.

nutrient A chemical element required by at least some living things.

old-growth forest A virgin forest that has never been cut, or a forest that has been undisturbed for a long time.

Oligocene epoch A slice of geological time spanning 35.4 million to 23.3 million years ago; the youngest unit of the Palaeogene period.

orobiomes Biomes associated with the altitudinal climatic zones on mountains.

overwinter To spend the winter in a particular place.

Palaeocene epoch A slice of geological time spanning 65 million to 56.5 million years ago; the oldest unit of the Palaeogene period.

Palaeogene period A slice of geological time between 65 million and 23.3 million years ago.

Palaeozoic era A slice of geological time spanning 570 million to 245 million years ago; comprises the Cambrian, Ordovician, Silurian, Devonian, Carboniferous, and Permian periods.

Pangaea The Triassic supercontinent comprising all the present continents.

pathogen An agent, such as bacterium or fungus, that causes a disease.

pedobiomes Areas of characteristic vegetation produced by distinctive soils.

pedosphere All the soils of the Earth.

Permian period A slice of geological time spanning 290 million to 245 million years ago.

photoperiod The duration of darkness and light.

photosynthesis The synthesis of carbohydrates from carbon dioxide and water by chlorophyll, using light as an energy source. Oxygen is a by-product.

phytomass The mass of living plant material in a community, or in a unit area; usually expressed as a dry weight.

phytoplankton Plant plankton: the plant community in marine and freshwater system which floats free in the water and contains many species of algae and diatoms.

phytotoxic Poisonous to plants.

plant formation The vegetational equivalent of a biome, that is, a community of plants of like physiognomy (life-form) occupying a climatically uniform area on a continental scale.

Pleistocene epoch A slice of geological time spanning 1.64 million to 10,000 years ago; the older unit of the Pleistogene period (or Quaternary sub-era).

Pleistogene period The most recent slice of geological time between 1.64 million years ago and the present; subdivided into Pleistocene and Holocene epochs.

Pliocene epoch A slice of geological time spanning 5.2 million to 1.64 million years ago; the younger unit of the Neogene period.

poikilotherm An organism whose body temperature is determined by the ambient temperature and who can control its body temperature only by taking advantage of sun and shade to heat up or to cool down; a 'cold-blooded' animal (cf. homeotherm).

prairie An extensive area of natural, dry grassland; equivalent to steppe in Eurasia.

profundal Deep water zone.

propagule The reproductive body of a plant.

protocooperation A non-obligatory interaction between two species where both benefit.

psychrophile Organisms that flourish at low temperatures.

race An animal or plant population that differs from other populations of the same species in one or more hereditary characters.

radioisotope A form of a chemical element that undergoes spontaneous radioactive decay.

rain shadow A region with relatively low rainfall that is sheltered from rain-bearing winds by high ground.

respiration A complex series of chemical reactions in all organisms by which energy is made available for use. The end products are carbon dioxide, water, and energy.

rhizome A root-like stem, usually horizontal, growing underground with roots emerging from its lower surface and leaves or shoots from its upper surface.

salinity Saltiness.

saltatorial Adapted for hopping and jumping.

savannah Tropical grassland.

saxicolous Growing on or living among rocks.

scavenger An organism feeding on dead animal flesh or other decaying organic material.

seasonality The degree of climatic contrast between summer and winter.

sessile Permanently attached or fixed; not free-moving.

social Darwinism Darwin's doctrine of the survival of the fittest applied to society.

soil Rock at, or near, the land surface that has been transformed by the biosphere.

solar radiation The total electromagnetic radiation emitted by the Sun. Electromagnetic radiation is energy propagated through space or through a material as an interaction between electric and magnetic waves; occurs at frequencies ranging from short wavelength, high frequency cosmic rays, through medium wavelength, medium frequency visible light, to long wavelength, low frequency radio waves.

speciation The process of species multiplication.

species A reproductively isolated collection of inter-breeding populations (demes).

steppe An extensive area of natural, dry grassland; equivalent to prairie in North American terminology.

stomata (singular **stoma**) Epidermal pores in a leaf or stem through which air and water vapour pass.

sweepstakes route A path along which organisms may disperse, but very few will do so owing to the difficulties involved.

symbiotic Pertaining to two or more organisms of different species living together. In a narrow sense equivalent to mutualism.

taxon (plural **taxa**) A population or group of populations (taxonomic group) that is distinct enough to be given a distinguishing name and to be ranked in a definite category.

territory An area defended by an animal against other members of its species (and occasionally other species).

Tertiary period A slice of geological time spanning 65 million to 1.64 million years ago and the youngest unit of the Cenozoic era; now designated a sub-era.

theory of island biogeography A model that explains the species diversity of islands chiefly as a function of distance from the mainland and island area.

thermophiles Organisms that love hot conditions.

toposequence The sequence of soils and vegetation found along a hillslope, from summit to valley bottom.

trace element A chemical element used by an organism in minute quantities and vital to its physiology.

tree-lines The altitudinal and latitudinal limits of tree growth.

tree-throw The toppling of trees by strong winds.

Triassic period A slice of geological time spanning 245 million to 208 million years ago; the oldest unit of the Mesozoic era.

trophic Of, or pertaining to, nourishment or feeding.

tsunamis The scientific (and Japanese) name for tidal waves.

ungulates Hoofed mammals.

vascular plant A plant containing vascular tissue, the conducting system which enables water and minerals to move through it.

vegetation Plant life collective; the plants of an area.

vernalization The exposure of some plants (or their seeds) to a period of cold that allows them to flower or to flower earlier than usual.

vicariance The splitting apart of a species distribution by a geological or environmental event.

vicars Species of different ancestry but with the same characteristics living in different biogeographical regions. Also called ecological equivalents.

vitalists Proponents of the idea that some kind of living force resides in organisms.

volatilization To convert to vapour.

wetlands All the places in the world that are wet for at least part of the year and support a characteristic soils and vegetation; they include marshes, swamps, and bogs.

xerophyte A plant adapted for life in a dry climate.

zonobiome One of nine, climatically defined, major community units of the Earth.

zoomass The mass of living animal material in a community, or in a unit area; usually expressed as a dry weight.

zooplankton Animal plankton; aquatic invertebrates living in sunlit waters of the hydrosphere.

FURTHER READING

General biogeography books

Cox, C. B. and Moore, P. D. (1993) *Biogeography: An Ecological and Evolutionary Approach*, 5th edn, Oxford: Blackwell.
Already a classic. A very popular textbook.

George, W. (1962) *Animal Geography*, London: Heinemann.
Written before the revival of continental drift, but well worth a look.

Pears, N. (1985) *Basic Biogeography*, 2nd edn, Harlow: Longman.
A good textbook on ecological biogeography.

Tivy, J. (1992) *Biogeography: A Study of Plants in the Ecosphere*, 3rd edn, Edinburgh: Oliver & Boyd.
A good introductory textbook.

Regional biogeography

Here is a small selection of regional biogeographical and ecological texts. A GEOBASE search will reveal many more.

Battistini, R. and Richard-Vindard, G. (eds) (1972) *Biogeography and Ecology in Madagascar* (Monographiae Biologicae, Vol. 21), The Hague: W. Junk.

Borhidi, A. (1991) *Phytogeography and Vegetation Ecology of Cuba*, Budapest: Akadémiai Kiadó.

Bowman, R. I. (ed.) (1966) *The Galápagos* (Proceedings of the Symposium of the Galápagos International Scientific Project), Berkeley and Los Angeles, California: University of California Press.

Fernando, C. H. (ed.) (1984) *Ecology and Biogeography in Sri Lanka* (Monographiae Biologicae, Vol. 57), The Hague: W. Junk.

Gressitt, J. L. (ed.) (1982) *Biogeography and Ecology of New Guinea*, 2 vols (Monographiae Biologicae, Vol. 42), The Hague: W. Junk.

Kuschel, G. (ed.) (1975) *Biogeography and Ecology in New Zealand* (Monographiae Biologicae, Vol. 27), The Hague: W. Junk.

Stoddart, D. R. (ed.) (1984) *Biogeography and Ecology of the Seychelles Islands* (Monographiae Biologicae, Vol. 55), The Hague: W. Junk.

Whitmore, T. C. (ed.) (1981) *Wallace's Line and Plate Tectonics* (Oxford Monographs in Biology, Vol. 1), Oxford: Clarendon Press.

Whitmore, T. C. (ed.) (1987) *Biogeographical Evolution of the Malay Archipelago* (Oxford Monographs in Biology, Vol. 4), Oxford: Clarendon Press.

Williams, W. D. (ed.) (1974) *Biogeography and Ecology in Tasmania* (Monographiae Biologicae, Vol. 25), The Hague: W. Junk.

Woods, C. A. (ed.) (1989) *Biogeography of the West Indies: Past, Present, and Future*, Gainesville, Florida: Sandhill Crane Press.

Ecological aspects

Dansereau, P. (1957) *Biogeography: An Ecological Perspective*, New York: Ronald Press.
Old but have a look.

Forman, R. T. T. (1995) *Land Mosaics: The Ecology of Landscapes and Regions*, Cambridge: Cambridge University Press.
A weighty tome on landscape ecology with some useful sections for biogeographers.

Huggett, R. J. (1995) *Geoecology: An Evolutionary Approach*, London: Routledge.
A survey of all environmental factors.

Kormondy, E. J. (1996) *Concepts of Ecology*, 4th edn, Englewood Cliffs, NJ: Prentice Hall.
An excellent basic ecology text.

Larcher, W. (1995) *Physiological Plant Ecology: Ecophysiology and Stress Physiology of Functional Groups*, 3rd edn, Berlin: Springer.
First-rate coverage of physiological aspects of plant ecology.

Stoutjesdijk, P. and Barkman, J. J. (1992) *Microclimate, Vegetation and Fauna*, Uppsala, Sweden: Opulus Press.
Unusual but highly interesting book.

Viles, H. A. (ed.) (1988) *Biogeomorphology*, Oxford: Basil Blackwell.
Links life to geomorphology.

Whelan, R. J. (1995) *The Ecology of Fire*, Cambridge: Cambridge University Press.
A clear introduction to the subject.

Historical aspects

Briggs, J. C. (1995) *Global Biogeography* (Developments in Palaeontology and Stratigraphy, 14), Amsterdam: Elsevier.
An advanced text but worth dipping into.

Cox, C. B. and Moore, P. D. (1993) *Biogeography: An Ecological and Evolutionary Approach*, 5th edn, Oxford: Blackwell.
Excellent basic coverage of historical aspects.

Cronk, Q. C. B. and Fuller, J. L. (1995) *Plant Invaders: The Threat to Natural Ecosystems*, London: Chapman & Hall.
Many examples.

Darlington, P. J., Jr (1957) *Zoogeography: The Geographical Distribution of Animals*, New York: John Wiley & Sons.
A classic in its day, which was before the revival of continental drift.

Elton, C. S. (1958) *The Ecology of Invasions by Animals and Plants*, London: Chapman & Hall.
A little gem.

Hallam, A. (1995) *An Outline of Phanerozoic Biogeography*, Oxford: Oxford University Press.
A palaeontological perspective.

Nelson, G. and Rosen, D. E. (eds) *Vicariance Biogeography: A Critique* (Symposium of the Systematics Discussion Group of the American Museum of Natural History, 2–4 May 1979), New York: Columbia University Press.
Only for the brave.

Udvardy, M. D. F. (1969) *Dynamic Zoogeography, with Special Reference to Land Animals*, New York: Van Nostrand Reinhold.
Rather old, but still deserves to be read.

Wallace, A. R. (1876) *The Geographical Distribution of Animals; with a Study of the Relations of Living and Extinct Faunas as Elucidating the Past Changes of the Earth's Surface*, 2 vols, London: Macmillan.
All biogeographers should read this great work.

Woods, C. A. (ed.) *Biogeography of the West Indies: Past, Present, and Future*, Gainesville, Florida: Sandhill Crane Press.
Caribbean case studies.

Single populations

Hanski, I. and Gilpin, M. E. (1997) *Metapopulation Biology: Ecology, Genetics, and Evolution*, New York: Academic Press.
Advanced but worth reading.

Kormondy, E. J. (1996) *Concepts of Ecology*, 4th edn, Englewood Cliffs, NJ: Prentice Hall.
Relevant sections worth reading.

Kruuk, H. (1995) *Wild Otters: Predation and Populations*, Oxford: Oxford University Press.
A good case study.

Perry, J. H., Woiwod, I. P., Smith, R. H., and Morse, D. (1997) *Chaos in Real Data: Analysis on Nonlinear Dynamics from Short Ecological Time Series*, London: Chapman & Hall.
If mathematics is not your strong point, forget it.

Taylor, V. J. and Dunstone, N. (eds) (1996) *The Exploitation of Mammal Populations*, London: Chapman & Hall.
Examples of exploitation.

Tuljapurkar, S. and Caswell, H. (1997) *Structured-Population Models*, New York: Chapman & Hall.
Somewhat demanding.

Interacting populations

Crawley, M. J. (1983) *Herbivory: The Dynamics of Animal–Plant Interactions* (Studies in Ecology, Vol. 10), Oxford: Blackwell Scientific Publications.
A superb book, despite the mathematics.

Grover, J. P. (1997) *Resource Competition*, New York: Chapman & Hall.
Up-to-date exposition.

Jackson, A. R. W. and Jackson, J. M. (1996) *Environmental Science: The Natural Environment and Human Impact*, Harlow: Longman.
Excellent text.

Kingsland, S. E. (1985) *Modeling Nature: Episodes in the History of Population Biology*, Chicago and London: University of Chicago Press.
A first-rate account of the history of population biology. More exciting than it might sound.

Kormondy, E. J. (1996) *Concepts of Ecology*, 4th edn, Englewood Cliffs, NJ: Prentice Hall.
Clear explanations of basic ideas.

MacDonald, D. (1992) *The Velvet Claw: A Natural History of the Carnivores*, London: BBC Books.
All about carnivores with beautiful photographs.

Communities

Archibold, O. W. (1994) *Ecology of World Vegetation*, New York: Chapman & Hall.
Covers ecological aspects of plant communities.

Chameides, W. L. and Perdue, E. M. (1997) *Biogeochemical Cycles: A Computer-Interactive Study of Earth System Science and Global Change*, New York: Oxford University Press.
If you demand much, this is the one for you.

Hochberg, M. E., Colbert, J., and Barbault, R. (eds) (1996) *Aspects of the Genesis and Maintenance of Biological Diversity*, Oxford: Oxford University Press.
Contains many case studies, though not an easy read.

Jeffries, M. L. (1997) *Biodiversity and Conservation*, London and New York: Routledge.
An excellent basic text.

Lawton, J. H. and May, R. M. (eds) (1995) *Extinction Rates*, Oxford: Oxford University Press.
Lots of recent figures on extinction rates.

Morgan, S. (1995) *Ecology and Environment: The Cycles of Life*, Oxford: Oxford University Press.
A good starting text.

Polis, G. A. and Winemiller, K. O. (1995) *Food Webs: Integration of Patterns and Dynamics*, New York: Chapman & Hall.
Not easy, but worth a look.

Quammen, D. (1996) *The Song of the Dodo: Island Biogeography in an Age of Extinctions*, London: Hutchinson.
Highly readable.

Reaka-Kudlka, M. L., Wilson, D. E., and Wilson, E. O. (1997) *Biodiversity II: Understanding and Protecting Our Biological Resources*, Washington, DC: National Academy Press.
Sure to become a classic.

Rosenzweig, M. L. (1995) *Species Diversity in Space and Time*, Cambridge: Cambridge University Press.
Fascinating, but not easy.

Schultz, J. (1995) *The Ecozones of the World: The Ecological Divisions of the Geosphere*, Hamburg: Springer.
A very good summary of the world's main ecosystems.

Szaro, R. C. and Johnston, D. W. (1996) *Biodiversity in Managed Landscapes: Theory and Practice*, New York: Oxford University Press.
Try it.

Wilson, E. O. (1992) *The Diversity of Life*, Cambridge, Massachusetts: Belknap Press of Harvard University Press.
A highly readable book.

Community change

Committee on Characterization of Wetlands, National Research Council (1995) *Wetlands: Characteristics and Boundaries*, Washington, DC: National Academy Press.
Readable case study.

Gates, D. M. (1993) *Climate Change and Its Biological Consequences*, Sunderland, Massachusetts: Sinauer Associates.
A clear account.

Huggett, R. J. (1993) *Modelling the Human Impact on Nature: Systems Analysis of Environmental Problems*, Oxford: Oxford University Press.
If you like modelling but are not too mathematically minded, this might be of interest.

Huggett, R. J. (1997) *Environmental Change: The Evolving Ecosphere*, London: Routledge.
Provides a broad perspective.

Jackson, A. R. W. and Jackson, J. M. (1996) *Environmental Science: The Natural Environment and Human Impact*, Harlow: Longman.
A basic textbook with several relevant sections.

Matthews, J. A. (1992) *The Ecology of Recently-Deglaciated Terrain: A Geoecological Approach to Glacier Forelands and Primary Succession*, Cambridge: Cambridge University Press.
An excellent case study and lots more.

Peters, R. L. and Lovejoy, T. E. (eds) (1992) *Global Warming and Biological Diversity*, New Haven, Connecticut and London: Yale University Press.
A host of examples in this one.

Schwartz, M. (1997) *Conservation in Highly Fragmented Landscapes*, New York: Chapman & Hall.
Topical and highly recommended.

Ethical issues and conservation

This is a selection from a long list.

Anderson, E. N. (1996) *Ecologies of the Heart: Emotion, Belief, and the Environment*, New York: Oxford University Press.

Botkin, D. B. (1990) *Discordant Harmonies: A New Ecology for the Twenty-First Century*, New York: Oxford University Press.

Committee on Scientific Issues in the Endangered Species Act, National Research Council (1995) *Science and the Endangered Species Act*, Washington, DC: National Academy Press.

Cooper, N. and Carling, R. C. J. (eds) (1996) *Ecologists and Ethical Judgements*, London: Chapman & Hall.

Norton, B. G. (1991) *Toward Unity Among Environmentalists*, New York: Oxford University Press.

Ryder, R. D. (ed.) (1993) *Animal Welfare and the Environment*, London: Duckworth and RSPCA.

Shafer, C. L. (1990) *Nature Reserves: Island Theory and Conservation Practice*, Washington, DC and London: Smithsonian Institution Press.

Sylvan, R. and Bennett, D. (1994) *The Greening of Ethics: From Human Chauvinism to Deep-Green Theory*, Cambridge: White Horse Press; Tucson: University of Arizona Press.

BIBLIOGRAPHY

Ables, E. D. (1969) 'Home range studies of red foxes (*Vulpes vulpes*)', *Journal of Mammalogy* 50: 108–20.

Agee, J. K. and Johnson, D. R. (1988) *Ecosystem Management for Parks and Wilderness*, Seattle, Washington: University of Washington Press.

Allen, R. (1980) *How to Save the World: Strategy for World Conservation*, Foreword by Sir Peter Scott, London: Kogan Page for IUCN, UNEP, and WWF.

Anderson, P. and Shimwell, D. W. (1981) *Wild Flowers and other Plants of the Peak District: An Ecological Study*, Ashbourne, Derbyshire: Moorland Publishing.

Arnold, G. W., Steven, D. E., Weeldenburg, J. R., and Smith, E. A. (1993) 'Influences of remnant size, spacing pattern and connectivity on population boundaries and demography in euros *Macropus robustus* living in a fragmented landscape', *Biological Conservation* 64: 219–30.

Arnold, H. R. (1993) *Atlas of Mammals In Britain* (Institute of Terrestrial Ecology Research Publication no. 6), London: Her Majesty's Stationery Office.

Atkinson, I. A. E. and Cameron, E. K. (1993) 'Human influence on the terrestrial biota and biotic communities of New Zealand', *Trends in Ecology and Evolution* 8: 447–51.

Bailey, R. G. (1995) *Description of the Ecoregions of the United States*, 2nd edn, revised and enlarged (Miscellaneous Publication no. 1391), Washington, DC: United States Department of Agriculture, Forest Service.

Bailey, R. G. (1996) *Ecosystem Geography*, Foreword by Jack Ward Thomas, Chief, USDA Forest Service, New York: Springer.

Baker, A. J. M., Proctor, J., and Reeves, R. D. (eds) (1992) *The Vegetation of Ultramafic (Serpentine) Soils*, Andover, Hants: Intercept.

Barfield, C. S. and Stimac, J. L. (1980) 'Pest management: an entomological perspective', *BioScience* 30: 683–8.

Bartholomew, G. A. (1968) 'Body temperature and energy metabolism', pp. 290–354 in M. S. Gordon, G. A. Bartholomew, A. D. Grinnell, C. B. Jorgensen, and F. N. White (eds) *Animal Function: Principles and Adaptations*, New York: Macmillan.

Bartlett, D. H. (1992) 'Microbial life at high pressures', *Science Progress* 76: 479–96.

Bascompte, J. and Sole, R. V. (1995) 'Rethinking complexity: modelling spatiotemporal dynamics in ecology', *Trends in Ecology and Evolution* 10: 361–6.

Bellamy, D. (1976) *Bellamy's Europe*, London: British Broadcasting Corporation.

Birks, J. (1990) 'Feral mink and nature conservation', *British Wildlife* 1: 313–23.

Bjärvall. A. and Ullström. S. (1986) *The Mammals of Britain and Europe*, Foreword by Ernest Neal, translated by David Christie, London and Sydney: Croom Helm.

Bloomer, J. P. and Bester, M. N. (1992) 'Control of feral cats on sub-Antarctic Marion Island, Indian Ocean', *Biological Conservation* 60: 211–19.

Botkin, D. B. (1990) *Discordant Harmonies: A New Ecology for the Twenty-First Century*, New York: Oxford University Press.

Botkin, D. B. and Keller, E. A. (1995) *Environmental*

Science: Earth as a Living Planet, New York: John Wiley & Sons.

Botkin, D. B., Woodby, D. A., and Nisbet, R. A. (1991) 'Kirtland's warbler habitats: a possible early indicator of climatic warming', *Biological Conservation* 56: 63–78.

Boulding, K. E. (1966) 'The economics of the coming Spaceship Earth', pp. 3–14 in Henry Jarrett (ed.) *Environmental Quality in a Growing Economy: Essays from the Sixth Resources of the Future Forum*, Baltimore, Maryland and London: Johns Hopkins University Press for Resources of the Future.

Boulding, K. E. (1970) *Economics as a Science*, New York: McGraw-Hill.

Boursot, P., Auffray, J. C., Britton-Davidian, J., and Bonhomme, F. (1993) 'The evolution of the house mouse', *Annual Review of Ecology and Systematics* 24: 119–52.

Box, E. O. and Meentemeyer, V. (1991) 'Geographic modeling and modern ecology', pp. 773–804 in G. Esser and D. Overdieck (eds) *Modern Ecology: Basic and Applied Aspects*, Amsterdam: Elsevier.

Box, E. O., Crumpacker, D. W., and Hardin, E. D. (1993) 'A climatic model for location of plant species in Florida, USA', *Journal of Biogeography* 20: 629–44.

Boyd, L. E. (1991) 'The behavior of Przewalski's horses and its importance to their management', *Applied Animal Behaviour Science* 29: 301–18.

Briggs, J. C. (1995) *Global Biogeography* (Developments in Palaeontology and Stratigraphy, 14), Amsterdam: Elsevier.

Bright, D. A., Dushenko, W. T., Grundy, S. L., and Reimer, K. J. (1995) 'Effects of local and distant contaminant sources: polychlorinated biphenyls and other organochlorines in bottom-dwelling animals from an Arctic estuary', *Science of the Total Environment* 160–1: 265–83.

Brooks, R. R. (1987) *Serpentine and its Vegetation: A Multidisciplinary Approach*, London and Sydney: Croom Helm.

Brothers, N. (1991) 'Albatross mortality and associated bait loss in the Japanese longline fishery in the Southern Ocean', *Biological Conservation* 55: 255–68.

Brower, L. P. (1969) 'Ecological chemistry', *Scientific American* 220: 22–9.

Brown, J. H. (1968) 'Adaptation to environmental temperature in two species of woodrats, *Neotoma cinerea and N. albigula*', *Miscellaneous Publications, Museum of Zoology, University of Michigan* 135: 1–48.

Brown, J. H. (1971) 'Mammals on mountaintops: non-

equilibrium insular biogeography', *The American Naturalist* 105: 467–78.

Brown, J. H. (1995) *Macroecology*, Chicago and London: University of Chicago Press.

Brown, J. H. and Heske, E. J. (1990) 'Control of a desert-grassland transition by a keystone rodent guild', *Science* 250: 1750–7.

Budiansky, S. (1995) *Nature's Keepers: The New Science of Nature Management*, London: Weidenfeld & Nicolson.

Buringh, P. and Dudal, R. (1987) 'Agricultural land use in space and time', pp. 9–44 in M. G. Wolman and F. G. A. Fournier (eds) *Land Transformation in Agriculture* (SCOPE 32), Chichester: John Wiley & Sons.

Burnham, C. P. and Mackney, D. (1968) 'Soils of Shropshire', *Field Studies* 2: 83–113.

Burt, W. H. (1943) 'Territoriality and home range concepts as applied to mammals', *Journal of Mammalogy* 24: 346–52.

Burton, R. (1896 edn) *The Anatomy of Melancholy*, edited by Revd A. R. Shilleto, MA, with an introduction by A. H. Bullen, 3 vols, London: George Bell & Sons.

Bush, M. B. and Whittaker, R. J. (1991) 'Krakatau: colonization patterns and hierarchies', *Journal of Biogeography* 18: 341–56.

Bush, M. B., Whittaker, R. J., and Partomihardjo, T. (1992) 'Forest development on Rakata, Panjang and Sertung: contemporary dynamics (1979–1989)', *GeoJournal* 28: 185–99.

Butler, D. R. (1992) 'The grizzly bear as an erosional agent in mountainous terrain', *Zeitschrift für Geomorphologie* NF 36: 179–89.

Butler, D. R. (1995) *Zoogeomorphology: Animals as Geomorphic Agents*, Cambridge: Cambridge University Press.

Callicott, J. B. (1985) 'Intrinsic value, quantum theory, and environmental ethics', *Environmental Ethics* 7: 257–75.

Carson, R. (1962) *Silent Spring*, Boston, Massachusetts: Houghton Mifflin.

Case, T. J. and Bolger, D. T. (1991) 'The role of introduced species in shaping the distribution and abundance of island reptiles', *Evolutionary Ecology* 5: 272–90.

Caughley, G. (1970) 'Eruption of ungulate populations, with emphasis on the Himalayan thar in New Zealand', *Ecology* 51: 53–72.

Chapman, N., Harris, S., and Stanford, A. (1994) 'Reeves' muntjac (*Muntiacus reevesi*) in Britain: their history, spread, habitat selection, and the role of human intervention in accelerating their dispersal', *Mammal Review* 24: 113–60.

Chapman, R. N. (1928) 'The quantitative analysis of environmental factors', *Ecology* 9: 111–22.

Choi, G. H. and Nuss, D. L. (1992) 'Hypovirulence of chestnut blight fungus conferred by an infectious viral cDNA', *Science* 257: 800–3.

Chown, S. L. and Smith, V. R. (1993) 'Climate change and the short-term impact of feral house mice at the sub-Antarctic Prince Edward Islands', *Oecologia* 96: 508–16.

Christian, K. A., Tracy, C. R., and Porter, W. P. (1983) 'Seasonal shifts in body temperature and use of micro-habitats by the Galápagos land iguana (*Conolophus pallidus*)', *Ecology* 64: 463–8.

Clements, F. E. (1916) *Plant Succession: An Analysis of the Development of Vegetation* (Carnegie Institute of Washington, Publication no. 242), Washington, DC: Carnegie Institute of Washington.

Clements, F. E. and Shelford, V. E. (1939) *Bio-Ecology*, New York: John Wiley & Sons.

Cloudsley-Thompson, J. L. (1975a) *Terrestrial Environments*, London: Croom Helm.

Cloudsley-Thompson, J. L. (1975b) *The Ecology of Oases*, Watford, Hertfordshire: Merrow.

Colbert, E. H. (1971) 'Tetrapods and continents', *The Quarterly Review of Biology* 46: 250–69.

Colinvaux, P. A. (1980) *Why Big Fierce Animals are Rare*: Harmondsworth, Middlesex: Pelican Books.

Colwell, T. (1987) 'The ethics of being part of Nature', *Environmental Ethics* 9: 99–113.

Connell, J. H. (1961) 'The influence of interspecific competition and other factors on the distribution of the barnacle *Chthamalus stellatus*', *Ecology* 42: 710–23.

Conway, G. (1981) 'Man versus pests', pp. 356–86 in R. M. May (ed.) *Theoretical Ecology: Principles and Applications*, 2nd edn, Oxford: Blackwell Scientific Publications.

Coope, G. R. (1994) 'The response of insect faunas to glacial–interglacial climatic fluctuations', *Philosophical Transactions of the Royal Society of London* 344B: 19–26.

Cooper, W. S. (1923) 'The recent ecological history of Glacier Bay, Alaska', *Ecology* 6: 197.

Cooper, W. S. (1931) 'A third expedition to Glacier Bay, Alaska', *Ecology* 12: 61–95.

Cooper, W. S. (1939) 'A fourth expedition to Glacier Bay, Alaska', *Ecology* 20: 130–59.

Corbett, A. L., Krannitz, P. G., and Aarssen, L. W. (1992) 'The influence of petals on reproductive success in the Arctic poppy (*Papaver radicatum*)', *Canadian Journal of Botany* 70: 200–4.

Cortner, H. J. and Moote, M. A. (1994) 'Trends and issues in land and water resources management: setting the agenda for change', *Environmental Management* 18: 167–73.

Costanza, R., Sklar, F. H., and White, M. L. (1990) 'Modeling coastal landscape dynamics', *BioScience* 40: 91–107.

Cottrell, A. (1978) *Environmental Economics: An Introduction for Students of the Resources and Environmental Sciences*, London: Edward Arnold.

Cox, C. B. (1990) 'New geological theories and old biogeographical problems', *Journal of Biogeography* 17: 117–30.

Cox, C. B. and Moore, P. D. (1993) *Biogeography: An Ecological and Evolutionary Approach*, 5th edn, Oxford: Blackwell.

Cox, W. T. (1936) 'Snowshoe rabbit migration, tick infestation, and weather cycles', *Journal of Mammalogy* 17: 216–21.

Cracraft, J. (1973) 'Continental drift, palaeoclimatology and the evolution and biogeography of birds', *Journal of Zoology*, London 169: 455–545.

Cracraft, J. (1974) 'Phylogeny and evolution of ratite birds', *Ibis* 116: 494–521.

Cramp, S. (ed.) (1988) *Handbook of the Birds of Europe, the Middle East and North Africa (The Birds of the Western Palaearctic). Volume 5: Tyrant Flycatchers to Thrushes*, Oxford: Oxford University Press.

Crawford, A. R. (1974) 'A greater Gondwanaland', *Science* 184: 1,179–81.

Crawley, M. J. (1983) *Herbivory: The Dynamics of Animal–Plant Interactions* (Studies in Ecology, Vol. 10), Oxford: Blackwell Scientific Publications.

Crocker, R. L. and Major, J. (1955) 'Soil development in relation to vegetation and surface age at Glacier Bay, Alaska', *Journal of Ecology* 43: 427–48.

Croizat, L. (1958) *Pangeography*, 2 vols, Caracas: published by the author.

Croizat, L. (1964) *Space, Time, Form: The Biological Synthesis*, Caracas: published by the author.

Crome, F. H. J. and Moore, L. A. (1990) 'Cassowaries in north-eastern Queensland: a report or a survey and a review and assessment of their status and conservation and management needs', *Australian Wildlife Management* 17: 369–85.

Dansereau, P. (1957) *Biogeography: An Ecological Perspective*, New York: Ronald Press.

Darlington, P. J., Jr (1957) *Zoogeography: The Geographical Distribution of Animals*, New York: John Wiley & Sons.

de Candolle, A.-P. (1820) 'Géographie botanique', pp. 359–436 in F. C. Levrault (ed.) *Dictionnaire des Sciences Naturelles*, Vol. 18, Paris: Levrault.

Delcourt, H. R. and Delcourt, P. A. (1988) 'Quaternary landscape ecology: relevant scales in space and time', *Landscape Ecology* 2: 23–44.

Delcourt, H. R. and Delcourt, P. A. (1994) 'Postglacial rise and decline of *Ostrya virginiana* (Mill.) K. Koch and *Carpinus caroliniana* Walt. in eastern North America: predictable responses of forest species to cyclic changes in seasonality of climate', *Journal of Biogeography* 21: 137–50.

Delcourt, P. A. and Delcourt, H. R. (1987) *Long-term Forest Dynamics of the Temperate Zone: A Case Study of Late-Quaternary Forests in Eastern North America* (Ecological Studies 63), New York: Springer.

Desponts, M. and Payette, S. (1993) 'The Holocene dynamics of jack pine at its northern range limit in Quebec', *Journal of Ecology* 81: 719–27.

Diamond, J. M (1974) 'Colonization of exploded volcanic islands by birds: the supertramp strategy', *Science* 184: 803–6.

Diamond, J. M. (1987) 'Human use of world resources', *Nature* 328: 479–80.

Dobler, G., Schneider, R., and Schweis, A. (1991) 'Die Invasion des Rauhfußbussards (*Buteo lagopus*) in Baden-Württemberg im Winter 1986/87', *Die Vogelwarte* 36: 1–18.

Dony, J. G. (1963) 'The expectation of plant records from prescribed areas', *Watsonia* 5: 377–85.

Drake, J. A. (1990) 'The mechanics of community assembly and succession', *Journal of Theoretical Biology* 147: 213–33.

Drury, W. H. and Nisbet, I. C. T. (1973) 'Succession', *Journal of the Arnold Arboretum* 54: 331–68.

Duggins, D. O. (1980) 'Kelp beds and sea otters: an experimental approach', *Ecology* 61: 447–53.

Ehleringer, J. R., Mooney, H. A., Rundel, P. W., Evans, R. D., Palma, B., and Delatorre, J. (1992) 'Lack of nitrogen cycling in the Atacama Desert', *Nature* 359: 316–18.

Elton, C. S. (1958) *The Ecology of Invasions by Animals and Plants*, London: Chapman & Hall.

Eyre, S. R. (1963) *Vegetation and Soils: A World Picture*, London: Edward Arnold.

Fastie, C. L. (1995) 'Causes and ecosystem consequences of multiple pathways of ecosystem succession at Glacier Bay, Alaska', *Ecology* 76: 1,899–916.

Feeny, P. P (1970) 'Seasonal changes in oak leaf tannins and nutrients as a cause of spring feeding by winter moth caterpillars', *Ecology* 51: 565–81.

Fenneman, N. M. (1916) 'Physiographic divisions of the United States', *Annals of the Association of American Geographers* 6: 19–98.

Fischer, S. F., Poschlod, P., and Beinlich, B. (1996) 'Experimental studies on the dispersal of plants and animals on sheep in calcareous grasslands', *Journal of Applied Ecology* 33: 1,206–22.

Fisher, J. A. (1987) 'Taking sympathy seriously: a defense of our moral philosophy toward animals', *Environmental Ethics* 9: 197–215.

Flannery, T. F., Rich, T. H., Turnbull, W. D., and Lundelius, E. L., Jr (1992) 'The Macropodoidea (Marsupialia) of the early Pliocene Hamilton local fauna, Victoria, Australia', *Fieldiana: Geology*, new series no. 25, Chicago, Illinois: Field Museum of Natural History.

Flux, J. E. C. and Fullagar, P. J. (1992) 'World distribution of the rabbit *Oryctolagus cuniculus* on islands', *Mammal Review* 22: 151–205.

Ford, M. J. (1982) *The Changing Climate: Responses of the Natural Fauna and Flora*, London: George Allen & Unwin.

Forman, R. T. T. (1995) *Land Mosaics: the Ecology of Landscapes and Regions*, Cambridge: Cambridge University Press.

Forman, R. T. T. and Godron, M. (1986) *Landscape Ecology*, New York: John Wiley & Sons.

Friday, L. E. (1992) 'Measuring investment in carnivory: seasonal and individual variation in trap number and biomass in *Utricularia vulgaris* L.', *New Phytologist* 121: 439–45.

Funch, P. and Kristensen, R. M. (1995) 'Cycliophora is a new phylum with affinities to Entoprocta and Ectoprocta', *Nature* 378: 711–14.

Gale, R. P. and Cordray, S. M. (1991) 'What should forests sustain? Eight answers', *Journal of Forestry* 89: 31–6.

Gates, D. M. (1980) *Biophysical Ecology*, New York: Springer.

Gates, D. M. (1993) *Climate Change and Its Biological Consequences*, Sunderland, Massachusetts: Sinauer Associates.

Gause, G. F. (1934) *The Struggle for Existence*, New York: Hafner.

Genelly, R. E. (1965) 'Ecology of the common mole-rat (*Cryptomys hottentotus*) in Rhodesia', *Journal of Mammalogy* 46: 647–65.

Gerlach, L. P. and Bengston, D. N. (1992) 'If ecosystem management is the solution, what is the problem?', *Journal of Forestry* 92: 18–21.

Goldspink, C. R. (1987) 'The growth, reproduction and mortality of an enclosed population of red deer (*Cervus elaphas*) in north-west England', *Journal of Zoology, London* 213: 23–44.

Gorman, M. L. (1979) *Island Ecology* (Outline Studies in Ecology), London: Chapman & Hall.

Gosling, L. M. and Baker, S. J. (1989) 'The eradication of muskrats and coypus from Britain', *Biological Journal of the Linnean Society* 38: 39–51.

Gottfried, R. R. (1992) 'The value of a watershed as a series of linked multiproduct assets', *Ecological Economics* 5: 145–61.

Goudriaan, J. and Ketner, P. (1984) 'A simulation study for the global carbon cycle, including Man's impact on the biosphere', *Climatic Change* 6: 167–92.

Grace, J. (1987) 'Climatic tolerance and the distribution of plants', *New Phytologist* 106 (Supplement): 113–30.

Graham, R. W. (1979) 'Paleoclimates and late Pleistocene faunal provinces in North America', pp. 46–69 in R. L. Humphrey and D. J. Stanford (eds) *Pre-Llano Cultures of the Americas: Paradoxes and Possibilities*, Washington, DC: Anthropological Society of Washington.

Graham, R. W. (1992) 'Late Pleistocene faunal changes as a guide to understanding effects of greenhouse warming on the mammalian fauna of North America', pp. 76–87 in R. L. Peters and T. E. Lovejoy (eds) *Global Warming and Biological Diversity*, New Haven, Connecticut and London: Yale University Press.

Graham, R. W. and Grimm, E. C. (1990) 'Effects of global climate change on the patterns of terrestrial biological communities', *Trends in Ecology and Evolution* 5: 289–92.

Graham, R. W. and Mead, J. I. (1987) 'Environmental fluctuations and evolution of mammalian faunas during the last deglaciation', pp. 371–402 in W. F. Ruddiman and H. E. Wright, Jr (eds) *North America and Adjacent Ocean during the Last Deglaciation* (The Geology of North America, Vol. K-3), Boulder, Colorado: The Geological Society of America.

Grime, J. P. (1977) 'Evidence for the existence of three primary strategies in plants and its relevance to ecological and evolutionary theory', *American Naturalist* 111: 1,169–94.

Grime, J. P. (1989) 'The stress debate: symptom of impending synthesis?', *Biological Journal of the Linnean Society* 37: 3–17.

Grime, J. P., Hodgson, J. G., and Hunt, R. (1988) *Comparative Plant Ecology: A Functional Approach to Common British Species*, London: Unwin Hyman.

Gulland, F. M. D. (1992) 'The role of nematode parasites in Soay sheep (*Ovis aries* L.) mortality during a population crash', *Parasitology* 105: 493–503.

Gutiérrez, R. J. and Harrison, S. (1996) 'Applications of metapopulation theory to spotted owl management: a history and critique', in D. McCullough (ed.) *Metapopulations and Wildlife Conservation Management*, Covelo, California: Island Press.

Hafner, D. J. (1994) 'Pikas and permafrost: post-Wisconsin historical zoogeography of *Ochotona* in the southern Rocky Mountains, U.S.A.', *Arctic and Alpine Research* 26: 375–82.

Hall, E. R. (1946) *Mammals of Nevada*, Berkeley, California: University of California Press.

Hanksi, I. and Henttonnen, H. (1996) 'Predation on competing rodent species: a simple explanation of complex patterns', *Journal of Animal Ecology* 65: 220–32.

Hanski, I. and Korpimäki, E. (1995) 'Microtine rodent dynamics in northern Europe: parameterized models for the predator–prey interaction', *Ecology* 76: 840–50.

Hanski, I., Turchin, P., Korpimäki, E., and Henttonnen, H. (1993) 'Population oscillations of boreal rodents: regulation by mustelid predators leads to chaos', *Nature* 364: 232–5.

Hanski, I., Pakkala, T., Kuussaari, M., and Guangchun Lei (1995) 'Metapopulation persistence of an endangered butterfly in a fragmented landscape', *Oikos* 72: 21–8.

Hanksi, I., Foley, P., and Hassell, M. (1996) 'Random walks in a metapopulation: how much density dependence is necessary for long-term persistence?', *Journal of Animal Ecology* 65: 274–82.

Hanski, I., Moilanen, A., Pakkala, T., and Kuussaari, M. (1996) 'The quantitative incidence function model and persistence of an endangered butterfly population', *Conservation Biology* 10: 587–90.

Hardin, G. (1960) 'The competitive exclusion principle', *Science* 131: 1,292–97.

Harper, J. L. (1961) 'Approaches to the study of plant competition', *Symposia of the Society for Experimental Biology* 15: 1–39.

Harris, P. (1993) 'Effects, constraints and the future of weed biocontrol', *Agriculture, Ecosystems and Environment* 46: 289–303.

Harrison, G. W. (1995) 'Comparing predator–prey models to Luckinbill's experiment with *Didinium* and *Paramecium*', *Ecology* 76: 357–74.

Harrison, S. (1994) 'Metapopulations and conservation', pp. 111–28 in P. J. Edwards, R. M. May, and N. R. Webb (eds) *Large-Scale Ecology and Conservation Biology* (The 35th Symposium of the British Ecological Society with the Society for Conservation Biology, University of Southampton, 1993), Oxford: Blackwell Scientific Publications.

Hartman, G. (1994) 'Long-term population development of a reintroduced beaver (*Castor fiber*) population in Sweden', *Conservation Biology* 8: 713–17.

Heard, D. C. and Ouellet, J. P. (1994) 'Dynamics of an introduced caribou population', *Arctic* 47: 88–95.

Heard, S. B. (1994) 'Pitcher-plant midges and mosquitoes: a processing chain commensalism', *Ecology* 75: 1,647–60.

Hedges, S. B. (1989) 'Evolution and biogeography of West Indian frogs of the genus *Eleutherodactylus*: slow-evolving loci and the major groups', pp. 305–70 in C. A. Woods (ed.) *Biogeography of the West Indies: Past, Present, and Future*, Gainesville, Florida: Sandhill Crane Press.

Heywood, V. H. (ed.) (1978) *Flowering Plants of the World*, Oxford: Oxford University Press.

Hill, J. M. and Knisley, C. B. (1992) 'Frugivory in the tiger beetle, *Cicindela repanda* (Coleoptera: Cicindelidae)', *Coleopterist's Bulletin* 46: 306–10.

Hilligardt, M. (1993) 'Durchsetzungs- und Reproduktionsstrategien bei *Trifolium pallescens* Schreb. und *Trifolium thalii* Vill. II. Untersuchungen zur Populationsbiologie', *Flora (Jena)*: 188: 175–95.

Horn, H. S. (1981) 'Succession', pp. 253–71 in R. M. May (ed.) *Theoretical Ecology: Principles and Applications*, 2nd edn, Oxford: Blackwell Scientific Publications.

Horsley, G. A. (1966) 'Trees and shrubs', pp. 34–51 in B. L. Sage (ed.) *Northaw Great Wood: Its History and Natural History*, Hertford: Education Department of the Hertfordshire Country Council.

Hosking, J. R., Sullivan, P. R., and Welsby, S. M. (1994) 'Biological control of *Opuntia stricta* (Haw.) Haw. var. *stricta* using *Dactylopius opuntiae* (Cockerell) in an area of New South Wales, Australia, where *Cactoblastis cactorum* (Berg) is not a successful biological control agent', *Agriculture, Ecosystems and Environment* 48: 241–55.

Houde, P. W. (1988) 'Paleognathous birds from the early Tertiary of the northern hemisphere', *Publications of the Nuttall Ornithological Club* 22: 1–148.

Huffaker, C. B. (1958) 'Experimental studies on predation: dispersion factors and predator–prey oscillations', *Hilgardia* 27: 343–83.

Huffaker, C. B., Shea, K. P., and Herman, S. G. (1963) 'Experimental studies on predation: complex dispersion and levels of food in an acarine predator–prey interaction', *Hilgardia* 34: 305–30.

Huggett, R. J. (1980) *Systems Analysis in Geography*, Oxford: Clarendon Press.

Huggett, R. J. (1995) *Geoecology: An Evolutionary Approach*, London: Routledge.

Huggett, R. J. (1997) *Environmental Change: The Evolving Ecosphere*, London: Routledge.

Hylander, L. D., Silva, E. C., Oliveira, L. J., Silva, S. A., Kuntze, E. K., and Silva, D. X. (1994) 'Mercury levels in Alto Pantanal: a screening study', *Ambio* 23: 478–84.

Illies, J. (1974) *Introduction to Zoogeography*, translated by W. D. Williams, London: Macmillan.

Irland, L. C. (1994) 'Getting from here to there: implementing ecosystem management on the ground', *Journal of Forestry* 92: 12–17.

Irving, L. (1966) 'Adaptations to cold', *Scientific American* 214: 94–101.

Iversen, J. (1944) '*Viscum*, *Hedera* and *Ilex* as climate indicators', *Geologiska föreningens i Stockholm förhandlinger* 66: 463–83.

Jackson, R. M. and Raw, F. (1966) *Life in the Soil* (The Institute of Biology's Studies in Biology no. 2), London: Edward Arnold.

Jacob, M. (1994) 'Sustainable development and deep ecology: an analysis of competing traditions', *Environmental Management* 18: 477–88.

Jaffe, K., Michelangeli, F., Gonzalez, J. M., Miras, B., and Ruiz, M. C. (1992) 'Carnivory in pitcher plants of the genus *Heliamphora* (Sarraceniaceae)', *New Phytologist* 122: 733–44.

Johnson, D. L. (1980) 'Problems in the land vertebrate zoogeography of certain islands and the swimming powers of elephants', *Journal of Biogeography* 7: 383–98.

Johnson, D. L. (1989) 'Subsurface stone lines, stone zones, artifact-manuport layers and biomantles produced by bioturbation via pocket gophers (*Thomomys bottae*)', *American Antiquity* 54: 370–89.

Johnson, D. L. (1990) 'Biomantle evolution and the redistribution of earth materials and artefacts', *Soil Science* 149: 84–102.

Johnson, M. P. and Simberloff, D. S. (1974) 'Environmental determinants of island species numbers in the British Isles', *Journal of Biogeography* 1: 149–54.

Joshi, S. (1991) 'Biological control of *Parthenium hysterophorus* L. (Asteraceae) by *Cassia uniflora* Mill (Leguminosae), in Bangalore, India', *Tropical Pest Management* 37: 182–4.

Karanth, K. U. and Sunquist, M. E. (1995) 'Prey selection by tiger, leopard and dhole in tropical forests', *Journal of Animal Ecology* 64: 439–50.

Kaufman, D. M. (1995) 'Diversity of New World mammals:

universality of the latitudinal gradients of species and bauplans', *Journal of Mammalogy* 76: 322–34.

Kaufmann, M. R., Graham, R. T., Boyce, D. A. Jr, Moir, W. H., Perry, L., Reynolds, R. T., Bassett, R. L., Mehlhop, P., Edminster, C. B., Block, W. M., and Corn, P. S. (1994) *An Ecological Basis for Ecosystem Management* (General Technical Report RM–246, United States Department of Agriculture, Forest Service, Rocky Mountain Forest and Range Experiment Station, Fort Collins, Colorado), Washington, DC: United States Government Printing Office.

Keiter, R. and Boyce, M. (1991) *The Greater Yellowstone Ecosystem*, New Haven, Connecticut: Yale University Press.

Kennerly, T. E., Jr (1964) 'Microenvironmental conditions of the pocket gopher burrow', *Texas Journal of Science* 14: 397–441.

Kenward, R. E. and Holm, J. L. (1993) 'On the replacement of the red squirrel in Britain: a phytotoxic explanation', *Proceedings of the Royal Society of London* 251B: 187–94.

Kessel, B. (1953) 'Distribution and migration of the European starling in North America', *Condor* 55: 49–67.

Kessler, W. B., Salwasser, H., Cartwright, C. Jr, and Caplan, J. (1992) 'New perspective for sustainable natural resources management', *Ecological Applications* 2: 221–5.

Kevan, P. G. (1975) 'Sun-tracking solar furnaces in high Arctic flowers: significance for pollination and insects', *Science* 189: 723–6.

King, C. M. (1990) 'Introduction', pp. 3–21 in C. M. King (ed.) *The Handbook of New Zealand Mammals*, Auckland: Oxford University Press.

King, D. W. (1966) 'The soils', pp. 26–33 in B. L. Sage (ed.) *Northaw Great Wood: Its History and Natural History*, Hertford: Education Department of the Hertfordshire Country Council.

Kitayama, K., Mueller-Dombois, D., and Vitousek, P. M. (1995) 'Primary succession of Hawaiian montane rain forest on a chronosequence of eight lava flows', *Journal of Vegetation Science* 6: 211–22.

Knapp, P. A. (1992) 'Secondary plant succession and vegetation recovery in two western Great Basin Desert ghost towns', *Biological Conservation* 6: 81–9.

Kohn, D. D. and Walsh, D. M. (1994) 'Plant species richness – the effect of island size and habitat diversity', *Journal of Ecology* 82: 367–77.

Kormondy, E. J. (1996) *Concepts of Ecology*, 4th edn, Englewood Cliffs, NJ: Prentice Hall.

Kurtén, B. (1969) 'Continental drift and evolution', *Scientific American* 220: 54–64.

Lack, D. (1933) 'Habitat selection in birds with special references to the effects of afforestation on the Breckland avifauna', *Journal of Animal Ecology* 2: 239–62.

Lack, D. (1947) *Darwin's Finches*, Cambridge: Cambridge University Press.

Lamberson, R. H., McKelvey, R., Noon, B. R., and Voss, C. (1992) 'A dynamic analysis of northern spotted owl viability in a fragmented forest landscape', *Conservation Biology* 6: 505–12.

Larcher, W. (1975) *Physiological Plant Ecology*, 1st edn, Berlin: Springer.

Larcher, W. (1995) *Physiological Plant Ecology: Ecophysiology and Stress Physiology of Functional Groups*, 3rd edn, Berlin: Springer.

Laurie, W. A. and Brown, D. (1990a) 'Population biology of marine iguanas (*Amblyrhynchus cristatus*). I. Changes in fecundity related to a population crash', *Journal of Animal Ecology* 59: 515–28.

Laurie, W. A. and Brown, D. (1990b) 'Population biology of marine iguanas (*Amblyrhynchus cristatus*). II. Changes in annual survival rates and the effects of size, sex, age and fecundity in a population crash', *Journal of Animal Ecology* 59: 529–44.

Laws, R. M. (1970) 'Elephants as agents of habitat and landscape change in East Africa', *Oikos* 21: 1–15.

Lean, G. and Hinrichsen, D. (1992) *Atlas of the Environment*, Oxford: Helicon.

Leggett, J. (1989) 'The biggest mass-extinction of them all', *New Scientist* 122 (no. 1,668): 62.

Lenihan, J. M. (1993) 'Ecological response surfaces for North American boreal tree species and their use in forest classification', *Journal of Vegetation Science* 4: 667–80.

Lenihan, J. M. and Neilson, R. P. (1993) 'A rule-based vegetation formation model for Canada', *Journal of Biogeography* 20: 615–28.

Leopold, A. S. (1949) *A Sand County Almanac, with Sketches Here and There*, Illustrated by Charles W. Schwartz, New York: Oxford University Press.

Leslie, P. H. (1945) 'The use of matrices in certain population mathematics', *Biometrika* 33: 183–212.

Leslie, P. H. (1948) 'Some further notes on the use of matrices in population mathematics', *Biometrika* 35: 213–45.

Lever, C. (1979) *The Naturalized Animals of the British Isles*, London: Granada.

Levins, R. (1970) 'Extinction', pp. 77–107 in M.

Gerstenhaber (ed.) *Some Mathematical Questions in Biology*, Providence, Rhode Island: American Mathematical Society.

Li, Y., Glime, J. M., and Liao, C. (1992) 'Responses of two interacting *Sphagnum* species to water level', *Journal of Bryology* 17: 59–70.

Liebig, J. (1840) *Organic Chemistry and its Application to Agriculture and Physiology*, English edn edited by L. Playfair and W. Gregory, London: Taylor & Walton.

Ligon, J. D. (1978) 'Reproductive interdependence of piñon jays and piñon pines', *Ecological Monographs* 48: 111–26.

Lindemann, R. L. (1942) 'The trophic-dynamic aspect of ecology', *Ecology* 23: 399–418.

Lindenmayer, D. B. and Lacy, R. C. (1995) 'Metapopulation viability of Leadbeater's possum, *Gymnobelideus leadbeateri*, in fragmented old-growth forests', *Ecological Applications* 5: 164–82.

Linton, D. L. (1949) 'The delimitation of morphological regions', *Transactions of the Institute of British Geographers, Publication* 14: 86–7.

Lodge, D. M. (1993) 'Biological invasions: lessons for ecology', *Trends in Ecology and Evolution* 8: 133–7.

Lotka, A. J. (1925) *Elements of Physical Biology*, Baltimore, Maryland: Williams & Wilkins. Reprinted with corrections and bibliography as *Elements of Mathematical Biology*, New York: Dover, 1956.

Loughlin, T. R. and Miller, R. V. (1989) 'Growth of the northern fur seal colony on Bogoslof Island, Alaska', *Arctic* 42: 368–72.

Lovelock, J. E. (1979) *Gaia: A New Look at Life on Earth*, Oxford: Oxford University Press.

Lovelock, J. E. (1989) 'Geophysiology', *Transactions of the Royal Society of Edinburgh: Earth Sciences* 80: 169–75.

Luckinbill, L. S. (1973) 'Coexistence in laboratory populations of *Paramecium aurelia* and its predator *Didinium nasutum*', *Ecology* 54: 1,320–7.

Luckinbill, L. S. (1974) 'The effects of space and enrichment on a predator–prey system', *Ecology* 55: 1,142–7.

Lundelius, E. L., Jr, Graham, R. W., Anderson, E., Guilday, J., Holman, J. A., Steadman, D., and Webb, S. D. (1983) 'Terrestrial vertebrate faunas', pp. 311–53 in S. C. Porter (ed.) *Late-Quaternary Environments of the United States. Vol. 1. The Late Pleistocene*, London: Longman.

MacArthur, R. H. (1958) 'Population ecology of some warblers of northeastern coniferous forests', *Ecology* 39: 599–619.

McDaniel, J. (1986) 'Christian spirituality as openness to fellow creatures', *Environmental Ethics* 8: 33–46.

MacFadden, B. J. (1980) 'Rafting mammals or drifting islands?: biogeography of the Greater Antillean insectivores', *Journal of Biogeography* 7: 11–22.

McKenna, M. C. (1973) 'Sweepstakes, filters, corridors, Noah's arks, and beached Viking funeral ships in palaeogeography', pp. 295–308 in D. H. Tarling and S. K. Runcorn (eds) *Implications of Continental Drift to the Earth Sciences, Volume 1*, London and New York, Academic Press.

MacLulich, D. A. (1937) *Fluctuations in the Numbers of the Varying Hare* (Lepus americanus) (University of Toronto Studies, Biological Series, no. 43), Toronto, Canada: University of Toronto.

Madigan, M. T. and Marrs, B. L. (1997) 'Extremophiles', *Scientific American* 276: 66–71.

Marsh, G. P. (1864) *Man and Nature; or, Physical Geography as Modified by Human Action*, New York: Charles Scribner.

Marsh, G. P. (1965 edn) *Man and Nature*, edited by David Lowenthal, Cambridge, Massachusetts: Belknap Press of Harvard University Press.

Marshall, J. K. (1978) 'Factors limiting the survival of *Corynephorus canescens* (L.) Beauv. in Great Britain at the northern edge of its distribution', *Oikos* 19: 206–16.

Marshall, L. G. (1980) 'Marsupial paleobiogeography', pp. 345–86 in L. L. Jacobs (ed.) *Aspects of Vertebrate History: Essays in Honor of Edwin Harris Colbert*, Flagstaff, Arizona: Museum of Northern Arizona Press.

Marshall, L. G. (1981a) 'The Great American Interchange – an invasion induced crisis for South American mammals', pp. 133–229 in M. H. Nitecki (ed.) *Biotic Crises in Ecological and Evolutionary Time*, New York: Academic Press.

Marshall, L. G. (1981b) 'The Argentine connection', *Field Museum of Natural History Bulletin* (Chicago) 52: 16–25

Marshall, L. G. (1994) 'The terror birds of South America', *Scientific American* 270: 64–9.

Martin, R., Rodriguez, A., and Delibes, M. (1995) 'Local feeding specialization by badgers (*Meles meles*) in a Mediterranean environment', *Oecologia* 101: 45–50.

May, R. M. (1976) 'Simple mathematical models with very complicated dynamics', *Nature* 261: 459–67.

May, R. M. (1981) 'Models for single populations', pp. 5–29 in R. M. May (ed.) *Theoretical Ecology: Principles and Applications*, 2nd edn, Oxford: Blackwell Scientific Publications.

Menge, B. A., Berlow, E. L., Blanchette, C. A., Navarrete, S. A., and Yamada, S. B. (1994) 'The keystone species concept: variation in interaction strength in a rocky intertidal habitat', *Ecological Monographs* 64: 249–86.

Moloney, C. L., Cooper, J., Ryan, P. G., and Siegfried, W. R. (1994) 'Use of a population model to assess the impact of longline fishing on wandering albatross *Diomedia exulans* populations', *Biological Conservation* 70: 195–203.

Morhardt, J. E. and Gates, D. M. (1974) 'Energy-exchange analysis of the Belding ground squirrel and its habitat', *Ecological Monographs* 44: 14–44.

Moyle, P. B. and Williams, J. E. (1990) 'Biodiversity loss in the temperate zone: decline of the native fish fauna of California', *Conservation Biology* 4: 275–84.

Murphy, D. D. and Freas, K. (1988) 'Habitat-based conservation: the case of the Amargosa vole', *Endangered Species Update* 5: 6.

Myers, N. (1987) *The Gaia Atlas of Planet Management*, London: Pan Books.

Myklestad, Å. and Birks, H. J. B. (1993) 'A numerical analysis of the distribution patterns of *Salix* L. species in Europe', *Journal of Biogeography* 20: 1–32.

Naess, A. (1973) 'The shallow and the deep, long-range ecology movement', *Inquiry* 16: 95–100.

Naess, A. (1986) 'The deep ecological movement: some philosophical aspects', *Philosophical Inquiry* 8: 10–31.

Naiman, R. J., Johnson, C. A., and Kelley, J. C. (1988) 'Alteration of North American streams by beaver', *BioScience* 38: 753–62.

Naiman, R. J., Pinay, G., Johnson, C. A., and Pastor, J. (1994) 'Beaver influences on the long-term biogeochemical characteristics of boreal forest drainage networks', *Ecology* 75: 905–21.

Naqvi, S. M., Howell, R. D., and Sholas, M. (1993) 'Cadmium and lead residues in field-collected red swamp crayfish (*Procambarus clarkii*) and uptake by alligator weed, *Alternanthera philoxiroides*', *Journal of Environmental Science and Health* B28: 473–85.

Neilson, R. P. (1993a) 'Vegetation redistribution: a possible biosphere source of CO_2 during climatic change', *Water, Air, and Soil Pollution* 70: 659–73.

Neilson, R. P. (1993b) 'Transient ecotone response to climatic change: some conceptual and modelling approaches', *Ecological Applications* 3: 385–95.

Nelson, T. C. (1955) 'Chestnut replacement in the Southern Highlands', *Ecology* 36: 352–3.

Newton, I. (1972) *Finches*, London: Collins.

Nicholls, A. O. and McKenzie, N. J. (1994) 'Environmental control of the local-scale distribution of funnel ants, *Aphaenogaster longiceps*', *Memoirs of the Queensland Museum* 36: 165–72.

O'Brien, P. (1990) 'Managing Australian wildlife', *Search* 21: 24–7.

Odum, E. P. (1989) *Ecology and Our Endangered Life-Support Systems*, Sunderland, Massachusetts: Sinauer Associates.

Odum, H. T. (1971) *Environment, Power, and Society*, New York: John Wiley & Sons.

O'Riordan, T. (1988) 'Future directions for environmental policy', pp. 168–98 in D. C. Pitt (ed.) *The Future of the Environment: The Social Dimensions of Conservation and Ecological Alternatives*, London: Routledge.

O'Riordan, T. (1996) 'Environmentalism on the move', pp. 449–76 in I. Douglas, R. J. Huggett, and M. E. Robinson (eds) *Companion Encyclopedia of Geography: The Environment and Humankind*, London: Routledge.

Orphanides, G. M. (1993) 'Control of *Saissetia oleae* (Hom.: Coccidae) in Cyprus through establishment of *Metaphycus bartletti* and *M. helvolus* (Hym.: Encyrtidae)', *Entomophaga* 38: 235–9.

Osburn, R. C., Dublin, L. I., Shimer, H. W., and Lull, R. S. (1903) 'Adaptation to aquatic, arboreal, fossorial, and cursorial habits in mammals', *American Naturalist* 37: 651–65; 731–6; 819–25; 38: 322–32.

Overpeck, J. T., Rhind, D., and Goldberg, R. (1990) 'Climate-induced changes in forest disturbance and vegetation', *Nature* 343: 51–3.

Owen-Smith, R. N. (1987) 'Pleistocene extinctions: the pivotal role of megaherbivores', *Paleobiology* 13: 351–62.

Owen-Smith, R. N. (1988) *Megaherbivores: The Influence of Very Large Body Size on Ecology*, Cambridge: Cambridge University Press.

Owen-Smith, R. N. (1989) 'Megafaunal extinctions: the conservation message from 11,000 years B.P.', *Conservation Biology* 3: 405–12.

Paine, R. T. (1974) 'Intertidal community structure. Experimental studies on the relationship between a dominant competitor and its principal predator', *Oecologia* 15: 93–120.

Parker, A., Holden, A. N. G., and Tomley, A. J. (1994) 'Host specificity testing and assessment of the pathogenicity of the rust, *Puccinia abrupta var. partheniicola*, as a biological control agent of parthenium weed (*Parthenium hysterophorus*)', *Plant Pathology* 43: 1–16.

Parkes, J. P. (1993) 'Feral goats: designing solutions for a designer pest', *New Zealand Journal of Ecology* 17: 71–83.

Parmesan, C. (1996) 'Climate and species' range', *Nature* 382: 765–6.

Pastor, J. R. and Post, W. M. (1988) 'Response of northern forests to CO_2-induced climatic change', *Nature* 334: 55–8.

Payette, S. (1993) 'The range limit of boreal tree species in Quebec–Labrador: an ecological and palaeoecological interpretation', *Review of Palaeobotany and Palynology* 79: 7–30.

Pearl, R. (1928) *The Rate of Living*, New York: Alfred Knopf.

Pearlstine, L., McKellar, H., and Kitchens, W. (1985) 'Modelling the impacts of a river diversion on bottom-land forest communities in the Santee River floodplain, South Carolina', *Ecological Modelling* 29: 283–302.

Pears, N. (1985) *Basic Biogeography*, 2nd edn, Harlow: Longman.

Perfit, M. R. and Williams, E. E. (1989) 'Geological constraints and biological reintroductions in the evolution of the Caribbean Sea and its islands', pp. 47–102 in C. A. Woods (ed.) *Biogeography of the West Indies: Past, Present, and Future*, Gainesville, Florida: Sandhill Crane Press.

Perring, F. H. and Walters, S. M. (1962) *Atlas of the British Flora*, London: Nelson.

Perrins, C. (ed.) (1990) *The Illustrated Encyclopaedia of Birds: The Definitive Guide to Birds of the World*, London: Headline Publishing.

Peters, R. L. (1992a) 'Conservation of biological diversity in the face of climatic change', pp. 15–30 in R. L. Peters and T. E. Lovejoy (eds) *Global Warming and Biological Diversity*, New Haven, Connecticut and London: Yale University Press.

Peters, R. L. (1992b) 'Introduction', pp. 3–14 in R. L. Peters and T. E. Lovejoy (eds) *Global Warming and Biological Diversity*, New Haven, Connecticut and London: Yale University Press.

Phillipson, J. (1966) *Ecological Energetics* (The Institute of Biology's Studies in Biology no. 1), London: Edward Arnold.

Pielou, E. C. (1979) *Biogeography*, New York: John Wiley & Sons.

Pigott, C. D. (1974) 'The response of plants to climate and climatic change', pp. 32–44 in F. H. Perring (ed.) *The Flora of a Changing Britain*, London: Classey.

Pigott, C. D. (1981) 'Nature of seed sterility and natural regeneration of *Tilia cordata* near its northern limit in Finland', *Annales Botanici Fennici* 18: 255–63.

Pigott, C. D. and Huntley, J. P. (1981) 'Factors controlling the distribution of *Tilia cordata* at the northern limits of its geographical range. III. Nature and causes of seed sterility', *New Phytologist* 87: 817–39.

Pimm, S. L. (1991) *Balance of Nature? Ecological Issues in the Conservation of Species and Communities*, Chicago: University of Chicago Press.

Poiani, K. A. and Johnson, W. C. (1991) 'Global warming and prairie wetlands', *BioScience* 41: 611–18.

Polischuk, S. C., Letcher, R. J., Norstrom, R. J., and Ramsay, M. A. (1995) 'Preliminary results of fasting on the kinetics of organochlorines in polar bears (*Ursus maritimus*)', *Science of the Total Environment* 160–1: 465–72.

Pond, D. (1992) 'Protective-commensal mutualism between the queen scallop *Chlamys opercularis* (Linnaeus) and the encrusting sponge *Suberites*', *Journal of Molluscan Studies* 58: 127–34.

Pounds, J. A. (1990) 'Disappearing gold', *BBC Wildlife* 8: 812–17.

Pounds, J. A. and Crump, M. L. (1994) 'Amphibian declines and climate disturbance: the case of the golden toad and the harlequin frog', *Conservation Biology* 8: 72–85.

Pregill, G. K. (1981) 'An appraisal of the vicariance hypothesis of Caribbean biogeography and its application to West Indian terrestrial vertebrates', *Systematic Zoology* 30: 147–55.

Priddel, D. and Wheeler, R. (1994) 'Mortality of captive-raised malleefowl, *Leipoa ocellata*, released into a mallee remnant within the wheat-belt of New South Wales', *Wildlife Research* 21: 543–52.

Prigogine, I. and Stengers, I. (1984) *Order out of Chaos: Man's New Dialogue with Nature*, London: William Heinemann.

Prins, H. H. T. and Van der Jeugd, H. P. (1993) 'Herbivore population crashes and woodland structure in East Africa', *Journal of Ecology* 81: 305–14.

Puigcerver, M., Gallego, S., Rodriguez-Teijeiro, J. D., and Senar, J. C. (1992) 'Survival and mean life span of the quail *Coturnix c. coturnix*', *Bird Study* 39: 120–3.

Rabb, G. B. (1994) 'The changing roles of zoological parks in conserving biological diversity', *American Zoologist* 34: 159–64.

Rabinowitz, A. R. and Walker, S. R. (1991) 'The carnivore community in a dry tropical forest mosaic in Huai Kha Khaeng Wildlife Sanctuary, Thailand', *Journal of Tropical Ecology* 7: 37–47.

Rapoport, E. H. (1982) *Aerography: Geographical Strategies of Species*, Oxford: Pergamon Press.

Raunkiaer, C. (1934) *The Life Forms of Plants and Statistical Plant Geography, Being the Collected Papers of C. Raunkiaer*, translated by H. Gilbert-Carter and A. G. Tansley, Oxford: Clarendon Press.

Raven, P. H. (1963) 'Amphitropical relationships in the floras of North and South America', *The Quarterly Review of Biology* 38: 151–77.

Ray, P. M. and Alexander, W. E. (1966) 'Photoperiodic

adaptation to latitude in *Xanthium strumasium*', *American Journal of Botany* 53: 806–16.

Reddish, P. (1996) *Spirits of the Jaguar: The Natural History and Ancient Civilizations of the Caribbean and Central America*, London: BBC Books.

Regan, T. (1983) *The Case for Animal Rights*, London: Routledge & Kegan Paul.

Reid, D. G., Herrero, S. M., and Code, T. E. (1988) 'River otters as agents of water loss from beaver ponds', *Journal of Mammalogy* 69: 100–7.

Revelle, R. (1984) 'The effects of population on renewable resources', *Population Studies* 90: 223–40.

Rhodes II, R. S. (1984) 'Paleoecological and regional paleo-climatic implications of the Farmdalian Craigmile and Woodfordian Waubonsie mammalian local faunas, southwestern Iowa', *Illinois State Museum Report of Investigations* 40: 1–51.

Ripple, W. J., Johnson, D. H., Hershey, K. T., and Meslow, E. C. (1991) 'Old-growth and mature forests near spotted owl nests in western Oregon', *Journal of Wildlife Management* 55: 316–18.

Rodríguez de la Fuente, F. (1975) *Animals of South America* (World of Wildlife Series), English language version by John Gilbert, London: Orbis Publishing.

Rohde, K. (1992) 'Latitudinal gradients in species diversity: the search for the primary cause', *Oikos* 65: 514–27.

Rolls, E. C. (1969) *They All Ran Wild*, Sydney: Angus & Robertson.

Romer, A. S. (1966) *Vertebrate Paleontology*, 3rd edn, Chicago and London: University of Chicago Press.

Root, R. B. (1967) 'The niche exploitation pattern of the blue-gray gnatcatcher', *Ecological Monographs* 37: 317–50.

Root, T. L. (1988a) 'Environmental factors associated with avian distributional boundaries', *Journal of Biogeography* 15: 489–505.

Root, T. L. (1988b) 'Energy constraints on avian distribu-tions and abundances', *Ecology* 69: 330–9.

Root, T. L. and Schneider, S. H. (1993) 'Can large-scale climatic models be linked with multiscale ecological studies?', *Conservation Biology* 7: 256–70.

Rorison, I. H., Sutton, F., and Hunt, R. (1986) 'Local climate, topography and plant growth in Lathkill Dale NNR. I. A twelve-year summary of solar radiation and temperature', *Plant, Cell, and Environment* 9: 49–56.

Rosenzweig, M. L. (1992) 'Species diversity gradients: we know more and less than we thought', *Journal of Mammalogy* 73: 715–30.

Rosenzweig, M. L. (1995) *Species Diversity in Space and Time*, Cambridge: Cambridge University Press.

Rudge, M. R. (1990) 'Feral goat', pp. 406–23 in C. M. King (ed.) *The Handbook of New Zealand Mammals*, Auckland: Oxford University Press.

Ryrholm, N. (1988) 'An extralimital population in a warm climatic outpost: the case of the moth *Idaea dilutaria* in Scandinavia', *International Journal of Biometeorology* 32: 205–16.

Sage, B. L. (ed.) (1966a) *Northaw Great Wood: Its History and Natural History*, Hertford: Education Department of the Hertfordshire Country Council.

Sage, B. L. (1966b) 'Geology', pp. 22–5 in B. L. Sage (ed.) *Northaw Great Wood: Its History and Natural History*, Hertford: Education Department of the Hertfordshire Country Council.

Sakai, A. (1970) 'Freezing resistance in willows from different climates', *Ecology* 51: 485–91.

Sakai, A. and Otsuka, K. (1970) 'Freezing resistance of alpine plants', *Ecology* 51: 665–71.

Salisbury, E. J. (1926) 'The geographical distribution of plants in relation to climatic factors', *Geographical Journal* 57: 312–35.

Sallabanks, R. (1992) 'Fruit fate, frugivory, and fruit charac-teristics: a study of the hawthorn, *Crataegus monogyna* (Rosaceae)', *Oecologia* 91: 296–304.

Samways, M. J. (1989) 'Climate diagrams and biological control: an example from the areography of the ladybird *Chilocorus nigritus* (Fabricius, 1798) (Insecta, Coleoptera, Coccinellidae)', *Journal of Biogeography* 16: 345–51.

Sarre, S. (1995) 'Size and structure of populations of *Oedura reticulata* (Reptilia: Gekkonidae) in woodland remnants: implications for the future regional distribution of a currently common species', *Australian Journal of Ecology* 20: 288–98.

Saunders, H. (1889) *An Illustrated Manual of British Birds*, London: Gurney & Jackson.

Schaetzl, R. J., Burns, S. F., Johnson, D. L., and Small, T. W. (1989a) 'Tree uprooting: a review of impacts on forest ecology', *Vegetatio* 79: 165–76.

Schaetzl, R. J., Johnson, D. L., Burns, S. F., and Small, T. W. (1989b) 'Tree uprooting: review of terminology, process, and environmental implications', *Canadian Journal of Forest Research* 19: 1–11.

Schennum, W. E. and Willey, R. B. (1979) 'A geographical analysis of quantitative morphological variation in the grasshopper *Arphia conspersa*', *Evolution* 33: 64–84.

Schindler, D. W., Kidd, K. A., Muir, D. C. G., and Lockhart, W. L. (1995) 'The effects of ecosystem characteristics on contaminant distribution in northern freshwater lakes', *Science of the Total Environment* 160–1: 1–17.

Schmidt-Nielsen, K. and Schmidt-Nielsen, B. (1953) 'The desert rat', *Scientific American* 189: 73–8.

Schultz, J. (1995) *The Ecozones of the World: The Ecological Divisions of the Geosphere*, Hamburg: Springer.

Schwalbe, C. P., Mastro, V. C., and Hansen, R. W. (1991) 'Prospects for genetic control of the gypsy moth', *Forest Ecology and Management* 39: 163–71.

Sclater, P. L. (1858) 'On the general geographical distribution of the members of the class Aves', *Journal of the Linnean Society, Zoology* 2: 130–45.

Sears, P. (1949) *Deserts on the March*, London: Routledge & Kegan Paul.

Sengonca, C., Uygun, N., Kersting, U., and Ulusoy, M. R. (1993) 'Successful colonization of *Eretmocerus debachi* (Hym.: Aphelinidae) in the eastern Mediterranean citrus region of Turkey', *Entomophaga* 38: 383–90.

Senut, B., Pickford, M., and Dauphin, Y. (1995) 'Découverte d'œufs de type "Aepyornithoïde" dans le Miocène inférieur de Namibie', *Comptes Rendus, Académie des Sciences, Série II: Sciences de la Terre et des Planètes* 320: 71–6.

Shelford, V. E. (1911) 'Physiological animal geography', *Journal of Morphology* 22: 551–618.

Sherlock, R. L. (1922) *Man as a Geological Agent: An Account of His Action on Inanimate Nature*, Foreword by A. S. Woodward, London: H. F. & G. Witherby.

Shorten, M. (1954) *Squirrels*, London: Collins.

Shrader-Frechette, K. S. and McCoy, E. D. (1994) 'How the tail wags the dog: how value judgements determine ecological science', *Environmental Values* 3: 107–20.

Simberloff, D. and Dayan, T. (1991) 'The guild concept and the structure of ecological communities', *Annual Review of Ecology and Systematics* 22: 115–43.

Simmons, I. G. (1979) *Biogeography: Natural and Cultural*, London: Edward Arnold.

Simpson, G. G. (1940) 'Mammals and land bridges', *Journal of the Washington Academy of Science* 30: 137–63.

Simpson, G. G. (1980) *Splendid Isolation: The Curious History of South American Mammals*, New Haven, Connecticut and London: Yale University Press.

Sinsch, U. (1992) 'Structure and dynamic of a natterjack toad metapopulation (*Bufo calamita*)', *Oecologia* 90: 489–99.

Sjögren, P. (1991) 'Extinction and isolation gradients in metapopulations: the case of the pool frog (*Rana lessonae*)', *Biological Journal of the Linnean Society* 42: 135–47.

Smith, A. G., Smith, D. G., and Funnell, B. M. (1994) *Atlas of Mesozoic and Cenozoic Coastlines*, Cambridge: Cambridge University Press.

Smith, A. T. (1974) 'The distribution and dispersal of pikas: consequences of insular population structure', *Ecology* 55: 1,112–19.

Smith, A. T. (1980) 'Temporal changes on insular populations of the pika (*Ochotona princeps*)', *Ecology* 61: 8–13.

Smith, C. H. (1983) 'A system of world mammal faunal regions. I. Logical and statistical derivation of the regions', *Journal of Biogeography* 10: 455–66.

Smith, F. D. M. (1994) 'Geographical ranges of Australian mammals', *Journal of Animal Ecology* 63: 441–50.

Smith, J. M. (1974) *Models in Ecology*, London: Cambridge University Press.

Solomon, A. M. (1986) 'Transient response of forests to CO_2-induced climatic change: simulation modeling experiments in eastern North America', *Oecologia* 68: 567–79.

Stanton, M. L. and Galen, C. (1989) 'Consequences of flower heliotropism for reproduction in an alpine buttercup (*Ranunculus adoneus*)', *Oecologia* 78: 477–85.

Stevens, G. C. (1989) 'The latitudinal gradient in geographical range: how so many species coexist on the tropics', *The American Naturalist* 133: 240–56.

Stevens, G. C. (1992) 'The elevational gradient in altitudinal range: an extension of Rapoport's latitudinal rule to altitude', *The American Naturalist* 140: 893–911.

Stewart, I. (1995) *Nature's Numbers: Discovering Order and Pattern in the Universe*, London: Weidenfeld & Nicolson.

Stokes, D. W. and Stokes, L. Q. (1996) *Stokes Field Guide to Birds: Eastern Region*, Boston: Little, Brown.

Stoltz, J. F., Botkin, D. B., and Dastoor, M. N. (1989) 'The integral biosphere', pp. 31–49 in M. B. Rambler, L. Margulis, and R. Fester (eds) *Global Ecology: Towards a Science of the Biosphere*, San Diego, California: Academic Press.

Stone, C. D. (1972) 'Should trees have standing? Toward legal rights for natural objects', *California Legal Review* 45: 450.

Storer, R. W. (1966) 'Sexual dimorphism and food habits in three North American accipiters', *Auk* 83: 423–6.

Stoutjesdijk, P. and Barkman, J. J. (1992) *Microclimate, Vegetation and Fauna*, Uppsala, Sweden: Opulus Press.

Strahan, R. (ed.) (1995) *Mammals of Australia*, 2nd edn, Washington, DC: Smithsonian Institution Press.

Sukachev, V. N. and Dylis, N. V. (1964) *Fundamentals of*

Forest Biogeocoenology, translated by J. M. Maclennan, Edinburgh: Oliver & Boyd.

Tagawa, H. (1992) 'Primary succession and the effect of first arrivals on subsequent development of forest types', *GeoJournal* 28: 175–83.

Tambussi, C. P., Noriega, J. I., Gaździcki, A., Tatur, A., Reguero, M. A., and Vizcaino, S. F. (1994) 'Ratite bird from the Paleogene La Meseta Formation, Seymour Island, Antarctica', *Polish Polar Research* 15: 15–20.

Tansley, A. G. (1935) 'The use and abuse of vegetational concepts and terms', *Ecology* 16: 284–307.

Tansley, A. G. (1939) *The British Isles and their Vegetation*, Cambridge: Cambridge University Press.

Taylor, C. R. (1969) 'The eland and the oryx', *Scientific American* 220: 89–95.

Terborgh, J. (1988) 'The big things that run the world – a sequel to E. O. Wilson', *Conservation Biology* 2: 402–3.

Thomas, C. D. and Jones, T. M. (1993) 'Partial recovery of a skipper butterfly (*Hesperia comma*) from population refuges: lessons for conservation in a fragmented landscape', *Journal of Animal Ecology* 62: 472–81.

Thomas, D. J., Tracey, B., Marshall, H., and Norstrom, R. J. (1992) 'Arctic terrestrial ecosystem contamination', *Science of the Total Environment* 122: 135–64.

Thoreau, H. D. (1893 edn) *The Writings of Henry David Thoreau. Volume 2: Walden; or, Life in the Woods*, Boston: Houghton Mifflin.

Tilman, D. (1988) *Plant Strategies and the Dynamics and Structure of Plant Communities* (Monographs in Population Biology 26), Princeton, NJ: Princeton University Press.

Tilman, D. (1994) 'Competition and biodiversity in spatially structured habitats', *Ecology* 75: 2–16.

Tilman, D. and Wedin, D. (1991) 'Oscillations and chaos in the dynamics of a perennial grass', *Nature* 353: 653–5.

Tivy, J. (1992) *Biogeography: A Study of Plants in the Ecosphere*, 3rd edn, Edinburgh: Oliver & Boyd.

Tudge, C. (1997) 'Rights and wrongs', *The Independent on Sunday*, 16 March, pp. 40–1.

Türk, A. and Arnold, W. (1988) 'Thermoregulation as a limit to habitat use in alpine marmots (*Marmota marmota*)', *Oecologia* 76: 544–8.

Usher, M. B. (1972) 'Developments in the Leslie matrix model', pp. 29–60 in J. N. R. Jeffers (ed.) *Mathematical Models in Ecology*, Oxford: Blackwell Scientific Publications.

Utida, S. (1957) 'Population fluctuation, an experimental and theoretical approach', *Cold Spring Harbor Symposium in Quantitative Biology* 22: 139–51.

van Wilgen, B. W., Richardson, D. M., Kruger, F. J., and van Hensbergen, H. J. (eds) (1992) *Fire in South African Mountain Fynbos: Ecosystem, Community and Species Response at Swartboskloof* (Ecological Studies, Vol. 93), New York: Springer.

Varley, G. C. (1970) 'The concept of energy flow applied to a woodland community', pp. 389–405 in A. Watson (ed.) *Animal Populations in Relation to their Food Resources* (A Symposium of the British Ecological Society, Aberdeen 24–28 March 1969), Oxford: Blackwell Scientific Publications.

Vaughan, T. A. (1978) *Mammalogy*, 2nd edn, Philadelphia: W. B. Saunders.

Veena, T. and Lokesha, R. (1993) 'Association of drongos with myna flocks: are drongos benefited?', *Journal of Biosciences, Indian Academy of Sciences* 18: 111–19.

Verboom, J., Schotman, A., Opdam, P. and Metz, J. A. J. (1991) 'European nuthatch metapopulations in a fragmented agricultural landscape', *Oikos* 61: 149–56.

Vitousek, P. M., Ehrlich, P. R., Ehrlich, A. H., and Matson, P. A. (1986) 'Human appropriation of the products of photosynthesis', *BioScience* 36: 368–73.

Volterra, V. (1926) 'Fluctuations in the abundance of a species considered mathematically', *Nature* 188: 558–60.

Volterra, V. (1928) 'Variations and fluctuations of the number of individuals in animal species living together', *Journal du Conseil International pour l'Éxploration de la Mer* 3: 3–51.

Volterra, V. (1931) *Leçons sur la Théorie mathématique de la Lutte pour la Vie*, edited by Marcel Brelot, Paris: Gauthier-Villars.

Wallace, A. R. (1876) *The Geographical Distribution of Animals; with a Study of the Relations of Living and Extinct Faunas as Elucidating the Past Changes of the Earth's Surface*, 2 vols, London: Macmillan.

Walter, H. (1985) *Vegetation of the Earth and Ecological Systems of the Geo-Biosphere*, 3rd revised and enlarged edn, translated from the 5th revised German edn by O. Muise, Berlin: Springer.

Walter, H. and Breckle, S.-W. (1985) *Ecological Systems of the Geobiosphere. Vol. 1. Ecological Principles in Global Perspective*, translated by S. Gruber, Berlin: Springer.

Walter, H. and Lieth, H. (1960–7) *Klimadiagramm-Weltatlas*, Jena: Gustav Fischer.

Ward, T. J. and Jacoby, C. A. (1992) 'A strategy for assessment and management of marine ecosystems: baseline and monitoring studies in Jervis Bay, a temperate Australian embayment', *Marine Pollution Bulletin* 25: 163–71.

Wardle, D. A., Nicholson, K. S., and Rahman, A. (1993) 'Influence of plant age on the allelopathic potential of nodding thistle (*Carduus nutans* L.) against pasture grasses and legumes', *Weed Research* 33: 69–78.

Watt, A. S. (1947) 'Pattern and process in the plant community', *Journal of Ecology* 35: 1–22.

Webb, N. R. and Thomas, J. A. (1994) 'Conserving insect habitats in heathland biotopes: a question of scale', pp. 129–51 in P. J. Edwards, R. M. May, and N. R. Webb (eds) *Large-Scale Ecology and Conservation Biology* (The 35th Symposium of the British Ecological Society with the Society for Conservation Biology, University of Southampton, 1993), Oxford: Blackwell Scientific Publications.

Webb, S. D. (1991) 'Ecogeography and the Great American Interchange', *Paleobiology* 17: 266–80.

Weiser, C. J. (1970) 'Cold resistance and injury in woody plants', *Science* 169: 1,269–78.

Whelan, R. J. (1995) *The Ecology of Fire*, Cambridge: Cambridge University Press.

Whitmore, T. C. (1975) *Tropical Rain Forests of the Far East*, Oxford: Oxford University Press.

Whittaker, R. H. (1953) 'A consideration of climax theory: the climax as a population and pattern', *Ecological Monographs* 23: 41–78.

Whittaker, R. H. (1954) 'The ecology of serpentine soils. IV. The vegetational response to serpentine soils', *Ecology* 35: 275–88.

Whittaker, R. J. and Bush, M. B. (1993) 'Dispersal and establishment of tropical forest assemblages, Krakatoa, Indonesia', pp. 147–60 in J. Miles and D. W. H. Walton (eds) *Primary Succession on Land* (The British Ecological Society, special publication no. 12), Oxford: Blackwell Scientific Publications.

Whittaker, R. J. and Jones, S. H. (1994) 'Structure in re-building insular ecosystems: an empirically derived model', *Oikos* 69: 524–30.

Whittaker, R. J., Bush, M. B., and Richards, K. (1989) 'Plant recolonization and vegetation succession on the Krakatau Islands, Indonesia', *Ecological Monographs* 59: 59–123.

Whittaker, R. J., Bush, M. B., Asquith, N. M., and Richards, K. (1992) 'Ecological aspects of plant colonisation of the Krakatau Islands', *GeoJournal* 28: 201–11.

Whittlesey, D. (1954) 'The regional concept and the regional method', pp. 19–68 in P. E. James and C. F. Jones (eds) *American Geography: Inventory and Prospect*, Syracuse, New York: Syracuse University Press.

Williams, E. E. (1989) 'Old problems and new opportunities in West Indian Biogeography', pp. 1–46 in C. A. Woods (ed.) *Biogeography of the West Indies: Past, Present, and Future*, Gainesville, Florida: Sandhill Crane Press.

Williams, M. (1996) 'European expansion and land cover transformation', pp. 182–205 in I. Douglas, R. J. Huggett, and M. E. Robinson (eds) *Companion Encyclopedia of Geography: The Environment and Humankind*, London and New York: Routledge.

Wilson, E. O. (1992) *The Diversity of Life*, Cambridge, Massachusetts: Belknap Press of Harvard University Press.

Wilson, R., Allen-Gil, S., Griffin, D., and Landers, D. (1995) 'Organochlorine contaminants in fish from an Arctic lake in Alaska, USA', *Science of the Total Environment* 160–1: 511–19.

Woods, C. A. (ed.) (1989) *Biogeography of the West Indies: Past, Present, and Future*, Gainesville, Florida: Sandhill Crane Press.

Woodward, F. I. (1992) 'A review of the effects of climate on vegetation: ranges, competition, and composition', pp. 105–23 in R. L. Peters and T. E. Lovejoy (eds) *Global Warming and Biological Diversity*, New Haven, Connecticut and London: Yale University Press.

Woodwell, G. M. (1967) 'Toxic substances and ecological cycles', *Scientific American* 216 (March): 24–31.

World Commission on Environment and Development (1987) *Our Common Future* (Brundtland Report), New York and Oxford: Oxford University Press.

Worster, D. (1994) *Nature's Economy: A History of Ecological Ideas*, 2nd edn, Cambridge: Cambridge University Press.

Wright, S. J., Gompper, M. E., and DeLeon, B. (1994) 'Are large predators keystone species in Neotropical forests? The evidence from Barro Colorado Island', *Oikos* 71: 279–94.

Zackrisson, O. and Nilsson, M. C. (1992) 'Allelopathic effects by *Empetrum hermaphroditum* on seed germination of two boreal tree species', *Canadian Journal of Forest Research* 22: 1310–19.

Zimmerman, M. E. (1988) 'Quantum theory, intrinsic value, and pantheism', *Environmental Ethics* 10: 3–30.

Ziswiler, V. (1967) *Extinct and Vanishing Animals: A Biology of Extinction and Survival*, revised English edn by F. and P. Bunnell (The Heidelberg Science Library, Vol. 2), London: English Universities Press.

INDEX